AF443275

ANNALS OF
THE NEW YORK ACADEMY
OF SCIENCES

Volume 824

EDITORIAL STAFF

Executive Editor
BILL BOLAND

Managing Editor
JUSTINE CULLINAN

Associate Editor
LINDA HOTCHKISS MEHTA

The New York Academy of Sciences
2 East 63rd Street
New York, New York 10021

CHALLENGES AND OPPORTUNITIES IN PEDIATRIC ONCOLOGY

ANNALS OF THE NEW YORK ACADEMY OF SCIENCES
Volume 824

CHALLENGES AND OPPORTUNITIES IN PEDIATRIC ONCOLOGY

Edited by Frederick F. Holmes, John J. Kepes, Tribhawan S. Vats,
Dezső Schuler, and István Nyáry

The New York Academy of Sciences
New York, New York
1997

Cover art on the softcover edition of this volume is *The Sick Child* by Eugène Carrière. © Photo RMN, Jean Schomans. The painting is in the collection of the Musee d' Orsay, Paris.

Library of Congress Cataloging-in-Publication Data

Challenges and opportunities in pediatric oncology/editors,
    Frederick F. Holmes . . . [et al.].
        p.   cm. — (Annals of the New York Academy of Sciences; v.
824)
    Includes bibliographical references and index.
    ISBN 1-57331-082-4 (cloth: alk. paper). — ISBN 1-57331-083-2
(pbk.: alk. paper)
    1. Tumors in children—Congresses.   I. Holmes, Frederick F.,
1932– . II. Series.
Q11.N5   vol. 824
[RC281.C4]
500 s—dc21
[618'.92994]
                                                            97-34035
                                                               CIP

BiC/PCP
Printed in the United States of America
ISBN 1-57331-082-4 (cloth)
ISBN 1-57331-083-2 (paper)
ISSN 0077-8923

ANNALS OF THE NEW YORK ACADEMY OF SCIENCES
Volume 824
September 17, 1997

CHALLENGES AND OPPORTUNITIES IN PEDIATRIC ONCOLOGY[a]

Editors
FREDERICK F. HOLMES, JOHN J. KEPES, TRIBHAWAN S. VATS,
DEZSŐ SCHULER, AND ISTVÁN NYÁRY

Conference Organizers
FREDERICK F. HOLMES, JOHN J. KEPES, TRIBHAWAN S. VATS, DEZSŐ SCHULER,
ISTVÁN NYÁRY, JOSEPH BORSI, AND HEDVIG BODÁNSZKY

CONTENTS

[a] This volume is the result of a conference entitled **Challenges and Opportunities in Pediatric Oncology,** which was sponsored by the New York Academy of Sciences, the Pediatric Oncology Outreach to Hungary program of the University of Kansas Medical Center, and the U.S. Agency for International Development, and held on October 6 through 9, 1996, in Budapest, Hungary.

Financial assistance was received from:

Major Funder
  • U.S. AGENCY FOR INTERNATIONAL DEVELOPMENT

Supporters
   • ASTA MEDICA KFT.
   • BIOTEST HUNGARIA KFT.
   • BECKMAN INSTRUMENTS, INC.

Preface:
Understanding the Challenges and Opportunities of Pediatric Oncology in Our Small World

FREDERICK F. HOLMES

Department of Medicine
University of Kansas Medical Center
39th and Rainbow Boulevard
Kansas City, Kansas 66160

This volume contains papers from physicians, scientists, and surgeons who attended the fifth annual Winnie the POOH Conference on pediatric malignancies held in Budapest, Hungary, in October of 1996. With funding from the U.S. Agency for International Development (USAID) and sponsorship of the University of Kansas Medical Center, there were 150 attendees from 23 countries, more than half from the nations of Central and Eastern Europe, still adjusting after decades of domination by the former Soviet Union.

Cancer is the second leading cause of death for children in the Western world. Rates of remission, survival, and cure of childhood malignancy may well represent the best measurements of the sophistication and efficiency of medical services provided to children in particular societies. The conference attempted to close the gaps between basic and clinical sciences in pediatric oncology and between countries with the most advanced treatment modalities and those striving to join them. All aspects of childhood malignancy, from basic research to clinical trials to supportive care were addressed and are presented in the papers of this volume. The multidisciplinary approach to treating the child with cancer was central to the presentations, with concentration on basic and clinical science information useful to the day-to-day practices of those who care for children with cancer. Opening the conference, His Excellency, Árpád Göncz, President of Hungary, reminded those present that its children are the most important resource any nation has in planning for its future.

The reader of this volume will find consideration of molecular genetics and cellular events such as apoptosis, with particular focus on bone marrow and stem cell transplantation in childhood leukemia. Direct clinical application of basic science knowledge to chemotherapy is considered in detail as well as the importance of biological markers in guiding and directing therapy.

Pediatric central nervous system tumors are considered in detail, from the immunobiology of gliomas generally to the specifics of surgical approaches to uncommon tumors and, of course, the remarkable advances of recent years in the multidisciplinary management of tumors such as medulloblastomas. In respect to quality of life for long-term survival for children cured of brain tumors, careful attention to the specifics of initial management has proved to be of the greatest importance.

In this regard, considering all pediatric malignancies, the conference carefully and thoroughly addressed the consequences of treatment, cure, and long survival. Supportive care, treatment toxicity, quality assurance, and the psychosocial aspects of pediatric cancer are presented in papers in this volume as well. Finally, the great challenge of helping children with pediatric malignancies in the developing world is described.

The Honorable Donald M. Blinken, Ambassador of the United States to Hungary, found particular hope in addressing the conference. Coming earlier from academia, he saw particular significance in scientists from all parts of the world seeking to level the playing field for the thousands of children who are diagnosed with cancer each year.

The expert or the casual reader will find this volume to be a wide-ranging review of the remarkable accomplishments of pediatric oncology in recent decades as well as a clear statement of the challenges remaining in all parts of the world for the diagnosis and treatment of childhood cancer. It was serendipity that named the USAID project that sponsored this conference, Pediatric Oncology Outreach to Hungary, the POOH project. It is to this redoubtable bear, his friend Christopher Robin, and the children of our small world that this volume is dedicated.

Molecular Genetic Basis of Cancer Development

LÁSZLÓ KOPPER,[a] ISTVÁN PETÁK, AND
ANNA SEBESTYÉN

*First Institute of Pathology and Experimental Cancer Research
Semmelweis University of Medicine
Budapest, H-1085, Hungary*

Life is based on the continuous flow of information among cells and inside a cell (inter- and intracellular communications). To maintain the structural and/or functional integrity of a multicellular organism, the cells should actively respond to physiological or pathological information by receiving and generating signals (e.g., the phosphorylation and dephosphorylation of hundreds of proteins by thousands of kinases and phosphatases) that can ultimately lead to a decision, that is, activation of a genetic program.

Although the signals are using seemingly independent pathways, there are many connections between them, forming a complex network, serving to fine-tune the signals (specificity, amplification, regulation), and, at the same time, making our effort to understand this complexity rather unrealistic. Nevertheless, these decision-making processes and genetic programs should be understood in order to identify the key genetic defects in human diseases, including cancer, and to design proper gene-based diagnosis and therapy.

The basic decisions at the cellular level are either do nothing (i.e., signals inhibit each other without activating genes) or (b) do something (i.e., activating programs for proliferation, cell death, or to exercise specific, differentiated functions).

If everything runs smoothly, the complexity mentioned above remains unnoticed. Problems can appear if the afferent or efferent components of the decision-making process, including the genes, acquire one (or more) defect, which could be caused either by exogenous (environmental), or endogenous factors (reflecting the extreme difficulty in completing these programs without mistakes as well as inherited defects). In reality the number of DNA defects per cell per day are about 80,000 (due to oxidation, depurination, depyrimidination, cytosine deamination, single-strand breaks, O6-methylguanine production). Most of these defects are without consequences (the defects are repaired, neutral, or recessive), but if not, their effect could be positive (contributing to the evolution of the species) or negative (contributing to the development of a disease in an individual).

Cancer is a failure of the regulated cell functions, mainly due to defects in the decision-making programs, resulting in a continuous accumulation of cells without obvious purpose, but with increasing autonomy by a selection of the most vital (or resistant) clone(s), and with the capacity to invade the host tissues. The continuous accumulation of cancerous cells clearly indicates that the balance between arising and dying cells has been destroyed by means of a multistep disorganization of the regulation of cell proliferation and/or cell death. The stepwise progression is caused

---

[a] Address for correspondence: László Kopper, M.D., Ph.D., D.Ac.S., Üllői út 26, Budapest, H-1085, Hungary; tel/fax (36)(1) 117 0891; e-mail: kopper@korb1.sote.hu

by the appearance of more and more defects in the genome with time (one defect, by creating genetic instability, can help the development of the next) and can be further supported by a decline in repair capacity. (Frequently, it is also the result of a gene defect.)

In cancer all kinds of genetic defects/errors have been observed: for example, DNA mutations (point mutations such as transition, transversion, and frame-shift, leading to missense, nonsense, or neutral mutations; deletion, insertion; duplication; inversion) and chromosomal mutations (accompanied by rearrangement due to deletion or translocation or as numerical aberrations). Obviously, many other factors can influence gene activity, for example, state of methylation.

These genetic defects have significant importance in tumor formation and progression especially when the targets are those genes that have the key role in the regulation of cell proliferation and cell death as well as in the repair. These well-known genes and their products are (didactic classification with much overlap): (a) activated (defected) proto-oncogenes, that is, oncogenes; (b) suppressor genes; (c) growth factors/cytokines and their receptors; (d) signal-transducers, for example, receptors, protein kinases, transcription factors; and (e) repair enzymes. It seems that the accumulation of gene defects is generally more important than their exact order. It will lead to the selection of those clones which in that particular microenvironment and essentially in that particular host have the greatest growth advantge. Further selection pressure(s) could be introduced by the therapy.

If one lists the frequent malignancies in adults and children the difference is obvious. In adults the most frequent tumor types are either epithelial (bronchial, colonic, gastric, oral) or hormone related (breast, prostate), whereas in childhood leukemias/lymphomas, brain tumors, quite unique types such as neuroblastoma, Wilms' tumor, retinoblastoma, or rhabdomyosarcoma, bone tumors are the most frequent. The lists suggest that the difference in tumor types is probably due to different etiology: the genetic defects in adults are caused mainly (not exclusively) by exogenous factors, whereas the inherited genetic errors play a more important (again not exclusive) role in pediatric tumors. Similarly, in adults chronic physical or chemical irritations, inflammations, hormonal influences are seen, whereas in children inherited diseases and malformations are the most significant risk factors.

ERRORS IN REGULATORY GENES

Oncogenes are the activated forms of proto-oncogenes, which are a mixture of genes (and gene products) and represent widely different components of signal transduction. They could be growth factors (e.g., *sis*); growth factor receptors (e.g., EGFR); membrane-bound tyrosine kinases (e.g., *src, abl*), G-proteins (e.g., *ras*); cytoplasmic serine/threonine kinases (e.g., *raf, mos*), or transcription factors (e.g., *myc, myb, fos, jun*). Most proto-oncogenes support cell proliferation; therefore, mutations, amplifications, and translocations are common (deletion of a proto-oncogene is not "useful" for cancer). The genetic error is usually dominant and leads to the production of a continuous signal for proliferation without the need for the usual regulatory factors (e.g., truncated EGFR can miss EGF, mutated *ras* remains activated, fused *abl/bcr* escapes from gene regulation, or translocated *myc* produces normal products but in an inappropriate time frame and quantity). Transcription factors, especially *myc*, can also act as cell death inducers (see below). Activated forms of known oncogenes were shown in about 15–20% of human malignancies, with a great variety among tumor types and oncogenes.

Suppressor genes represent the negative arm in the regulation of cell proliferation. Therefore, those defects are especially important that result in the loss of the activity (deletion, mutation, translocation; here, amplification provides no advantage for cancer). The defect is usually recessive, except the dominant p53 defect; that is, both alleles should be damaged. The loss of the activity of the second, still-working allele causes loss of heterozygosity (LOH). p53 is a very common suppressor gene, although it has been discovered to be an oncogene, because of the capacity of certain viral proteins to block suppressive gene products by complex formation. The main duty of p53 is to guard the genome integrity and stop the proliferation in case of DNA mutations, allowing time for the repair, usually via activating p21[WAF1] and therefore inhibiting cyclin-dependent kinases. But if the defect is unrepairable, p53 should trigger apoptosis. If p53 is not working properly (the gene is mutated, deleted, or the normal p53 protein is inhibited by other proteins without regulatory reasons, e.g., by mutated mdm2 or viral proteins), the most severe consequence is the appearance of the gene defects in the daughter cells. (Mdm2 expression can also be induced by p53, suggesting the possibility of self-regulation.) The defect in p53 activity is one of the most common genetic errors (together with Rb) in human malignancies.[1,2] The retinoblastoma (Rb) gene is a prototype of the suppressor genes and normally is the main supervisor of the R (restriction) point in G1. pRb inhibits the traverse of cells through the cell cycle by complexing with the E2F transcription factor. Phosphorylation of pRb (e.g., by cyclinD/cdk4) releases E2F, and the genes of the proliferation program are activated. The inactivation or loss of pRb can do the same, leaving the cell with uncontrolled proliferation.[3] (The regulation of p53 and Rb is much more complicated than described, with many exceptions such as p53- or pRb-independent pathways in different cell types.) There are many other suppressors with more or less defined functions: inhibitors of cyclin-dependent kinases (CKI),[4] such as p21 (*WAF1*), p16 (*MTS1*; inhibitor of cdk4/6), p15 (*MST2*), p27 (*kip1*; TGFβ responsive), p57 (*kip2*); neurofibromatosis genes (*NF1*, G-protein modulator; and *NF2*), Wilm's tumor gene (*WT1*), *APC*, *MCC*, *DCC*, (mainly in colorectal cc), *BRCA* 1 and 2 (mainly in breast and ovarian cc), *VHL*, etc.

Transcription factors (TF; there are more than a hundred) form usually homo- or heterodimers (e.g., *myc* should dimerize with *max* to be activated; interestingly, *max* has no oncogenic form) and can bind to a specific segment of DNA and modulate gene expression. Certain TFs have a general function; they are needed for the expression of many genes, for example, TATA-factor, AP-1, and CREB proteins. Others are more cell type–specific: for example, steroid receptors, WT1, hepatocyte nuclear factors, MyoD, and myogenin. TFs frequently appear as part of translocations, especially in leukemias and lymphomas[5] (TABLE 1). The new—either non-fused or fused—gene product has usually TF activity, too.

It has been suggested, more than 30 years ago, that cell loss rate has a profound effect on tumor growth. Obviously, cells could be killed "passively" by hypoxia or by other, partly therapeutic, cytotoxic effects, slowing down the increase in tumor volume (tumor-doubling time). However, the emphasis here is on active cell death, which is not an effective contributor to cell loss in tumors. On the contrary, the defect in the active cell death machinery (*apoptosis*) can lead to longer cell survival or even to the continuous accumulation of cells. The first clear example was given by showing the role of overexpression of *bcl*-2 in follicular non-Hodgkin's lymphomas, due to translocation: t[14;18].[6] Since then, apoptosis research has become very popular, proving that almost anything can switch on the apoptotic cascade, and if one trigger failed in a certain cell type, another can do the job.[7,8]

TABLE 1. Transcription Factors in Translocations of Hemopoietic Malignancies (Examples)

Translocation	Involved Genes	Diseases[a]
t[8q24;14q32]	*myc*; Ig heavy chain	
t[8q24;22q11]	*myc*; Ig λ light chain	BL, B-ALL (L3)
t[2p12;14q32]	Ig κ light chain; *myc*	
t[9p21-22;14q11]	p16INK4/MTS1 and TCRα/δ	B-ALL
t[1q23;19p13]	PBX1;E12/47(E2A)	pre-B-ALL
t[17q22;19p13]	HLF;E12/47	preB-ALL
t[1p32;14q11]	tal-1;TCRα/δ	T-ALL
t[7q35;19p13]	TCRβ;lyl-1	T-ALL
t[10q24;14q11]	HOX11;TCRδ	T-ALL
t[11q23;19p13.3]	MLL/ENL (mainly ALL) or MLL/ELL (mainly AML)	
t[15q22;17q12-21]	PML;RARa	APL (M3)
t[6p23;9q34]	DEK;CAN	AML
t[8q22;21q22]	AML1/ETO	AML

[a] Abbreviations: BL, Burkitt's lymphoma; ALL, acute lymphocytic leukemia; B, B-cell origin; T, T-cell origin; APL, acute promyelocytic leukemia; AML, acute myelogenous leukemia.

There is no question that apoptosis is an important, evolutionarily conserved process and that it has many distinct morphological and biochemical features. Strictly speaking, apoptosis is not equivalent to programmed cell death, which does not evoke apoptosis, for example, during physiological spermatogenesis; apoptosis is also not always triggered by nuclear activities but may be started by cytoplasmic constituents, for example, fas, UV, ceramide. It would be impossible in this space to count all apoptosis-inducing agents, so only a few of them will be mentioned.

The Bcl-2 family has several members, including inhibiting (bcl-2, bcl-x_L, mcl-1, and bcl-w) and some viral components (adenovirus E1B19K, EBV BHRF1), as well as stimulating (bax, nbk/bik1, bad, bak, bcl-x_S) apoptosis. Members of the bcl-2 family can form dimers, which usually require their BH1 and BH2 domains. In most eukaryotic cells, there is substantial redundancy in the expression of the bcl-2 family members, and the final decision depends on the regulation of the actual balance between anti- and pro-apoptotic molecules. However, some cells prefer to use one particular member as the survival factor (e.g., peripheral B cells: mcl-1; Reed-Sternberg cells: bcl-x_L). In many studies the progression and prognosis of cancer patients has shown correlation with increased expression of bcl-2 or decreased expression of bax. The function of bcl-2 is still unclear, but its cellular localization (mitochondrial membrane) suggests it has a role in the regulation of oxygen tension. As previously mentioned, wild-type p53 can activate the apoptotic program (triggered by DNA mutations, caused by agents such as irradiation and cytotoxic drugs), but other stimuli do not require the presence of functional p53 to induce apoptosis, or even disregard whether p53 is wild-type or mutant. *Myc*, a nuclear transcriptional activator and powerful stimulator of cell proliferation, in certain circumstances can induce apoptosis, especially when the cells are confused by opposite signals, and when proliferative signal arrives but the survival factor is actually missing. Ligands like FasL and TNF can induce apoptosis by binding to the receptors *Fas/APO1* (CD95) and *TNFR1*, respectively. The signal is transmitted through the "death domain" of the cytoplasmic part of the receptors. The next step is, probably, the activation of the cysteine proteases, especially the ICE family (interleukin-1β-converting enzymes), whose potential targets besides IL1β are PARP (poli[ADP]ribose polymerase), lamins, U1snRNP, and the proteases themselves. PARP is also a guard of genomic integrity and at the same time is a potent regulator of the Ca^{2+}, Mg^{2+}-dependent endonucleases. If the proteases inactivate PARP, it is unable to work for the integrity of the cell, and apoptosis can start. ICE overproduction enhances, whereas inhibition prevents apoptosis.[9]

ERRORS IN OTHER GENES

The errors (mismatches) in the newly synthetized DNA should be repaired by the mismatch repair system (MMR). The insufficient repair capacity can be recognized by studying microsatellites (tandem-repeated short sequences, which are very polymorphous in a population but identical in one person). Errors in microsatellites are described as microsatellite instability (MIN), which reflects the decreased potential of the whole genom to handle DNA errors (mutator phenotype). A defect in the MMR system is understood today to be related to the errors of four genes: hMSH2, hMLH1, hPMS1, and hPMS2. In HNPCC (hereditary non-polyposis colon cancer) families, the mutation of these genes appears in germ cells. Although MIN is a characteristic lesion in HNPCC, it occurs in many other tumor types. The lack of MMR activity can contribute to the resistance against alkylating agents.

It has been recognized that in certain genes the expression of the alleles is related to their maternal or paternal origin (genetic imprinting), which is probably regulated by the methylation state of the allele.[10] The loss of imprinting (LOI) could play a role in tumorigenesis. For example, in certain pediatric tumors—Wilms' tumor, rhabdomyosarcoma, hepatoblastoma—deletion appears in maternal chromosome 11; in AML in the paternal chromosome 7; in osteosarcoma in maternal chromosome 13; and in CML the translocation occurs between paternal chromosome 9 and maternal chromosome 22.

SOME RECENT RESULTS ON PEDIATRIC TUMORS

Pediatric tumors can carry the prototype of certain, sometimes specific, genetic errors. Insulin-like growth factor 2 (IGF2) has recently been demonstrated to be maternally imprinted in both mice and humans. LOI of IGF2 in rhabdomyosarcoma (RMS) has been reported, where IGF2 acts as an autocrine growth factor. (In Ewing sarcoma, however, LOI of IGF2 was not associated with increased expression of IGF2 mRNA, suggesting that LOI may not be involved in the regulation of IGF2 expression.) Embryonal and alveolar RMS showed *c-met* expression, in some cases amplification. In another study alveolar RMS cells had t[2;13][PAX3;FKHR], resulting in the tumor-specific expression of a transcriptional factor. PAX3 can stimulate *c-met* oncogene expression; the *c-met* oncogene is a receptor for the ligand HGF (hepatocyte growth factor/scatter factor). HGF controls cell motility and invasion.

The Wilms' tumor 1 gene (WT1) encodes a transcription factor of the zinc finger family. WT1 represses transcription of several growth factors and growth factor receptors as well as transcription of bcl-2 and c-myc.[11] WT1-induced apoptosis in an osteosarcoma cell line was accompanied by decreased synthesis of EGFR, but not of other postulated WT1-targeted genes. This effect, which was differentially mediated by the alternative splicing variants of WT1, was independent of p53.

In 29% of medulloblastomas LOH of 17p (loss in the telomeric region) was observed without the involvement of the p53 gene and indicated a worse prognosis.[12] In the pathogenesis of ependymomas a—probably suppressor—gene on chromosome 22 could be involved (it is not EWS or NF2). However, NF2 showed responsibility for autosomal, dominant appearance of multiple CNS tumors, including schwannomas, meningiomas, and ependymomas.

Neuroblastoma (NBL) is a tumor derived from cells of the neural crest, with a widely variable outcome. NBL was among the first tumors for which the relationship between the expression of an oncogene (*N-myc*) and the prognosis was discovered. Besides amplification of *N-myc* (increased expression without amplification was not correlated with poor prognosis), LOH of several chromosomal loci occur in NBL, representing genetic instability. However, microsatellite instability is infrequent. The *N-myc* replicon is rather large (350 kb-1 Mb); therefore presence of co-amplified genes was suggested. The co-amplification and concomitant high level of expression of a DEAD box gene with *N-myc* was observed in human NBL. DDXI belongs to a family of genes that encode DEAD (Asp-Glu-Ala-Asp) box proteins, which are putative ATP-dependent RNA helicases implicated in several cellular processes involving alterations of RNA secondary structures.[13] It was also found that NBLs express the neurotrophin receptors TrkA and TrkB. Expression of TrkA and TrkC correlates with favorable outcome, whereas expression of full-length TrkB is associated with unfavorable, more aggressive NBL.[14] Recently, a high level of nucleoside diphosphate kinase A (NDPK A/nm23-H1) in NBL is associated with advanced stage disease.[15] There is no question that NBLs demonstrate both clinical and biological heterogeneity. Three genetically distinct subtypes can be separated[16]: (1) hyperdiploid or triploid karyotypes; 1p LOH and *N-myc* amplification absent, *TrkA* expression is high; usually infants are involved; stage is I, II, or IVS; prognosis is good (>90% cure); (2) near diploid or tetraploid karyotypes, no *N-myc* amplification; 1p allelic loss, 14q allelic loss or other structural changes are present; *TrkA* expression is low; patients are older; in stage III or IV, prognosis is bad (25–30% cure); (3) near diploid or tetraploid karyotype; *N-myc* amplification and 1p allelic loss are present; *TrkA* expression is low or absent; age 1–5 years; advanced stages (III, IV); very poor prognosis (<5% cure).

These are just examples. The molecular genetics of malignant tumors is a rapidly expanding area with a huge amount of new, sometimes conflicting information. We are now at the learning phase, and our real challenge is figuring out how to take advantage of the daily increasing knowledge and the opportunities offered by the molecular studies.

REFERENCES

1. GREENBLATT, M. S., W. P. BENNETT, M. HOLLSTEIN & C. C. HARRIS. 1994. Mutations in the p53 tumor suppressor gene: Clues to cancer etiology and molecular pathogenesis. Cancer Res. **54:** 4855–4878.
2. GOTTLIEB, T. M. & M. OREN. 1996. P53 in growth control and neoplasia. Biochim. Biophys. Acta **1287:** 77–102.
3. BEISERSBERGEN, R. L. & R. BERNARDS. 1996. Cell cycle regulation by the retinoblastoma: A family of growth inhibitory proteins. BBA **1287:** 103–120.
4. HALL, M., S. BATES & G. PETERS. 1995. Evidence for different modes of action of cyclin-dependent kinase inhibitors: p15 and p16 bind to kinases; p21 and p27 bind to cyclins. Oncogene **11:** 1581–1588.

5. RABBITS, T. H. 1994. Chromosomal translocation in human cancer. Nature **372:** 143–148.
6. KORSMEYER, S. J. 1992. Bcl-2: An antidote to programmed cell death. Cancer Surv. **15:** 105–118.
7. KERR, J. F. R., A. H. WYLLIE & A. R. CURRIE. 1972. Apoptosis: A basic biological phenomenon with wide-ranging implications in tissue kinetics. Br. J. Cancer **26:** 239–257.
8. HALE, A. J., C. A. SMITH, L. C. SUTHERLAND, V. E. STONEMAN, V. L. LOGTHORNE, A. L. CULHAME & G. T. WILLIAMS. 1996. Apoptosis: Molecular regulation of cell death. Eur. J. Biochem. **236:** 1–26.
9. PATEL, T., G. J. GORES & S. H. KAUFMANN. 1996. The role of proteases during apoptosis. FASEB J. **10:** 587–597.
10. LATHAM, K. E., J. MCGRATH & D. SOLTER. 1995. Mechanistic and developmental aspects of genetic imprinting in mammals. Int. Rev. Cytol. **160:** 53–98.
11. HEWITT, S. M., M. C. HAMADAS, T. J. DONNELL, F. J. RAUSCHER III & G. F. SAUNDERS. Regulation of the proto-oncogenes bcl-2 and c-myc by the Wilms' tumor suppressor gene WT1. Cancer Res. **55:** 5386–5389.
12. BATRA, S. K., R. E. MCLENDON, J. S. KOO, S. CASTELINO-PRABHU, H. E. FUCHS, J. P. KRISCHER, H. S. FRIEDMAN, D. D. BIGNER & S. BIGNER. 1995. Prognostic implications of chromosome 17p deletions in human medulloblastomas. J. Neurooncol. **24:** 39–45.
13. MANOHAR, C. F., H. R. SALWEN, G. M. BRODEUR & S. L. COHN. 1995. Co-amplification and concomitant high levels of expression of a DEAD box gene with *Myc-N* in human neuroblastoma. Genes Chromosomes Cancer **14:** 196–203.
14. YAMASHIRO, D. J., A. NAKAGAWARA, N. IKEGAKI, A. G. LIU & G. M. BRODEUR. 1996. Expression of TrkC in favorable human neuroblastoma. Oncogenes **12:** 37–41.
15. CHANG, C. L., J. R. STRAHLER, D. H. PHORAVAL, M. QIAN, R. HIDERER & S. M. HANASH. 1996. A nucleoside diphosphate kinase A (nm23-H1) serine 120→glycine substitution in advanced stage neuroblastoma affects enzyme stability and alters protein–protein interactions. Oncogene **12:** 659–667.
16. BRODEUR, G. M. 1995. Molecular basis for heterogenity in human neuroblastomas. Eur. J. Cancer **31:** 505–510.

Diagnostic and Prognostic Significance of Chromosome Abnormalities in Childhood Acute Lymphoblastic Leukemia[a]

É. OLÁH,[b,c] E. BALOGH,[c] P. KAJTÁR,[d] L. PAJOR,[d]
Z. JAKAB,[c] AND C. KISS[c]

[c]Department of Pediatrics
University Medical School of Debrecen
H-4012 Debrecen, P.O. Box 32
Hungary

[d]Department of Pediatrics and
Institute of Pathology
Medical School of Pécs
Pécs, Hungary

INTRODUCTION

Current intensive chemotherapy cures about 70% of children with acute lymphocytic leukemia (ALL). Although this percentage is relatively high, it means that a major proportion of the children with ALL is still not cured despite intensive treatment. At the same time some highly curable patients are treated too intensively and suffer from unnecessary side effects of the chemo- and radiotherapy applied, for example, anthracyline-related cardiomyopathy, infertility, and second malignancies associated with alkylating drugs and epipodophyllotoxins. In order to further improve the therapeutic results and reduce the therapy-related toxicity in childhood ALL, we have to distinguish between the cases with a better and those with a worse prognosis at the time of determining the diagnosis. The initial cytogenetic changes (both structural and numerical abnormalities) proved to be one of the most reliable prognostic parameters leading to the suggestion of developing genotype-specific therapy.[1-11]

In the spontaneously dividing blast cells of 70–75% of ALL children, acquired clonal chromosome abnormalities can be detected at the time of the diagnosis.[10-14] About half of these abnormalities seem to be specific to the tumor type involved. The cell specificity in this disease means that there is a close correlation between the aberrations and the immunophenotype of the blasts. Among the karyotype changes structural aberrations and numerical disturbances can be distinguished.

Structural abnormalities, mostly translocations and deletions involving specific breakpoints, are present in two-thirds of patients with abnormal karyotypes de-

[a] Supported by grants from Hungarian Ministry of Welfare (ETT/220).

[b] Address for correspondence: Éva Oláh, M.D., D.Sc., Department of Pediatrics, University Medical School of Debrecen, Debrecen H-4012, P.O. Box 32, Hungary. Telephone/Fax: (36) 52 414 992; e-mail: olah_e@gyermek.dote.hu

scribed to date. At present more than 15 different, consistently occurring structural rearrangements are known in ALL.[11,15–17] Most of them are specifically associated with distinctive immunophenotypes and are of particular clinical (diagnostic and prognostic) importance (TABLE 1).[5,8,11,12,18,19] A molecular analysis of the breakpoints involved in these aberrations provides an insight into the molecular mechanisms of the pathogenesis of the disease.[12,20–22]

Patients without ALL-specific structural abnormalities can be classified in terms of the modal chromosome number into five subgroups: *diploid* (46 chromosomes), *pseudodiploid* (46 chromosomes with structural or numerical abnormalities), *hypo-diploid* (less than 46 chromosomes), *hyperdiploid A* (47 to 50 chromosomes) and *hyperdiploid B* (more than 50 chromosomes).[8,9,12,14] The latter group can be further subdivided into the *hyperdiploid* (50–57 chromosomes), the *near-triploid* (58–80 chromosomes), and the *near-tetraploid* subgroup (81–103 chromosomes).[23]

Hyperdiploidy with a chromosome number between 50 and 57 with a mean peak at 55 chromosomes can be considered as a distinct karyotype pattern occurring in about 25–30% of childhood ALL.[10–12,14,23] In Sweden Heim and his co-workers found more than 50 chromosomes in 49% of ALL patients with abnormal karyotype.[24] The main feature of this group is the presence of nonrandom numerical abnormalities due to a gain of chromosomes 4, 6, 10, 14, 17, 18, 21, and X.[5,10,14] In contrast, chromosomes 1, 2, 3, 12, and 16 are rarely involved.[10,14]

Near-triploid and near-tetraploid cases are rare and account only for 1–1.5% of all childhood ALL cases.[23] Although near-triploid cases do not appear to differ from the general ALL population in clinical features, the near-tetraploid patients more often show a T-cell phenotype and have a poorer prognosis.[23]

Because these structural and numerical abnormalities together with genetic changes appearing at the molecular level determine the biological features of the blasts and the morphological, cytochemical, and immunological features as well as

TABLE 1. Most Common Structural Abnormalities Related to Typical Immunophenotype

Typical Immunophenotype	Rearrangement
B-cell ALL	t(8;14)(q24;q32)
	t(2;8)(p12;q24)
	t(8;22)(q24;q11)
Pre-B-cell ALL	t(1;11)(p32;q23)
	t(1;19)(q23;p13)
	t(9;22)(q34;q11)
Early B precursor	t(4;11)(q21;q23)
	t(9;22)(q34;q11)
	del/t(9)(p21)
	dic(9;20)(p1?3;q11)
Common ALL	t(9;22)(q34;q11)
	del/t(12p)
T-cell ALL	t(8;14)(q24;q11)
	t(10;14)(q24;q11)
	t(11;14)(p13;q11)
	del/t(9)(p21)
	other t/del(14)(q11)
Mixed phenotype	t(4;11)(q21;q23)
	t(9;22)(q34;q11)

NOTE: Due to cell specificity, the aberrations are of diagnostic significance.

TABLE 2. Cytogenetic Subgroups with Different Prognosis in ALL at Diagnosis

Karyotype		
> 50 chromosomes	< 46	t(4;11)
normal	t(9;22)	t(8;14), including variants
del(6q)	14q+	t(1;19)
46 abnormal		t(12;21)
47–50		

NOTE: Complete remission rates and complete remission durations are different in these cytogenetic subgroups.[12]

the tumor's invasivity, metastatic capacity, and response to therapy, it is evident that these genetic changes are of clinical significance. Their detection gives invaluable assistance to clinicians in confirming the diagnosis, subclassification of the disease, predicting the prognosis, and making the therapeutic decision.

From a clinical and therapeutic point of view, the prognostic implication of the initial karyotype is of utmost importance. Several cytogenetic subgroups with different prognoses can be distinguished (TABLE 2). Of these, three main subgroups should be pointed out (TABLE 3). The t(9;22), and t(4;11) subgroups are associated with an unfavorable prognosis and serve as an indication for bone marrow transplantation. In contrast, the hyperdiploidy (with >50 chromosomes and/or >1.16 DNA index) proved to be the strongest predictor of a favorable course of the disease and of a good response to therapy.[1,6,12,25] These patients can be characterized by favorable clinical and laboratory parameters, by a lower mean age, a usually moderate tumor load, relatively low initial white blood cell count, and an early B-cell precursor immunphenotype.[26] They can achieve remission on a metabolite-based therapy; therefore the toxic effect of a more intensive treatment with anthracycline or genotoxic agents may be avoided. There are data available currently suggesting that the prognostic value of hyperdiploidy may further be related to the particular karyotype composition, namely the chromosomes involved in the aberrations and/or the structural abnormalities associated.[9] For example, trisomy 6, or the combined trisomy 4 and 10, has proved to indicate an extremely favorable prognosis.[27,28] In contrast, recent data suggested that there is a small subset of patients with hyperdiploidy associated with chromosome 7 aberrations that has a relatively high relapse rate.[29] Structural aberrations added to numerical ones makes

TABLE 3. Three Main Prognostically Different Cytogenetic Subgroups of ALL Patients

Unfavorable	Favorable	Others
t(9;22)	Hyperdiploidy	Diploidy
t(4;11)	t(12;21)	Translocations
		Deletions
↓	↓	↓
Bone marrow transplantation	Less aggressive chemotherapy	Genotype-specific chemotherapy

NOTE: Presence of t(9;22) and t(4;11) serves as indication for bone marrow transplantation. Hyperdiploid karyotype with >50 chromosomes seems to be associated with a favorable prognosis. All the other patients, with various cytogenetic changes of different prognoses, belong to the third group. These patients need genotype-specific therapy.

the prognosis worse.[5] All other patients left with various cytogenetic abnormalities belong to the third group. These patients require a genotype-specific, individually chosen therapy.[30,31] Based on these observations, the initial cytogenetic data may serve as a guide for the stratification of therapy, for the choice of the genotype-specific, effective but not too agressive chemotherapy for each individual. Foremost, a reliable and accurate determination of patients with hyperdiploid karyotype and of their particular karyotype composition is becoming an increasingly important task.

Several laboratory methods have been proved to be useful for determination of leukemia-associated initial genetic alterations, first of all of hyperdiploidy in childhood ALL. To identify the patients with more than 50 chromosomes, three standard methods can be used: the classical cytogenetic analysis, flow cytometry, and interphase fluorescence *in situ* hybridization (FISH). A recently developed FISH technique, the so-called comparative genomic hybridization (CGH) offers a new possibility to determine all chromosomal gains and losses in tissues that are otherwise not suitable for traditional cytogenetic analysis.[32] Unfortunately, none of these methods give complete information about all genetic changes. Each of them has both advantages and disadvantages, which are explained as follows.

Obviously, the chromosome analysis provides the most complete information about the karyotype changes, about the numerical and structural aberrations, the chromosomes involved in the aberrations, the different cell lines, and so forth. On the other hand, it requires a great number of dividing cells and the commonly poor quality of abnormal metaphases often hampers accurate evaluation. In addition, it is occasionally difficult to decide whether the abnormalities detected in the few analyzable metaphases are representative of the whole leukemic clone. A further disadvantage of this method is that it requires a lot of expertise and is a rather time-consuming procedure.

To identify the relative DNA content of the blasts and ploidy level, a simple and rapid technique, flow cytometry, can be used. However, it gives information only about the gross quantitative deviations of DNA content.[25,33,34] No information about structural chromosome aberrations, the individual chromosomes, and the different cell lines is provided. Using this technique alone, diploidy and pseudodiploidy cannot be distinguished. The main disadvantage of the method is that in cases of trisomies of the smallest chromosomes the hyperdiploid chromosome number (>50 chromosomes) may be associated with a DNA index less than 1.16. As a result of this, about 30% of hyperdiploid patients are lost when we use flow cytometry as the only method.[5,35]

The two methods mentioned can be complemented by the fluorescence *in situ* hybridization (FISH) techniques offering a strong possibility of identifying some cytogenetic abnormalities even in interphase nuclei.[36–41] The method is suitable for screening a large number of cells. Applying centromere-specific probes it makes possible for us to identify the extra chromosomes in hyperdiploid ALL patients. However, the number of simultaneously applicable probes is limited, and some of the structural abnormalities cannot be detected in this way.

In order to identify the initial genetic abnormalities in childhood ALL as accurately as possible, the three methods—conventional chromosome analysis, flow cytometry, and occasionally the complementary FISH technique—should be used simultaneously.

In 1993 it was decided to develop a comprehensive, nationwide project in order to perform an initial genetic analysis on all ALL children diagnosed in the hematological/oncological centers of Hungary. Before starting with this project, it seemed to be important to evaluate the results of all genetic investigations performed on ALL children in Hungary until that time and, on the basis of the results obtained, prepare a proper strategy for our future work. Chromosome findings of 165 previously investigated patients were collected and evaluated. Comparing the results to those in the literature we come to the following conclusions:

Cytogenetic analyses were carried out only in the minority of ALL children (30%). Clonal chromosome aberrations were found in 40% of patients studied. The proportion of hyperdiploid patients with more than 50 chromosomes proved to be 12% instead of 30% of the expected proportion based on a large series.[10–12,23,40] The frequent involvement of chromosomes 8, 11, and 13 in the trisomies differed from data of other authors.[5,10,14,29] The survival data of patients with hyperdiploidy did not demonstrate a favorable prognosis.

All these clinical and cytogenetic differences were suggested to be due to technical problems, as well as the small number of patients investigated, but the role of real geographic differences of prognostic importance was also considered. That is why it was important to study the genetic characteristics of ALL patients in our country and compare them to observations made by other authors. The aim of our study was to answer the following questions:

- What is the frequency of clonal chromosome aberrations in ALL children at diagnosis in Hungary?
- What is the distribution of the leukemic children in terms of the modal chromosome number?
- What is the frequency of hyperdiploid (>50 chromosomes) cases at diagnosis?
- Which chromosomes are most frequently involved in the numerical and structural aberrations?
- Is there any difference in the frequency of hyperdiploid cases and in the spectrum of chromosomes involved as compared with the data in the literature?
- What is the frequency of additional structural aberrations in hyperdiploid patients?
- How can the patients with more than 50 chromosomes be characterized from a clinical and immunological point of view?
- Is there any correlation between the cytogenetic findings and the prognosis?
- From a methodological point of view, we planned to study the reliability of the three methods in order to determine a more effective strategy for future investigations.

In spite of the fact that the present situation in health service in Hungary is not favorable either in terms of financial support or instrumental and personnel facilities of genetic laboratories for the introduction of any new investigations, great efforts have been made in order to perform as many genetic analyses on ALL patients as possible in Hungary. Here we summarize our results, which were obtained during the period from 1993 to 1995, and we will also try to outline some guidelines for our future work.

PATIENTS AND METHODS

Diagnosis and treatment of children with malignant diseases have been performed for more than two decades at 10 hematological/oncological centers of the

Hungarian Pediatric Oncology Working Group established in 1971.[e] The diagnostic and therapeutic work of these centers is coordinated by the main center in the Department of Pediatrics No. 2. of Semmelweis Medical School, Budapest; and the work is based on standardized diagnostic and therapeutic principles.

Patients

Each ALL patient diagnosed in one of the 10 centers is registered, and the diagnosis of each case is confirmed by the coordinating center in Budapest. Cytochemical, immunologic, and genetic analyses are performed at the local laboratories of hematological centers. The treatment of patients included in this study was carried out in accordance with the "BFM 90" regimen recommended by the German Berlin–Frankfurt–Münster (BFM) Study Group.

Regarding the genetic analysis, the following strategy was recommended: To identify the hyperdiploid patients, cytogenetic analysis and flow cytometric evaluation should be performed simultaneously. In cases of successful chromosome analyses, no further investigation is needed. However, if the cytogenetic investigation fails to provide accurate information about the karyotypic changes and the DNA index is above 1.0, FISH technique is recommended for the identification of supernumerary chromosomes. Considering the chromosomes mostly involved in numerical aberrations, the application of probes specific for the centromeric region of chromosomes 4, 6, 10, 17, 18, 21, and X seems to be necessary. Besides, based on our earlier experience, it seemed worthwhile to complete this spectrum of chromosomes with probes for chromosome 8, 11, and 13.

Clinical, hematological, and genetic data were collected and evaluated by the hematological/oncological center and the genetic working team of the Department of Pediatrics of the University Medical School, Debrecen.

Methods

Cytogenetic Analysis

Unstimulated, isolated bone marrow and/or peripheral blood cells were cultured for 24 hours. Chromosomes were prepared according to standard procedures. Slides were G-banded with trypsin-Giemsa. Karyotyping was performed on a KaryoASK version 4.11 chromosome analysis system. In evaluating the karyotype changes, the International System for Human Cytogenetic Nomenclature (ISCN) was followed.[42]

---

[e] The hematological/oncological centers are the following: Department of Pediatrics No. 2, Semmelweis University Medical School, Budapest; Department of Pediatrics No. 1, Semmelweis University Medical School, Budapest; Heim Pal Children's Hospital, Budapest; Bethesda Children's Hospital, Budapest, Madarasz Children's Hospital, Budapest; Department of Pediatrics, University Medical School of Debrecen, Debrecen; County Children's Hospital, Miskolc; Department of Pediatrics, University Medical School of Pecs, Pecs; Department of Pediatrics, Albert Szent-Gyorgyi University Medical School, Szeged; County Children's Hospital, Szombathely.

Flow Cytometry

For the analysis of DNA content, freshly isolated, ethanol-fixed cell pellets were prepared. Fixed cells were resuspended in a solution containing 3% citric acid and 0.5% Tween 20 and incubated for 20 min at room temperature with slow agitation on a shaker board. Nuclei were passed through a 30-μm nylon mesh to remove cell clumps and centrifuged at 1200 rpm for 8 min. The cell pellet was resuspended in 70% ethanol. After two hours at 20°C, the material was centrifuged again, and the pellet suspended in 0.3 ml 0.5% Tween 20 for 10 min at room temperature. Cells were stained with propidium iodide for one hour at room temperature and simultaneously digested with RNAse. The samples were analyzed with a Beckton Dickinson Fascan flow cytometer. As a control, fresh samples from Ficoll-separated peripheral blood lymphocytes from healthy donors were used (CV < 2.5%). At least 20,000 nuclei were analyzed and plotted. A DNA index < 0.95 is considered as hypodiploidy, and a DNA index > 1.05 means hyperdiploidy. The hyperdiploidy with more than 50 chromosomes equals a DNA index 1.16.[33]

Fluorescence in situ *Hybridization (FISH)*

Numerical abnormalities were analyzed in interphase cells with probes specific for the centromeric regions of chromosomes 1, 8, 17, 18, 13/21, and X according to standard methods described by Berger *et al.*[38] In some patients a total panel of centromere-specific probes could be applied.

RESULTS

From 1993 until 1995, a total of 187 ALL children were diagnosed in 9 of the 10 hematological/oncological centers in Hungary. In 74% of newly diagnosed cases (140/187), chromosome analysis was performed. The proportion of successful cytogenetic investigations proved to be 55.7% (78/140). In 1994 the ratio of patients studied cytogenetically was rather high (92%), but this value was associated with the lowest ratio of successful investigations (TABLE 4). Flow cytometric evaluation was performed in 31 of 187 patients (16.5%), mostly together with a simultaneous chromosome analysis. FISH technique was applied in 12 cases (6.4%) when chromo-

TABLE 4. Ratio of Cytogenetically Successfully Investigated ALL Cases

	1993	1994	1995	Total
Total No. of ALL Patients	57	52	78	187
No. of cytogenetically investigated cases	39	48	53	140
	(68.4%)	(92.4%)	(67.9%)	(74.8%)
No. of successful analyses	22	21	35	78
	(56.4%)	(43.7%)	(65%)	(55.7%)

NOTE: Number of patients diagnosed and cytogenetically investigated in hematological centers of the Hungarian Pediatric Oncology Study Group in 1993–95. The ratio of successful investigations indicates an improving tendency from 1994 to 1995.

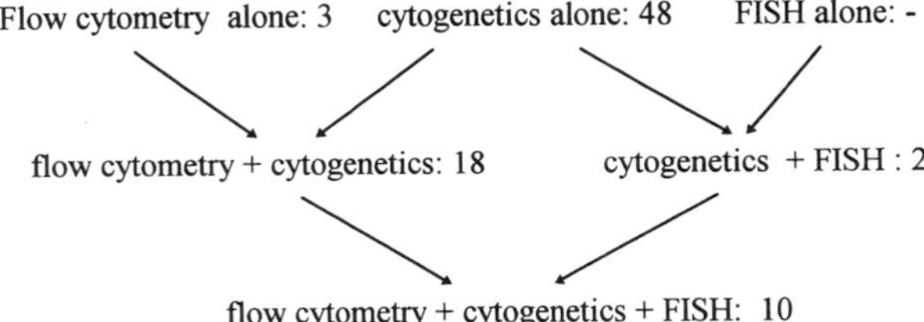

FIGURE 1. Distribution of patients studied genetically according to the methods applied. Total number of chromosome analyses: 78; total number of flow cytometry evaluations: 31; total number of FISH methods: 12. Although the number of chromosome analyses, flow cytometric evaluations, and FISH methods are separately rather low; in 80 of 187 (42.7%) newly diagnosed patients, evaluable information about the genotype could be obtained.

some study failed to offer full information about the genetic changes (FIG. 1). Evaluable information about the genetic abnormality is available in 80 of 187 patients (42.7%).

Chromosome Analysis

Out of 78 cytogenetically investigated patients, 42 (53.8%) proved to have a diploid chromosome set; 11 (14.1%) showed pseudodiploid; four (5.1%) hypodiploid; 8 (10.2%) hyperdiploid A (47–50 chromosomes); and 13 (17.1%) hyperdiploid B (>50 chromosomes) karyotype. The ratio of children with abnormal karyotype is 36/78 (46.1%). (FIG. 2).

Flow Cytometric Evaluation

In 21 of 31 patients (67.7%), the DNA index showed a diploid DNA content. Only one patient had a DNA index less than 1.0, and two between 1.05 and 1.16.

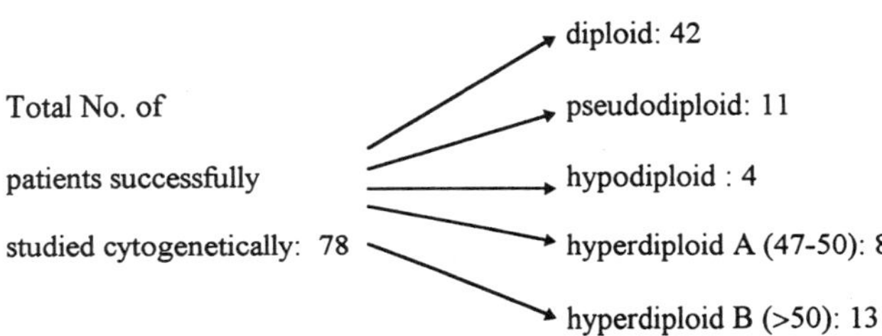

FIGURE 2. Distribution of patients according to modal chromosome number (ploidy). The ratio of patients with abnormal karyotype is rather low: 36 out of 78 (46.1%). Seven of 78 patients studied cytogenetically (16.6%) proved to have >50 chromosomes.

On the basis of flow cytometry, seven patients (7/31, 22.5%) could be classified as hyperdiploid (>50 chromosomes) (FIG. 3).

Taken together, 16 of 80 patients (20.0%) proved to have hyperdiploid (>50 chromosomes and/or DNA index above 1.16) karyotype: eight patients were assessed by chromosome analysis alone, three by flow cytometry alone, and five by using the two methods simultaneously (TABLE 5).

The hyperdiploid A subgroup (47–50 chromosomes) includes eight patients with various karyotype abnormalities (TABLE 6). The supernumerary chromosomes were as follows: +1,+6, +12, +D, +18, + 20. In three of the eight patients structural aberrations could also be detected (TABLE 6, No. 3, 6, 7). In a patient with B-cell ALL a t(8;22) was identified associated with kappa and lambda chain rearrangement. The patient died within three weeks. Median survival of the subgroup is 14.5 months. Three of eight children were lost.

Pseudodiploidy was identified in 11 patients. In 9 of the 11 children, ALL-specific translocations or structural aberrations involving specific breakpoints could be detected. Translocations were associated with the expected immunophenotype: t(1;19) and t(2;12) with pre-B cell ALL, t(9;22) and t(4;11) with biphenotypic (lymphoid/myeloid) diseases, and t(14;11) with a T-cell disease involving one patient each (TABLE 7). Although the exact breakpoints of aberrations were not given, the involvement of the well-known specific regions can be supposed: 7q+: 7q35, del(2p) : 2p11, del(11q): 11q23, del(6q): 6q21, 6q23.

In some patients (TABLE 7, No. 1, 8) combined structural and numerical aberrations were found.

The prognosis of pseudodiploid patients proved to be unfavorable. Median survival was 15 months. Two of them died.

Interphase Cytogenetics

FISH was performed in 12 patients (TABLE 8). In 10 cases centromere-specific probes for all chromosomes were used, whereas in two additional patients (No. 21, 24) probes only for chromosomes 8, 17, 18, 13/21, and X could be applied. In three of six patients with normal karyotype and diploid DNA index (1.0) (No. 11, 12, 13) and in a Down's syndrome case besides the 21-trisomy (No. 19), FISH revealed trisomies and monosomies of a few chromosomes in a small proportion of the cells. In two patients with >50 chromosomes and with a DNA index >1.16 (TABLE 8, No. 23, 25) accurate information about the chromosomes involved could be obtained only by FISH. In two additional cases, when detailed cytogenetic data based on

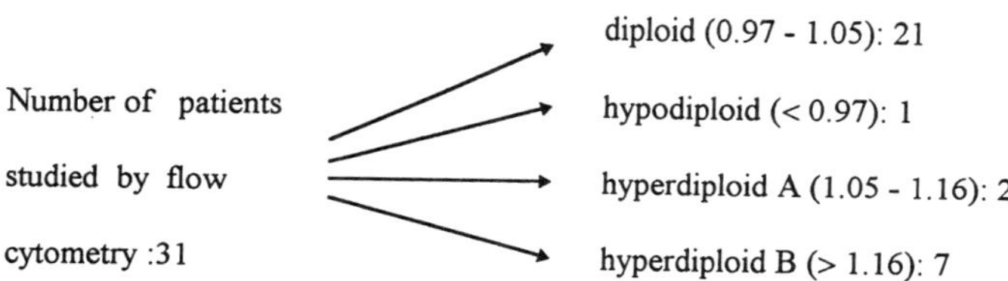

FIGURE 3. Subgroups of patients with different ploidy as indicated by DNA flow cytometry. The ratio of patients with hyperdiploidy (DNA index is above 1.16) detected by flow cytometry proved to be higher than that by chromosome analysis (7/31: 22.5%). The difference indicates that flow cytometry is a useful method for screening patients for ploidy.

TABLE 5. Genetic Data of Patients with Hyperdiploid (>50 Chromosomes/>1.16 DNA Index) Karyotype

No.	Ratio of Hyperdiploid Cells	Numerical Aberrations	Structural Aberrations	DNA Index	FISH
1	55 (53–58) (100%)	+6,+8,+11,+17, +19,+20,+21, +22	–	–	+17 (4%) +21 (35%)
2	51 (51–54)	n.i.	i(17q)	–	–
3	51 (80%)	?+8,+11,+19, +21,+X	–	–	–
4	–	–	–	1.77 (56%) /1.162 (44%)	–
5	–	–	–	1.269	–
6	55 (55–57) (75%)	?+6,+8,+10, +11,+13,+17, +18,+21,+22	–	1.24	–
7	–	–	–	1.28	53,XX,+4,+6,+8, +10,+18, +13/21,+X
8	53 (100%)	+6,+8,+11,+19, +20,+21,+22	–	–	+17 (20%) +18 (30%)
9	54 (20%)	+1,+4,+11,+19, +19,+21,+21, +22	–	–	–
10	50 (50–55) (40%)	–	–	1.19	52,XX,+10,+17, +13/21,+18, +X,+X
11	56 (56–60) (20%)	–	del(11q)	1.13	–
12	58 (5%)	+3,+8,+10,+11, +13,+14,+15, +16,+17,+19, +20,+22		1.0	–
13	52 (50%)	+21 Down	–	–	–
14	53 (55%)	–	–	1.203	–
15	53 (53–54) (30%)	–	–	–	–
16	55 (55–57) (60%)	–	–	–	–

NOTE: Hyperdiploidy in 8 of 16 patients was detected by chromosome analysis alone, in three cases by flow cytometry only, and in five children by using the two methods simultaneously. In some cases FISH provided more accurate information about the individual chromosomes involved in numerical aberrations than cytogenetic analysis.

FISH became available, a revision of some of the chromosomes became necessary (TABLE 8, No. 21, 24). In 20% of the cells from patient No. 24 trisomy 17 was identified by FISH; in 30% of the cells trisomy 18 was identified, also by FISH. These trisomies were not detected by the cytogenetic method. The karyotypic changes described by cytogenetics were found in 70% of the cells.

Based on the analysis of karyotypes and the FISH results of a total 16 patients, a predominant involvement of chromosomes 8, 10, 11, 13, 14, 17,18, 19, 21, 22, and X could be detected. Chromosomes 19, 21, and 22 appeared in tetrasomies. No pentasomy of any of the chromosomes was found. Structural aberrations added to the numerical ones were observed in only two of 16 patients (see TABLE 5, TABLE 8, FIG. 4).

Comparison of the Results Obtained by Using the Three Methods Simultaneously

We had an opportunity to compare the results of cytogenetic analysis, DNA content, and FISH performed (two or all three) simultaneously in 29 patients. Chromosome analysis and flow cytometry were compared in 17 cases, chromosome and FISH in two, flow cytometry and FISH in one, and the three methods were applied simultaneously in nine patients (see TABLE 8).

Cytogenetics and flow cytometry: In 18 cases a diploid karyotype (46,XX /46,XY) (TABLE 8, No. 1–18) and in another patient with Down's syndrome (No. 19), the 21-trisomy was associated with a diploid DNA index of 1.0. In an additional patient (No. 20) a normal karyotype was found together with a DNA index of 1.11. In three patients with >50 chromosomes (No. 22, 25, 28), the DNA index proved to be above 1.16. In two more cases (No. 26, 27) where the modal chromosome number

TABLE 6. Cytogenetic Data and Survival of Patients with Hyperdiploid A (47–50 Chromosomes)[a]

No.	Modal Chromosome Number	Karyotype	Numerical Aberrations	Structural Aberrations	DNA Index	Survival (months)
1	47	46,XX/47,XX,+20 (65%)	+20	−	−	41
2	47	46,XX/47,XX,+D (50%)	+D	−	−	+4
3	47	46,XX,t(16;17) (q24;q11)(30%)/ 47,XX,+18 (70%)	+18	t(16;17)	−	24
4	47	46,XY(30%)/ 47,XY,+mar (70%)	?	−	−	+13
5	46	46,XY/48-51,XY (10%)	−	−	1.0	16
6	49	46,XX/49-50,XX, +mar,Ph+(90%)	−	Ph	−	19
7	48 (47–50)	47,XY,+(7q), t(8;22)(q22;q11)/ 48,XY,+i(7q), +6/49,XY, +i(7q),+6, +12/50,XY,+1, +6,+12,+i(7q)	+1,+6,+12	i(7q) t(8;22)	−	+0.7
8	47	46,XX/47,XX,+20	+20	−	−	10

[a] Patients who died are signed by +. In patient No. 7, a complex karyotype change with approximately equal proportions of metaphases with 47, 48, 49, and 50 chromosomes could be identified. A B-cell-specific t(8;22) was detected in cells with 47 chromosomes, and an i(17q) occurred in each cell line. Additional supernumerary chromosomes representing a clonal evolution were also identified. The patient died in three weeks.

TABLE 7. Karyotype Changes, Immunophenotype, and Survival of Patients with Pseudodiploid Karyotype

No.	Karyotype	Immunophenotype	Survival (months)
1	46,XX(4)/46,XX,7q+(5)/48,XX,del(1q), +mar1,+mar2(1)	N.D.	34
2	46,XX,del(2p)	CALL	+1
3	46,XY/46,XY,t(1;19)(q21;p13)	Pre–B cell	N.D.
4	46,XY/46,XY,t(14;11)	T cell	15
5[a]	47,XY,+21,del(5)(p12)	N.D.	15
6	46,XX,t(9;11)	Pre–B cell	15
7	46,XY,,t(9;22)	Bipheno	15
8	46,XX(2)/46,XX,t(4;11)(6)/46,XX,t(4;11), del(6q)(5)/45,XX,t(4;11),−17(3)	Bipheno	+9
9	46,XY/46,,XY,1p+	CALL	16
10	46,XY/46,XY,del(11q)	Bipheno	1
11	46,XY,t(2;12)	Pre–B cell	8

NOTE: In 9 of 11 patients, ALL-specific translocations or aberrations with specific breakpoints could be detected. The pattern of aberrations observed together with the typical immunophenotype may explain the unfavorable prognosis of patients belonging to this group. N.D., no data.

[a] Down's syndrome.

was more than 50, and in another one with 47–48 chromosomes (No. 29), the DNA index remained below 1.16. In these latter cases the proportion of hyperdiploid cells was rather low (5–20%).

Conventional cytogenetics and FISH: Chromosome analysis and FISH technique without flow cytometry were carried out in only two cases (No. 21, 24). On the basis of FISH, some revision of the karyotype was needed. In patient No. 21 FISH analysis detected trisomy 8 in one-third and trisomy X in one-fifth of the cells. We think that trisomy 6 of the karyotype should be revised as trisomy X and trisomy 12 as trisomy 8. In the other case (No. 24) trisomy 17 and trisomy 18 were revealed by FISH. In our opinion, not instead of but in addition to trisomies of the other chromosomes identified by chromosome analysis.

Flow cytometry and FISH: These two methods, without cytogenetic analysis, were performed simultaneously in one patient only (No. 23). In accordance with a hyperdiploid DNA index (1.28), FISH showed trisomy 4, 6, 8, 10, 18, 13/21, and X in 80% of the cells.

Conventional cytogenetics, flow cytometry, and FISH: The three methods were simultaneously applied in a total of ten cases. In seven of 18 patients with diploid karyotype and diploid DNA content, FISH analysis was also performed. Random trisomies and monosomies were found in three of seven cases (No. 11, 12, 13). In a patient with Down's syndrome (No. 19), besides trisomy 21, trisomy X could also be detected in a few cells by FISH. The three methods were applied simultaneously in one hyperdiploid patient (No. 25). Using chromosome analysis, only the hyperdiploid modal chromosome number (55) could be determined in accordance with a DNA index 1.19. Applying the FISH method, the extra chromosomes could be clarified.

Using one of the three methods or their combinations, 16 of the 80 patients (20.0%) proved to have hyperdiploid karyotype with more than 50 chromosomes and/or a >1.16 DNA index. These cases were associated with favorable clinical,

TABLE 8. Comparison of Results Obtained by Cytogenetic, Flow Cytometric, and FISH Analysis

No.	Modal Chromosome Number	Range	Percent with Modal Chromosome Number	Karyotype	DNA	FISH	Revision
1–7	46	46	100	4X46,XX/3X46,XY	1.0	—	
8–10	46	46	100	46,XY/2x 46,XX	1.0	46,XY/2x 46XX	
11	46	46	100	46,XX	1.0	46,XX,+1,+3,−6,−X (5%)	+1,+3,−6,−X
12	46	46	100	46,XY	1.0	44,XY,−7,−20 (5%)	−7,−20
13	46	46	100	46,XY	1.0	46,XY,−4,+15 (10%)	−4,+15
14	46	46	100	46,XY	1.023	—	
15	46	46	100	46,XY	1.03	—	
16	46	46	100	46,XY	1.023	—	
17	46	46	100	46,XY	1.11	—	
18	46	45–46	75	46,XY/45,X,−C,−D	1.0	—	
19	47	47	100	47,XY,+21	1.0	46,XY,+21,+X (5%)	+X
20	46	46	100	46,XY,Ph+	1.0	46,XY,Ph+	
21	55	53–58	50	53,XX,+6,+12,+17,+19,+20,+21,+22	—	+8 (35%), +X (20%), +21 (80%)	6 = X, 12 = 8
22	55	55–57	75	55,XX,+6,+8,+10,+11,+13,+17,+18,+21,+22	1.24	—	
23	—	—	—	—	1.28	53,XX,+4,+6,+8,+10,+18,+13/21,+X	
24	53	51–54	70	53,XY,+6,+11,+12,+19,+20,+21,+22	—	+17 (20%) +18 (30%)	? 19 = 17 ? 20 = 18 or: +17,+18
25	46	46–55	40	—	1.19	52,XX,+10,+17,+18,+13/21,+X,+X	
26	46	46–60	80	46,XX,del(11q)	1.13	—	
27	46	46–59	95	58,XX,+3,+10,+11,+13,+14,+15,+16,+17,+19,+19,+20,+22	1.0	—	
28	46	46–56	40	—	1.203	—	
29	46	46–51	95	—	1.0	—	

NOTE: Comparison of the results obtained by chromosome analysis and flow cytometry, the latter one and FISH, or by the three methods simultaneously demonstrates that these methods are complementary to each other and their simultaneous application is essential. In some cases the FISH method provided new pieces of information about the chromosomes involved. In patients No. 22 and 25, a revision of the karyotype was needed. (In patient No. 22, chromosome 6 = X and 12 = 8; in patient No. 25, 19 = 17, 20 = 18?)

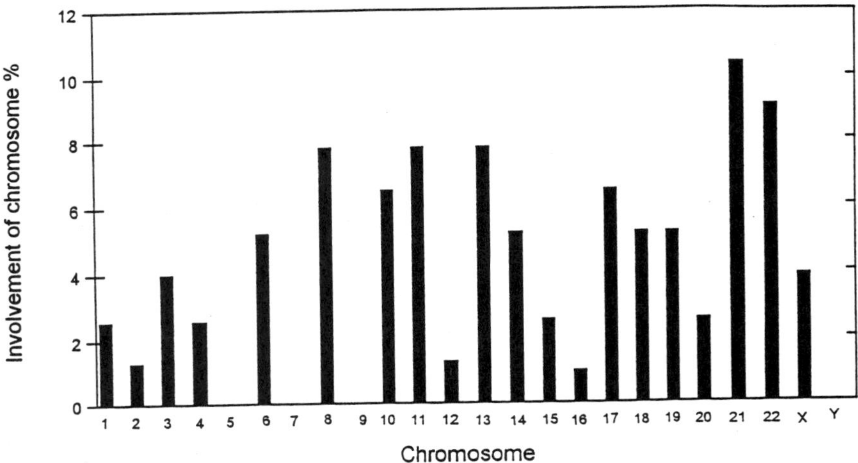

FIGURE 4. Percentage of aberrations of individual chromosomes. Although the absolute number of aberrations is low (76), the frequent involvement of chromosomes 8, 11, and 13 can be seen. Tetrasomy of chromosomes 19, 21, and X is considered to be a single aberration.

hematological, and immunological features (TABLE 9). With the exception of one patient (No. 7), all of them had a relatively low white blood cell count, favorable risk factors, and in the majority of cases a B-cell precursor immunophenotype could be found. In three of 16 patients (No. 8, 10, 16) biphenotypic leukemia was detected.

TABLE 9. Clinical, Hematological, and Immunological Features of Patients with Hyperdiploid Karyotype

No.	Sex/Age (months)	WBC (G/l)	Immunophenotype	Risk Factor	Survival (months)
1	F/51	1.4	cALL	0.7	35
2	M/83	8.1	Pre-pre B	1.8	41
3	M/24	4.1	ND	1.17	43
4	M/32	8.8	Pre B	1.35	44
5	M/30	1.9	cALL	1.0	36
6	F/32	3.8	Pre B	1.09	22
7	F/75	56.0	Pre B	1.09	27
8	M/15	10.0	Bipheno(lymphoid/M7)	—	+16
9	M/48	8.4	Pre B	0.72	26
10	F/90	13.0	Bipheno	0.75	9
11	F/198	9.0	Pre-pre B	1.0	16
12	M/120	6.2	cALL	1.08	15
13	M/30	8.4	Pre-pre B	1.09	36
14	F/188	28.0	Pre B	1.07	28
15	F/96	5.0	Pre B	1.0	14
16	F/32	6.0	Bipheno(lymphoid/M7)	0.08	+0.1

NOTE: With the exception of three patients with mixed immunophenotype hyperdiploidy proved to be associated with favorable clinical, hematological and immunological parameters. ND, no data.

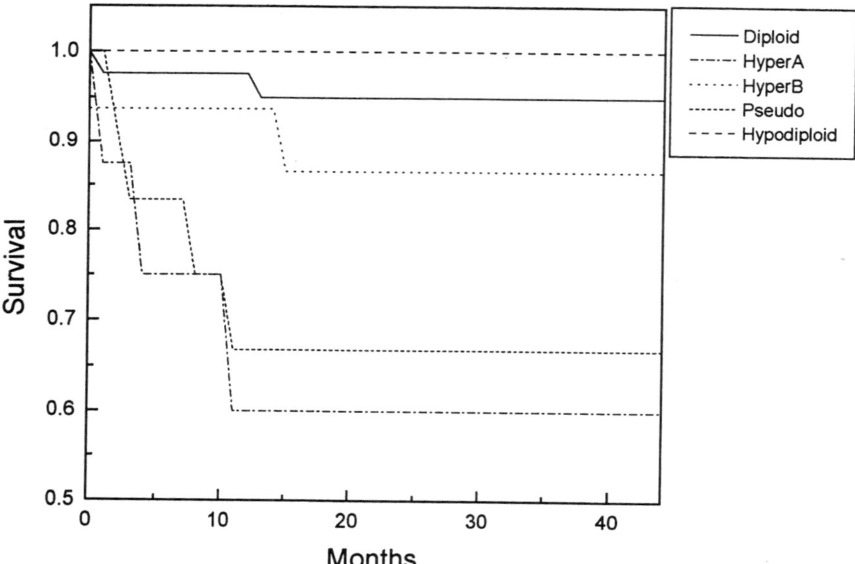

FIGURE 5. Survival curves of 80 patients with different modal chromosome numbers. Calculation of the curves was made based on the Kaplan-Meyer life table analysis. The number of cases analyzed: diploid, 42; pseudodiploid, 11; hypodiploid, 4; hyperdiploid A, 8; and hyperdiploid B, 16. A significant difference was found between diploid and pseudodiploid, and diploid and hyperdiploid A subgroups ($p < 0.01$). Survival curve of patients with hyperdiploid B karyotype remained under that of diploid patients.

Comparing the overall survival curves of subgroups with various modal chromosome number, an unfavorable prognosis of patients with a pseudodiploid and with a hyperdiploid A karyotype could be observed as compared to that of patients with a diploid chromosome set ($p < 0.01$). No significant difference between the other subgroups could be observed. The favorable prognosis of patients with more than 50 chromosomes could not be proved (FIG. 5). The number of patients in each group, however, is too small to avoid a possible statistical error. Thus, it is not possible to make any conclusions related to prognostic significance of hyperdiploidy in this study.

DISCUSSION

During the last three years ALL was diagnosed in 187 patients. In 140 of 187 patients, chromosome analysis was attempted, in 78 cases successfully. Although the ratio of successful investigations is rather low, it is important to point out that efforts have been made to perform these initial genetic analyses in various genetic laboratories working in collaboration with the hematological centers of the country. It represents a step forward that, in 80 of a total of 187 ALL patients (42.7%), the genetic diagnosis was available for the clinicians. This number will hopefully increase in the future if the staffing and instrumental facilities as well as the costs of the analyses will be provided.

Although chromosome analysis was performed in 78 patients, the DNA analysis and FISH were carried out in many fewer cases. Although chromosome analysis is the most informative method indicating the accurate numerical and structural aberrations, its value is decreased by the great number of unsuccessful cases. This is confirmed by our results presented here. Clonal chromosome aberrations were detected only in 35 of 78 successfully investigated patients (44.8%), and the ratio of cases with >50 chromosomes proved to be only 13:78 (16.6%). The low proportion of cases with abnormal karyotype may be the consequence of technical difficulties. When few analyzable metaphases are available, only a small number of cells are evaluated, and the normal metaphases of better quality are considered first of all. Furthermore, because of the poor quality of chromosome morphology, a great majority of the fine structural abnormalities remains unidentified. In addition, there are "hidden" structural abnormalities that cannot be detected by the traditional cytogenetic methods. For this reason it is essential to improve the standard of cytogenetic analyses and to complete them with molecular techniques.

The ratio of patients with more than 50 chromosomes also remains far behind the data in the literature. A frequency of 30% is usually mentioned,[10–12,14,23] but a much higher ratio (49%) was also found.[24] The low proportion of hyperdiploid cases in our material may also result from technical difficulties mentioned above, for example, selection of normal metaphases of better quality, or from the small number of patients studied, but the fact that our experience is characteristic of this geographic area cannot be excluded. On the other hand, the fact that the proportion of hyperdiploid cases proved to be higher when we used flow cytometry (7/31: 22.5%) may contradict the above hypothesis. These results call attention to the significance of simultaneous application of the two methods mentioned.

In evaluating the involvement of individual chromosomes in numerical aberrations of hyperdiploid patients, we have to point out the question of geographic factors anew. In accordance with our previous results, a difference in the spectrum of chromosomes involved compared to the data in the literature could be observed. In addition to the trisomies of chromosomes 4, 6, 10, 17, 18, 21, 22, and X, a supernumerary occurrence of chromosomes 3, 8, 11, and 13 was also found with a high frequency. This is not a unique observation. A geographic heterogeneity in the occurrence and frequency of chromosome aberrations was also observed in other diseases like that of t(8;21) in M2-type acute nonlymphocytic leukemia, and of t(15;17) in acute promyelocytic leukemia, respectively.[43] A difference in the frequency of the secondary aberrations of blastic-phase Ph-positive chronic myeloid leukemia was also proved.[44] The geographic differences in the occurrence and frequency of tumor-specific aberrations may reflect the role of different environmental mutagen/carcinogenic effects.[41] In order to determine whether these are or not real differences characteristic of our geographic area, further investigations are needed.

From a clinical point of view, the prognostic significance of the initial karyotype is of particular significance. Because the genetic changes determine the biological behavior of tumor cells, cell-to-cell interaction, invasivity, metastatic capacity, and response to therapy, the exact determination of the initial genetic alteration makes it possible to develop a genotype-specific therapy.[30,31] While certain translocations, t(9;22), t(4;11) suggest the necessity of a bone marrow transplantation, patients with hyperdiploidy of a favorable prognosis need only a standard therapy. In our study, however, the small number of patients in each group is too small to avoid a possible statistical error; therefore, it is not possible to make any conclusion related to prognostic significance of karyotype changes. Despite this, our observations might be indicative of probable geographic characteristics and might call our attention to a few avenues of study. In our study the comparison of survival curves of patients with different modal chromosome numbers proved a significantly worse prognosis for patients with pseudodiploid and hyperdiploid A karyotypes compared to that of the diploid group. The favorable prognosis of hyperdiploid patients (>50 chromosomes) could not be proved. The lack of the expected good prognosis of hyperdiploid patients may result from the small number of patients, perhaps the unfavorable prognosis of three biphenotypic cases (two of them had lymphoid/M7 phenotype), but may also be related to the different spectrum of chromosomes involved, such as the high frequency of trisomy 8 and 11. The unfavorable effect of chromosome 8 on the course of the disease is well known.[12,43] Similarly, the aberrations of chromosome 11 in acute leukemias are associated with a rather severe disease.[12,16,43] Besides, there are data regarding the special prognostic value of the particular extra chromosomes in these patients, for example, cases with trisomy 6 and with combined trisomy of 4 and 10 seem to have a particularly good prognosis.[27,28] Our findings and the latter observations call attention to the necessity for a more detailed analysis of involvement of chromosomes in numerical aberrations. An additional possible explanation for the lack of prognostic difference between the diploid and hyperdiploid groups may be that the diploid group also includes the cases with t(12;21), an aberration that is associated with a favorable prognosis.[22] This translocation cannot be detected by standard cytogenetic methods, only by molecular analysis. In this way this hidden aberration may improve the prognosis of the diploid group and decrease the prognostic difference compared to the hyperdiploid one. This aspect points to the importance of the molecular analysis of cases with a virtual normal karyotype.

Based on the data of patients studied in the last three years, we can say that genetic findings of ALL patients in Hungary seem to be different from those in the literature. The proportion of hyperdiploid cases is lower, the spectrum of chromosomes involved in numerical aberrations is different, and the favorable prognosis of the hyperdiploid group could not be proved. To clarify the significance of these observations, it seems to be essential to determine whether or not these clinical and genetic differences really exist, and if they do we have to reveal the causes leading to these differences. For this purpose we have to extend the genetic investigations to all ALL cases in Hungary to improve the proportion of successful investigations and complete them with molecular techniques, probably with comparative genomic hybridization. Besides, a detailed analysis of the individual chromosomes and a nationwide prospective study of the clinical–genetic correlation seem to be essential.

SUMMARY

Current intensive chemotherapies cure about 70% of the children with ALL. On the other hand a significant number of the children are not cured despite intensive treatment. At the same time some highly curable patients are treated too intensively and suffer from unnecessary side effects of the chemo- and radiotherapy applied. In order to further improve the therapeutic results in this disease, we have to distinguish between the cases with a better and a worse prognosis. The initial karyotype (both numerical and structural chromosome abnormalities) proved to be one of the most reliable prognostic parameters, leading to the suggestion of developing genotype-specific therapies. Although the prognosis in patients with pseudodiploid karyotype is usually unfavorable, a significantly better prognosis can be observed in those with more than 50 chromosomes. Because the latter patients can achieve remission on a metabolite-based therapy, the toxic effects of more aggressive chemotherapy with anthracyclines and genotoxic agents can be avoided; thus, the reliable and accurate identification of patients with >50 chromosomes is of particular importance. For this purpose three methods: chromosome analysis, DNA flow cytometry, and fluorescence *in situ* hybridization can be used.

In 1993 it was decided to develop a comprehensive nationwide project in order to perform the initial genetic analysis of all ALL children diagnosed in the hematological/oncological centers of Hungary. Here the data obtained on 187 ALL patients diagnosed in the period from 1993 to 1995 are presented. In about 75% of patients (in 140 of 187) chromosome analysis was performed, in 78 cases (55.7%) successfully. The proportion of patients with abnormal karyotype was 36 of 78 (46.1%), and hyperdiploidy with more than 50 chromosomes was detected in 13 of 78 (16.6%) children. The lower ratio of hyperdiploid cases in our patients as compared to the data in the literature may be due to technical difficulties and the small number of patients studied, but it may reflect real geographic characteristics. Using flow cytometry, seven of 31 patients investigated (22.5%) proved to be hyperdiploid with a DNA index above 1.16. A higher ratio of hyperdiploid patients in this study calls attention to the significance of simultaneous application of the two methods. Taken together, 16 of 80 (20.0%) succesfully studied patients proved to be hyperdiploid (>50 chromosomes and/or DNA index above 1.16).

The pattern of chromosome involvement in our study determined by chromosome analysis and/or FISH technique proved also to be different from the data of large international series. In addition to trisomies of chromosomes 4, 6, 10, 14, 17, 18, 21, and X, which are known to be the most frequently involved chromosomes, trisomies of chromosomes 3, 8, 11, and 13 were also observed with a high frequency. Comparison of survival curves of various cytogenetic subgroups showed a significant difference between diploid–pseudodiploid and diploid–hyperdiploid A (with 47–50 chromosomes) subgroups. No favorable prognosis of hyperdiploid patients (>50 chromosomes) could be proved. Because of the small number of patients studied, prognostic differences of cytogenetic subgroups need further confirmation. The clinical and genetic differences observed, however, call attention to the necessity for further genetic studies of ALL patients in Hungary, because these differences may reflect real geographic characteristics and may be related to different environmental mutagen/carcinogen effects of the given geographic area. It is essential to determine whether or not these differences really exist and if they do to reveal the causes leading to these differences. In our view this is one of the routes by which the therapeutic results in childhood ALL can be further improved simultaneously with the avoidance of early and late toxicity of chemotherapy.

ACKNOWLEDGMENT

We thank all clinicians of the Hematological/Oncological Centers of the Hungarian Pediatric Oncology Study Group for providing the patient data and the geneticists and technicians working at different laboratories for their valuable contribution.

REFERENCES

1. BLOOMFIELD, C. D., *et al.* 1990. Six-year follow up of the clinical significance of karyotype in acute lymphoblastic leukemia. Cancer Genet. Cytogenet. **40:** 171.

2. CASSANO, W. F., *et al.* 1993. Therapy for childhood acute lymphoblastic leukemia. Curr. Opin. Oncol. **5:** 42.
3. CHESSELS, J. M. 1986. Acute leukaemia in children. Clin. Haematol. **15:** 727.
4. CRIST, W., *et al.* 1986. Clinical and biological features predict a poor prognosis in acute lymphoid leukemias in infants. A. Pediatric Oncology Group Study. Blood **67:** 135.
5. HARBOTT, J., *et al.* 1993. Clinical significance of cytogenetic studies in childhood acute lymphoblastic leukemia. Experience of the BFM trials. Rec. Res. Cancer Res. **131:** 123.
6. MILLER, D. R. 1988. Childhood acute lymphoblastic leukemia. 1. Biological features and their use in predicting outcome of treatment. Am. J. Ped. Hematol. Oncol. **10:** 163.
7. POPLACK, D. G. & G. REAMAN. 1988. Acute lymphoblastic leukemia in childhood. Pediatr. Clin. N. Am. **35:** 903.
8. PUI, C. H., *et al.* 1988. Correlation of karyotype and immunophenotype in childhood acute lymphoblastic leukemia. J. Clin. Oncol. **6:** 56.
9. PUI, C. H., *et al.* 1989. Prognostic importance of structural chromosomal abnormalities in children with hyperdiploid (>50 chromosomes) acute lymphoblastic leukemia. Blood **73:** 1963.
10. PUI, C. H. & W. M. CRIST. 1992. Cytogenetic abnormalities in childhood acute lymphoblastic leukemia correlates with clinical features and treatment outcome. Leuk. Lymphoma **7:** 259.
11. SECKER-WALKER, L. M. 1990. Prognostic and biological importance of chromosome findings in acute lymphoblastic leukemia. Cancer Genet. Cytogenet. **49:** 1.
12. HEIM, S. & F. MITELMAN. 1987. Acute lymphoblastic leukemia. *In* Cancer Cytogenetics. Chapt. 8.: 141–173. Alan R. Liss, Inc. New York.
13. PUI, C. H., *et al.* 1990. Biology and clinical significance of cytogenetic abnormalities in childhood acute lymphoblastic leukemia. Blood **76:** 1449.
14. RAIMONDI, S. C. 1993. Current status of cytogenetic research in childhood acute lymphoblastic leukemia. Blood **81:** 2237.
15. MURPHY, S. B., *et al.* 1989. Nonrandom abnormalities of chromosome 9p in childhood acute lymphoblastic leukemia: Association with high risk clinical features. Blood **74:** 409.
16. RAIMONDI, S. C., *et al.* 1989. Childhood acute lymphoblastic leukemia with chromosomal breakpoints at 11q23. Blood **73:** 1627.
17. SLATER, R., *et al.* 1995. A nonrandom chromosome abnormality found in precursor-B lineage acute lymphoblastic leukemia: dic(9;20)(p1?3;q11). Leukemia **9:** 1613.
18. CRIST, W., *et al.* 1989. Prognostic importance of the pre-B-cell immunophenotype and other presenting features in B lineage childhood acute lymphoblastic leukemia. A. Pediatric Oncology Group Study. Blood **74:** 1252.
19. PUI, C. H., *et al.* 1993. Clinical and biological relevance of immunologic marker studies in childhood acute lymphoblastic leukemia. Blood **82:** 343.
20. BOROWITZ, M. J., *et al.* 1993. Predictability of the t(1;19)(q23;p13) from surface antigen phenotype: Implications for screening cases of childhood acute lymphoblastic leukemia for molecular analysis: A Pediatric Oncology Group Study. Blood **82:** 1086.
21. DEVARAJ, P. E., *et al.* 1995. Expression of the E2A-PBx1 fusion transcripts in t(1;19)(q23;p13) and leukaemia with different B lineage immunophenotypes. Leukaemia **9:** 821.
22. ROMANA, S. P. 1995. The t(12;21) of acute lymphoblastic leukemia results in a tel/AML1 gene fusion. Blood **85:** 3662.
23. PUI, C. H., *et al.* 1990. Near-triploid and near-tetraploid acute lymphoblastic leukemia of childhood. Blood **76:** 590.
24. HEIM, S., *et al.* 1990. Bone marrow karyotypes in 94 children with acute leukemia. Eur. J. Haematol. **44**(4): 227.
25. TRUEWORTHY, R., *et al.* 1992. Ploidy of lymphoblasts is the strongest predictor of treatment outcome in B-progenitor-cell acute lymphoblastic leukemia of childhood: A Pediatric Oncology Group Study. J. Clin. Oncol. **10:** 606.
26. HAMMOND, D., *et al.* 1986. Analysis of prognostic factors in acute lymphoblastic leukemia. Med. Ped. Oncol. **14:** 124.

27. HARRIS, M. B., *et al.* 1992. Trisomy of leukemic cell chromosomes 4 and 10 identifies children with B progenitor cell acute lymphoblastic leukemia with a very low risk of treatment failure: A Pediatric Oncology Group Study. Blood **79:** 3316.

28. JACKSON, J. F., *et al.* 1990. Favorable prognosis associated with hyperdiploidy in children with acute lymphocytic leukemia correlates with extra chromosome 6. A Pediatric Oncology Group Study. Cancer **66:** 1183.

29. HARBOTT, J., *et al.* 1994. Cytogenetics and clonal evolution in childhood acute lymphoblastic leukemia. Ann. Haematol. **68**(Suppl. I): A13.

30. PINKEL, D. 1987. Curing children of leukemia. Cancer **10:** 1683.

31. PINKEL, D. 1989. Species-specific therapy of acute lymphoid leukemia. Haematol. Blood Transf. **32:** 27.

32. DU MANOIR, S., *et al.* 1993. Detection of complete and partial chromosome gains and losses by comparative genomic in situ hybridization. Hum. Genet. **90:** 590.

33. ANDREEFF, M. 1990. Flow cytometry of leukaemia. *In* Flow Cytometry and Sorting. 2nd edit. M. R. Melamed, T. Lindmo, M. L. Mendelsohn, (Eds.: 697–724. Willy-Liss, Inc. New York.

34. HIDDEMANN, W., *et al.* 1987. Nachweis von aberrationen des Karyotyps bei Kindern mit akuten Leukämien: Eine vergleichende Analyse von Zytogenetik und Durchflubzytophotometrie. Klinische Pädiatrie **199/3:** 161.

35. LOOK, T. A., *et al.* 1985. Prognostic importance of blast cell DNA content in childhood acute lymphocytic leukaemia. Blood **65:** 1079.

36. ANASTASI, J. 1993. Fluorescence in situ hybridization in leukemia. Applications in diagnosis, subclassification and monitoring the response to therapy. Ann. N. Y. Acad. Sci. **677:** 214.

37. ANASTASI, J., *et al.* 1990. Detection of numerical chromosomal abnormalities in neoplastic hematopoietic cells by in situ hybridization with a chromosome-specific probe. Am. J. Pathol. **136:** 131.

38. BERGER, A., *et al.* 1994. Interphase cytogenetic study of childhood acute lymphoblastic leukemia. Med. Ped. Oncol. **23:** 413.

39. PINKEL, D., *et al.* 1986. Cytogenetic analysis using quantitaive high sensitivity fluorescence hybridization. Proc. Natl. Acad. Sci. USA **83:** 2934.

40. PODDIGHE, P. J., *et al.* 1991. Interphase cytogenetics of hematological cancer: Comparison of classical karyotyping and in situ hybridization using a panel of eleven chromosome specific DNA probes. Cancer Res. **51:** 1959.

41. TOSSI, S., *et al.* 1994. Double target in situ hybridization applied to the study of numerical aberrations in childhood acute lymphoblastic leukemia. Cancer Genet. Cytogenet. **73:** 10.

42. ISCN. 1995. An International System for Human Cytogenetic Nomenclature. S. Karger. Basel, Switzerland.

43. HEIM, S. & F. MITELMAN. 1987. Acute nonlymphocytic leukemia. Cancer Cytogenetics. Chapt. 5.: 65–110. Alan R. Liss, Inc. New York.

44. HEIM, S. & F. MITELMAN. 1987. Chronic myeloid leukemia. *In* Cancer Cytogenetics. Chapt. 4.: 41–46. Alan R. Liss, Inc. New York.

Apoptosis and Acute Lymphocytic Leukemia in Children[a]

DEZSŐ SCHULER[b] AND BÉLA SZENDE[c]

[b]Semmelweis Medical University
2nd Department of Pediatrics
1094 Budapest, Tûzoltó u.7-9, Hungary

[c]Semmelweis Medical University
1st Institute of Pathology and Experimental Cancer Research
1085 Budapest, Üllôi út 26, Hungary

Apoptosis, or programed cell death (PCD), is a physiologic phenomenon that ensures the balance between cell proliferation and cell death, similar to the balance between anabolism and catabolism within the cells. The physiologic function of apoptosis is to remove the cells that are not necessary for the organism during the different phases of organogenesis. Apoptosis has its role in normal function of the immune system, and it is needed for the homeostatic regulation of normal tissue mass.[1] Apoptosis removes the senescent, damaged, abnormal cells interfering with the normal function of the organism. The abnormalities of apoptosis play an important part in proliferative neoplastic diseases, degenerative disorders, immune diseases, and autoimmune diseases. Apoptosis has been studied intensively since the study of Kerr *et al.*[2] in 1972 concerning its basic biologic significance. It affects scattered, individual cells rather than all the cells in a particular area.

MORPHOLOGIC AND BIOCHEMICAL CHANGES IN APOPTOSIS

Several morphologic changes occur in apoptosis.[1] There is a reduction of nuclear size into crescentic caps at the nuclear periphery, resulting in nuclear disintegration with dissociation of the transcriptional complexes. The cell volume shrinks, cytoplasmic organelles become compacted but are preserved, and the smooth endoplasmic reticulum dilates. The mitochondria do not show swelling.[2] Both nucleus and cytoplasm may split into fragments.

The fragmentation of the chromatin into oligonucleosomal-length fragments by a nonlysosomal Ca^{2+}/Mg^{2+}-dependent endonuclease, laminin phosphorylation, solubilization, and breakdown of the nuclear envelope are early events in apoptosis. The early rise of cytosolic ionized calcium can activate proteases and transglutaminase, which cross-links cytosolic proteins. Then the cells become a cluster of round, smooth, membrane-bound apoptotic bodies, which either are shed from epithelial surfaces or are phagocytosed by neighboring cells or macrophages.[3] In cell cultures the apoptotic cells undergo lysis instead of phagocytosis. The ingestion of apoptotic cells by macrophages does not induce the release of proteolytic enzymes or toxic singlet oxygen. An essential characteristic of apoptosis is that there is no leakage

[a] This work was supported by OTKA T6335, 17849, 17722.

28

of cellular contents into the extracellular space; removal of apoptotic cells does not cause an inflammatory response.[4]

There are several metabolic pathways for apoptosis. It was formerly supposed that the fragmentation of DNA is an early and essential part of apoptosis. Recently, it was found that DNA is not indispensable to programed cell death; even anucleate cells can undergo apoptosis. In these cases cytosplasmic structures, especially the mitochondria, participate in the induction of the metabolic events in apoptosis. The mitochondrial transmembrane potential is reduced early during the apoptotic process, even before the nuclear and chromatin structures are altered (cell shrinkage, DNA fragmentation), and the plasma integrity is lost.[5] The role of mitochondria in apoptosis is independent of their DNA content, because cell lines lacking mitochondrial DNA die of apoptosis when stimulated with a phosphotyrosine kinase inhibitor.[6] This is also shown by the observation that the cell lines lacking mitochondrial DNA can also induce signs of nuclear apoptotic chromatin condensation and DNA fragmentation in isolated nuclei *in vitro*.[7] These observations indicate that the apoptotic functions of mitochondria are products of nuclear genes. It is postulated that mitochondria can release a factor causing isolated nuclei to undergo apoptotic chromatin condensation; these proteins are encoded by nuclear genes. According to experiments in the cell-free system based on cytosols of normally growing cells, the mitochondria trigger apoptosis by releasing cytochrome c.[8]

INITIATION OF APOPTOSIS

The cells "commit suicide" unless signaled by other cells except individual blastomers. They do not need the synthesis of killer proteins to PCD, because they already contain all the proteins needed for it.[9] Many endogenous and environmental influences induce apoptosis: It is a common occurrence in physiologic conditions; for example, some hormones (corticosteroids, thyroxin), or withdrawal of hormones from its target tissue, induce apoptosis.

Several pathologic or environmental factors are also known to start the apoptotic procedure. The c-*myc* oncogen[10] and chemotherapeutic agents,[11] inhibition of tyrosine kinase,[12] pokeweed mitogen,[13] tumor necrosis factor,[14] and other compounds can induce apoptosis in a given cell type either in cell culture or *in vivo*. It seems that PCD is triggered by diverse noxic stimuli, when they are present in concentrations, that do not precipitate necrosis.

PCD is initiated by binding of ligands to their respective cell surface receptors in the appropriate cell type. The receptors transduce the apoptotic signal into the susceptible cells. The signal pathways (Ca^{2+}, c-AMP, protein kinase C, other kinases, phosphatases, and the molecular details of the transcriptional response of cells to receptor stimulation) vary from one cell to another. The maturation and differentiation status of the cell determine the extent of their competence and how much they are primed for apoptosis.

Apoptosis is an active metabolic procedure that needs energy in the form of ATP. Depending on the cell type, the apoptotic stimulus may require the active synthesis of RNA and proteins. For example in long-lived lymphocytes, the inhibition of RNA or protein synthesis usually blocks apoptosis, whereas in short-lived neutrophils it increases apoptosis.[15] It may be that there are regulatory proteins of apoptosis in the cells and that the comparative rate of catabolism of the inhibitors and promoters in a particular cell type determines whether stopping RNA and protein synthesis prevents or induces apoptosis.

REGULATION OF APOPTOSIS

The regulation of apoptosis depends on the expression of other genes, especially *bcl*-2 and p53. In many cell types, proliferation requires at least two external stimuli, a competence signal, and a progression signal. The competence or induction signal starts metabolic events common to both replication and apoptosis, whereas the progression signal diverts the cell either toward replication or apoptosis. A decision point is in the late G1 phase, in which the protein of the p53 gene can delay the cell cycle progression before DNA replication, or from which the cell can proceed toward PCD.

The *bcl*-2 gene is a cell-death suppressor gene that directly or through blocking p53 regulates apoptosis. The proteins from the apoptosis-regulatory *bcl*-2 family are localized to the outer mitochondrial membrane (mainly to so-called contact sites with the inner membrane), as well as in the nucleus and the endoplasmic reticulum.[16] *Bcl*-2 can also protect the anucleated cells from apoptosis, indicating that in some cases at least, the nucleus is not required for *bcl*-2 protection.[17]

However, Strasser *et al.*[18] observed apoptosis in a lymphoid cell line without the expression of p53, and PCD exists without alteration of *bcl*-2 gene,[19] proving that there exist multiple pathways of apoptosis.

APOPTOSIS IN MALIGNANT DISEASES

In malignant tumors the basic pathologic procedure is the lack of balance between cell proliferation and cell death, which results in an anomalous accumulation of cells. This process may be due to increased cell division, prolonged survival of tumor cells, or decreased death of cells. Because of the regulatory role of apoptosis in cell proliferation, PCD has enormous significance. Most of the carcinogens damage DNA or interfere with enzymes necessary for the physiologic route of DNA replication. After damage of DNA, a cell can delay division until the damage is repaired, or it can undergo apoptosis or progress without interruption through the cell growth cycle, thus promoting cancer development. Hence, inducing apoptosis is an efficient method of preventing malignant transformation and also decreasing the tumor load by removing cells with genetic lesions. The p53 gene is an inhibitor of cell division, and in other situations even acts as a direct apoptogen. When a mutation blocks p53 activity, an uncontrolled proliferation takes over.[4]

All hemopoietic and lymphoid cells, many epithelial cells, and neurons contain bcl-2 protein, which protects the cells from apoptosis. Some chromosomal translocations, Epstein-Barr virus proteins, increase the expression of bcl-2 protein. The lymphocytes with overexpression of bcl-2 protein, which survive longer in culture after growth factor deprivation, are resistant to both ionizing radiation and corticosteroids.[4]

The significance of bcl-2 expression on the T-cell acute lymphoblastic leukemia blasts was proved by the transfection with bcl-2 alpha and beta elements, which added survival advantages to already malignant cells.[20] The effect of bcl-2 on the presenting clinical features and response to therapy could not be shown. In 52 children with newly diagnosed ALL, the bcl-2 expression in leukemic blasts ranged widely. Unexpectedly, the bcl-2 expression was lower in leukemic blasts with chromosomal translocations. The higher level of bcl-2 expression corresponded to higher leukemic cell recovery in culture in serum-free medium, but not after culture on stromal layers. However, it showed no correlation either with *in vitro* resistance to several antileukemic drugs (vincristine, dexamethasone, 6-thioguanine, cytarabine, teniposid, daunorubicin, methotrexate) or with the presenting clinical features. Higher levels of bcl-2 did not show correlation with the *in vivo* response to therapy either among 33 uniformly treated patients or in three patients at relapse.[21]

Most anticancer drugs produce apoptosis in sensitive cells and also influence the response to chemotherapy. In one animal study of slowly growing and nonmetastatic tumors, apoptosis was conspicuous at 14 days in mice injected with *c-myc* transfectants. The mice injected with cells expressing activated T24-*ras* developed aggressive, large, metastasizing tumors, in which apoptosis was rare.[3]

Apoptosis occurs spontaneously in solid tumors of various types, and apoptotic cell death predicts tumor size and development. However, the measurement of apoptotic cells in tumors is difficult because the half-life of apoptotic cells is short, although they can be histologically recognized, and the cell samples are often heterogeneous. Matsubara *et al.*[22] demonstrated that apoptosis occurs spontaneously in c-ALL, but not at all or little in AML cells *in vitro*. However, both cell types undergo apoptosis when incubated with chemotherapeutic agents. Spontaneous and induced apoptosis affect leukemia cells differently; childhood T-ALL cells are most susceptible to apoptosis. Patients with autonomous blast cell growth *in vitro* have both a lower remission rate and a higher relapse rate than otherwise comparable patients whose blasts exhibit nonautonomous growth. In a group of 50 patients, the actual disease-free survival for the autonomous growth group was 11% at 5 years, as compared with greater than 50% for the nonautonomous growth group.[23] The autonomous group of patients proved to be relatively resistant to apoptosis induced by cytotoxic drugs.

The differing susceptibility to apoptosis is related to the production of autocrine growth factors. The *bcl-2* expression was found unregulated by both exogenous and autocrine GM-CSF. It is supposed that the negative prognostic effect of autonomous growth on treatment outcome in AML may be in part due to the regulator effect of autocrine GM-CSF on *bcl-2* expression.

There is also an interaction between acute leukemia cells and the microenvironment of the bone marrow. Stromal cells are able to inhibit PCD in a proportion of cases of both acute myelocytic leukemia (AML) and ALL.[24]

In acute leukemia of children, the balance between cell proliferation and cell death is disturbed. The abnormal rate of apoptosis is of great importance in these diseases. We were interested in finding whether the incidence of spontaneous apoptosis in peripheral blast cells is elevated before the therapy. Other questions related to the *in vivo* effect of prednisolone on the apoptosis of leukemic blasts in children and the correlation, if any, between the *in vivo* decrease of blasts in children with ALL during the first 8 days of prednisolone chemotherapy and apoptosis. The *bcl-2* and p53 gene expression was also examined as a pilot study in 10 cases from the same blast samples, as well as the number of glucocorticoid receptors in six cases.

METHODS

Heparinized peripheral blood samples of 23 children with acute lymphoblastic leukemia (ALL) were taken for morphologic and flow cytometric studies before the onset of prednisolone therapy and 6, 24, and 48 hours after oral prednisolone administration. Blood samples were prepared for flow cytometric and morphological analysis within 30 minutes.

The methods of preparation for light microscopy and flow cytometry were identical with those described by us earlier.[25]

The number of glucocorticoid receptors (gcR) was evaluated by the determination of specific mRNA. We used reverse transcriptase polymerase chain reaction (RT-PCR) as described earlier by Falus *et al.*[26,27]

Method of Molecular Genetic Studies

DNA Extraction

DNA was extracted by standard methods using phenol and chloroform.

Assessment of Allele Loss

Highly polymorphic markers for p53 were used to determine the allelic status. To type the p53 marker, we used a two-step PCR amplification protocol. PCR products were run on sequencing gel. After fixation and drying, gels were autographed. Because of the highly informative capability of the markers when one allele was seen, this can be evaluated as loss of heterozygosity.

RNA Preparation, Determination of Gene Expression (Northern Analysis)

Total cellular RNA was harvested and electrophoresed in formaldehyde gels. Blots containing 10 μg RNA per sample were prepared on Hybond-N membrane (Amersham). Membranes were hybridized with ^{32}P-labeled genomic probes of p53 and *bcl*-2, then reprobed with glycerol aldehyde phosphate dehydrogenase (the latter was used as a control for lane loading). Quantitative densitometry of X-ray films was performed with a Molecular Dynamics laser densitometer.

Clinical Follow-up

Morphologic, immunologic, cytochemical, and cyto- and molecular genetic examination of the blast cells were carried out by the oncologic team of our department. The absolute blast number of the peripheral blood was determined before beginning the therapy and on day 8. Hematological remission was controlled by bone-marrow examination on day 33. Patients were regularly controlled by physical examination and by analysis of peripheral blood and bone marrow.

RESULTS AND DISCUSSION

A total of 23 children with ALL were examined. The diagnosis was made by morphologic and immunologic methods. The FAB classification was used for the morphologic diagnosis. Most cases belonged to group L1; in two cases some cells

TABLE 1. The Histological Type of Our Cases According to the FAB Classification

FAB Classification	Number of Cases
L1	15/23
L1-2	2/23
L2	5/23
L2-3	1/23

showed the characteristics of both L1 and L2 (TABLE 1). The c-ALL immunophenotype occurred in most cases; however, biphenotypic leukemia was relatively frequent (TABLE 2). *In vivo* apoptosis before the therapy was found in 15 of 23 cases by morphologic examination, and in all cases by flow cytometry. According to our pilot study published in 1994, the increase in the number of apoptotic cells was highest 6 hours after the beginning of prednisolone monotherapy.[25] The ratio of apoptotic cells is shown in TABLE 3. In 17 of 23 cases, the morphologic signs of increased apoptosis could be found; the results of flow cytometry showed increased DNA fragmentation in 15 of the 23 patients. By microscopic examination pyknosis, fragmentation of nuclei, and apoptotic bodies were seen. On the curve of flow cytometry, either a second peak or a shoulder on the preexisting peak indicated apoptosis. In 17 of 23 cases the results of the morphologic and flow cytometric examinations were congruent. In six cases only one method showed apoptosis. It is therefore necessary to use both methods in each case. It is remarkable that in two cases (Nos.14. and 16), considerably fewer apoptotic cells were found by flow cytometry 6 hours after the beginning of therapy than before the treatment. Such a finding has never been observed by morphologic examination. Our results do not support the observation of some authors, who could show neither spontaneous nor induced apoptosis *in vivo* in leukemic patients.[28,22] The explanation for these different findings may lie in a different method and timing of examination, because it is known that apoptotic cells may be removed rapidly from the circulation by phagocytosis.

The therapeutic results of ALL in children have improved considerably during the last two decades. However, in cases with poor prognoses, more intensive chemotherapy is necessary but with the possibility of more side effects and therapy-related mortality. The determination of prognosis at the onset of therapy is necessary. An important prognostic factor is the decrease of leukemic blast cells below $1.0 \times 10^9/l$ after 8 days of prednisolone monotherapy. We compared the prednisolone-induced *in vivo* apoptosis after the first 6 hours of prednisolone treatment with the decrease of blast cells over 8 days. One case was not evaluable because the number of blast cells was $1.0 \times 10^9/l$ on day 8, and the patient died of sepsis during the first month of therapy. The apoptosis correlated with the decrease of blast cells in

TABLE 2. Immunophenotype of Our Patients with ALL

Immunology	Number of Cases
cALL	15/23
Pre-pre-B	1/23
B-cell	1/23
T-cell	2/23
Biphenotypic	4/23

TABLE 3. The Ratio of Apoptotic Cells of Our Patients Examined by Morphologic and Flow Cytometric Methods before the Onset of Prednisolone Monotherapy and 6 Hours Later

No.	Name	Age/Sex	Morphology Apoptotic Ratio (%)			Flow Cytometry Apoptotic Ratio (%)		
			0 hr	6 hr		0 hr	6 hr	
1	C.Z.	15/m	0	8	+	2.5	2.9	−
2	C.H.	6/f	3	3	−	4.8	7.0	+
3	N.I.	4/f	1	5	+	1.7	2.5	+
4	S.D.	17/m	1	8	+	5.0	6.1	+
5	K.S.	6/m	0	3	+	2.9	5.5	+
6	N.A.	15/m	1	6	+	4.7	8.7	+
7	M.R.	4/m	0	2	+	4.6	5.0	+
8	H.O.	11/f	1	3	+	6.9	17.8	+
9	S.Z.	7/f	0	20	+	3.5	6.0	+
10	K.E.	14/f	0	1	−	5.6	4.5	−
11	B.I.	2/f	2	15	+	5.0	16.0	+
12	B.A.	11/m	0	3	+	5.0	13.0	+
13	B.P.	4/m	1	3	+	2.3	6.9	+
14	R.J.	4/f	1	1	−	3.4	0.2	−
15	M.P.	4/m	1	1	−	1.1	0.8	−
16	D.F.	9/m	2	6	+	4.5	1.5	−
17	N.K.	8/f	2	6	+	5.7	4.9	−
18	N.K.	4/f	1	6[a]	+	4.4	17.5[a]	+
19	K.G.	11/m	—	1	−	2.6	3.8	+
20	R.J.	6/f	1	1	−	1.6	1.5	−
21	B.L.	6/f	0	6	+	1.6	1.2	−
22	K.D.	6/m	3	8	+	1.72	2.92	+
23	K.I.	8/m	3	6	+	3.71	5.35	+

[a]Measurement at hour 24.

18 of 22 cases (TABLE 4). In one case, in which we did not find correlation between the blast number and apoptosis, the child did not achieve remission at day 33; it therefore became necessary to use more intensive chemotherapy (No. 20). In one child (No. 14), in whom the blast cells decreased below $1.0 \times 10^9/l$ in 8 days and we found no apoptosis, the patient needed intensive therapy because of 9:22 translocation, and relapsed early. There were only two cases out of the total of 22 in whom the clinical outcome did not correlate with apoptosis (Nos. 15, 17). It seems, therefore, that the degree of apoptosis before prednisolone monotherapy and 6 hours later is a useful prognostic factor. These measures also give information to clinicians about the prognosis one week earlier than the decrease of blast cells. More information is needed, however, to reach any conclusions.

Because of the regulatory effect of *bcl*-2 and p53 genes in apoptosis, we were interested in finding whether there was any abnormal expression of these genes in our blast cell samples. We performed a pilot study in 10 patients. The heterozygosity of p53 alleles was observed in six patients; one allele was seen for two patients. The expression of p53 gene was elevated in two patients. The *in vivo*–induced apoptosis did not correlate with the expression of p53. This is congruent with the observation that the presence of wild-type p53 protein is not required for apoptosis induced by glucocorticoids in *in vitro* studies of murine cells.[29] No elevation in the expression of *bcl*-2 gene was observed in our patients.

We also were interested in seeing whether a correlation between the number of glycocorticoid receptors and the extent of apoptosis could be found. We observed that gcRmRNA were present at high levels in three of the examined six cases, all responders, with respect to clinical parameters and apoptosis.

Our knowledge about apoptosis and its regulation may be useful in the future not only in relation to prognosis but also to therapy. Protein kinases participate in signal cascades of apoptosis. The inhibition of protein kinases by B43-Gen (Genistein) can induce apoptosis in B-lineage leukemia cells. Targeting this compound to the B-cell receptor CD19 with the monoclonal antibody B43 could kill the B-cell precursor leukemia cells in a severe combined immunodeficient mouse model.[12] Pokeweed antiviral protein (PAP) is also effective in inducing apoptosis in radiation-resistant primary B-cell precursor leukemic progenitor cells. Targeting to CD19-positive cells, B-43 (anti-CD19)–PAP immunotoxin killed more than 99%

TABLE 4. The Ratio of Apoptosis 6 Hours after Prednisolone Monotherapy and the Clinical Response to the Therapy in Our Patients

Case	Name	Morphology (hour 6)	Flow Cytometry (hour 6)	Absolute Blast Number at Day 8 $(1 \times 10^9/l)$	Remission at Day 33	Now
1	C.Z.	+	−	<1	+	Remission
2	C.H.	−	+	<1	+	Remission
3	N.I.	+	+	<1	+	Remission
4	S.D.	+	+	<1	+	Early relapse, died
5	K.S.	+	+	<1	+	Died after 6 month (sepsis)
6	N.A.	+	+	<1	+	Early relapse, died
7	M.R.	+	+	<1	+	Died after 4 month (sepsis)
8	H.O.	+	+	<1	+	Early relapse, died
9	S.Z.	+	+	<1	+	Remission
10	K.E.	−	−	=1	−	Died before day 33 (sepsis)
11	B.I.	+	+	<1	+	Remission
12	B.A.	+	+	<1	+	Remission
13	B.P.	+	+	<1	+	Remission
14	R.J.	−	−	<1	+	Early relapse, remission during therapy
15	M.P.	−	−	<1	+	Remission
16	D.F.	+	−	<1	+	Early relapse, died
17	N.K.	+	−	>1	+	Early relapse, died
18	N.K.	+ (24)[a]	+ (24)[a]	<1	+	Remission
19	K.G.	−	+	<1	+	Remission
20	R.J.	−	−	<1	−	Remission
21	B.L.	+	−	<1	+	Remission
22	K.D.	+	+	<1	+	Died after 2 month of therapy (sepsis)
23	K.I.	+	+	<1	+	Remission

[a] For this patient, the apoptosis was examined at 24 hours.

of radiation-resistant blast cells isolated from precursor B-cell leukemic patients, and prolonged the survival of severe combined immunodeficient mice xenografted with radiation-resistant human B-cell precursor leukemia.[30] A similar therapeutic effect could not be achieved by standard or investigational chemotherapeutic agents.

SUMMARY

We examined *in vivo* spontaneous and prednisolone-induced apoptosis in peripheral blood samples of 23 children with ALL by flow cytometric and morphologic methods. There was no significant spontaneous apoptosis before the therapy. Six hours after prednisolone therapy, increased apoptosis was found in 19 of 23 cases. In one case, the apoptosis of blast cells could not be compared with the clinical data, because the patient died in sepsis during the induction therapy. In 18 of 22 evaluable cases, the *in vivo* apoptosis correlated with the decrease of leukemic blasts during the first 8 days of prednisolone monotherapy. In 20 of 22 children, a correlation was found between *in vivo* prednisolone-induced apoptosis and clinical outcome.

The p53 gene expression was elevated in 2 of 10 patients. No elevation in the expression of *bcl*-2 gene was observed. In 6 of 23 cases, the glucocorticoid receptors were measured. The correlation of clinical responsiveness and gcR mRNA shows less parallelism than does the apoptosis correlation.

ACKNOWLEDGMENTS

For the glucocorticoid assay, the authors express their gratitude to Prof. A. Falus (Institute of Biology, Semmelweis University of Medicine, Budapest). The authors thank Prof. E. Oláh (National Institute of Oncology, Budapest) for the p53 and *bcl*-2 assays and J. Bocsi and M. Csóka for their collaboration.

REFERENCES

1. LYNCH, M. P., *et al.* 1986. Evidence for soluble factors regulating cell death and cell proliferation in primary cultures of rabbit endometrial cells grown on collagen. Proc. Natl. Acad. Sci. USA **83:** 4784.
2. KERR, J. F., *et al.* 1972. Apoptosis a basic biological phenomenon with wide ranging implications in tissue kinetics. Br. J. Cancer **26:** 239–257.
3. ARENDS, M. J. & A. H. WYLLIE. 1991. Apoptosis: Mechanism and roles in pathology. Int. Rev. Exp. Pathol. **32:** 223–253.
4. CARSON, D. A. & J. M. RIBEIRO. 1993. Apoptosis and disease. Lancet **341:** 1251–1254.
5. SCHULZE-OSTHOFF, K., *et al.* 1994. Cell nucleus and DNA fragmentation are not required for apoptosis. J. Cell Biol. **127:** 15–20.
6. JACOBSON, M. D., *et al.* 1994. Programmed cell death and bcl-2 protection in the absence of a nucleus. 1994. EMBO J. **13:** 1899–1910.
7. MARCHETTI, P., *et al.* 1996. Apoptosis-associated derangement of mitochondrial function in cells lacking mitochondrial DNA. Cancer Res. **56:** 2033–2038.
8. LIN, X., *et al.* 1996. Induction of apoptotic program in cell-free extracts: Requirement for dATP and cytochrome c. Cell **86:** 147–157.
9. RAFF, M. C. 1996. Death wish. Sciences **36**(4): 36–40.
10. WYLLIE, A. H. 1991. Glucocorticoid-induced thymocyte apoptosis is associated with endogenous endonuclease activation. J. NIH Res. **3:** 75–79.

11. DIVE, C. & A. HICKMAN. 1991. Drug-target interactions: Only the first step in the commitment to a programmed cell death? Br. J. Cancer **64:** 192–196.
12. UCKUN, F. M., *et al.* 1995. Biotherapy of B-cell precursor leukemia by targeting genistein to CD19-associated tyrosine kinases. Science **267:** 886–891.
13. WADDICK, K. G., *et al.* 1995. In vitro and in vivo antileukemic activity of B-43 pokeweed antiviral protein against radiation-resistant human B-cell precursor leukemia cells. Blood **86:** 4228–4233.
14. GENG, Y. J., *et al.* 1996. Apoptotic death of human leukemic cells induced by vascular cells expressing nitric oxide synthase in response to interferon-γ and tumor necrosis factor-α. Cancer Res. **56:** 866–874.
15. BRACH, M. A., *et al.* 1992. Prolongation of survival of human polymorphonuclear neutrophils by granulocyte-macrophage colony-stimulating factor is caused by inhibition of progammed cell death. Blood **80:** 2920–2924.
16. KRAJEWSKI, S., *et al.* 1993. Investigation of the subcellular distribution of the bcl-2 oncoprotein: Residence in the nuclear envelope, endoplasmic reticulum and outer mitochondrial membranes. Cancer Res. **53:** 4701–4714.
17. JACOBSON, M. D., *et al.* 1993. Bcl-2 blocks apoptosis in cells lacking mitochondrial DNA. Nature **361:** 365
18. STRASSER, A., *et al.* 1994. DNA Damage can induce apoptosis in proliferating lymphoid cells via p53-independent mechanisms inhibitable by bcl-2. Cell **79:** 329–339.
19. KAUFMANN, Y., *et al.* 1995. Brief report: Lymphoma with recurrent cycles of spontaneous remission and relapse—possible role of apoptosis. N. Engl. J. Med. **332:** 507–510.
20. ZHENG, C. Y., *et al.* 1994. Charasteristics of a cell line established from a T cell acute leukemia by transfection with the bcl-2 gene. Acta-Haematol. **91:** 181–186.
21. COUSTAN-SMITH, E., *et al.* 1996. Clinical relevance of bcl-2 overexpression in childhood acute lymphoblastic leukemia. Blood **87:** 1140–1146.
22. MATSUBARA, K., *et al.* 1994. Induction of apoptosis in childhood acute leukemia by chemotherapeutic agents: Failure to detect evidence of apoptosis in vivo. Eur. J. Haematol. **52:** 47–52.
23. RUSSELL, N. H., *et al.* 1995. Biological features of leukemic cells associated with autonomous growth and reduced survival in acute myeloblastic leukemia. Leuk-Lymphoma **16:** 223–229.
24. BRADSTOCK, K. F. & D. J. GOTTLIEB. 1995. Interaction of acute leukemia cells with the bone marrow microenvironment: Implications for control of minimal residual disease. Leuk-Lymphoma **18:** 1–16.
25. SCHULER, D., *et al.* 1994. Apoptosis as a possible way of destruction of lymphoblasts after glucocorticoid treatment of children with acute lymphoblastic leukemia. Ped. Hem. Oncol. **11:** 641–649.
26. FALUS, A., *et al.* 1994. Separate regulation of a membrane protein, gp130 present in receptor complex specific for interleukin-6 and funtionally related other cytokines. Mol. Recog. **7:** 277–281.
27. FALUS, A., *et al.* 1995. Cytokine networks and corticosteroid receptors. Ann. N. Y. Acad. Sci. **762:** 71–78.
28. JOHNSTON, J. B., *et al.* 1992. Induction of apoptosis in CD4 prolymphocytic leukemia by deoxyadenosine and 2-deoxycoformycin. Leuk. Res. **16:** 781–788.
29. BENEDETTI DE, V., *et al.* 1996. p53 tumor suppressor gene: Implications for iatrogenic cancer and cancer therapy. Med. Ped. Oncol. Suppl. **1:** 2–11.
30. WADDICK, K. G., *et al.* 1995. In vitro and in vivo antileukemic activity of B43 (anti-CD19) pokeweed antiviral protein immunotoxin against radiation-resistant human pre-B acute lymphoblastic leukemia. Blood **86**(11): 4228–4233.

The Present Role of Bone Marrow and Stem Cell Transplantation in the Therapy of Children with Acute Leukemia

RUTH LADENSTEIN, CHRISTINA PETERS,
AND HELMUT GADNER[a]

*St. Anna Children's Hospital
Kinderspitalgasse 6
A-1090 Vienna, Austria*

INTRODUCTION

Stem cell transplantation (SCT) in childhood leukemia in recent years has been widely accepted as a therapeutic option in an otherwise incurable disease. The antileukemia efficacy of transplants results from high-dose chemotherapy and/or radiation given pretransplant and from immune-mediated effects of the graft. Five-year leukemia-free survival ranges from approximately 25% for children transplanted with advanced leukemia, to >60% in those transplanted in first remission of acute leukemia.[1] This therapy's effectiveness, however, is diminished by transplant-related complications, including graft rejection, the problems of acute and chronic graft-versus-host disease (GVHD), and hazards of early and late infections due to a prolonged period of disturbed immune function and relapse. Also severe long-term sequelae, such as growth impairment, gonadal, and thyroid dysfunction, cardiomyopathy, and problems with learning, concentration, and development as well as the risk of second neoplasms, pose a particular problem for pediatric patients.

Worldwide, the number of children receiving bone marrow transplants has been steadily growing. Nevertheless, even after 20 years of experience in this field and a lot of improvements, the issue of defining the candidate who can really benefit from this procedure remains controversial. In general, children failing conventional therapy and, possibly, those with acute leukemia characterized by features predicting a poor response to conventional therapy are candidates for transplants. The field is constantly moving, because more effective antileukemic preparative regimens or post-transplant antileukemic strategies are under investigation. In particular, efforts are made to make preparative regimens less toxic, to decrease the incidence of early and late effects, to develop more effective means of preventing and treating GVHD, and to perform safely mismatched and/or unrelated marrow transplantation when no HLA-identical sibling is available. Even the chapter of autologous stem cell reinfusion (ASCR) is not closed, because with more effective means of purging leukemic cells from remission bone marrow and induction of a graft versus leukemia (GVL) effect even in the autologous setting, it still could overrule alternative stem cell resources with higher associated risks in first remission patients.[2]

---

[a] Address for correspondence: Prof. Dr. Helmut Gadner, FRCP (Glas.), St. Anna Children's Hospital, Kinderspitalgasse 6, A-1090 Vienna, Austria.

Problems of evaluation of SCT in childhood leukemia are related to the rather small numbers of patients, variations in primary therapy, reinduction protocols, pretreatments before BMT, purging and risk groups, and finally reports mixing children with adults.[3,4] This review aims to redefine the indications for SCT according to the various leukemic entities in children on the bases of the literature and recent strategies of large cooperative groups.

ACUTE LYMPHOBLASTIC LEUKEMIA: TREATMENT, ANALYSIS, AND RELAPSE

Conventional Dose Chemotherapy

Modern treatment for ALL achieves long-term survival in two-thirds of children, many of whom are truly cured.[5-11] Recent improvements have occurred as a result of progressive intensification of treatment, particularly during the first 6 months from diagnosis. Standard ALL treatment generally involves the following treatment periods: induction of remission as quickly as possible using a multi-drug schedule, followed by a more aggressive consolidation therapy to reduce tumor burden to a minimum, and finally a prolonged period of continuing or maintenance-phase therapy.

Immunophenotyping and Cytogenetic Analysis

Immunophenotyping and cytogenetic analysis have contributed to a more rational classification of ALL.[12,13] Immunophenotype, various prognostic features, and associated prognosis of ALL are summarized in TABLE 1. Cytogenetic abnormalities can now be detected in about 90 to 95% of patients with ALL.[14-16] Numerical abnormalities are more common than structural abnormalities, the latter occurring in about 40% of patients. Structural abnormalities are often associated with other unfavorable prognostic features: older age group, high total white blood cell counts at presentation, and immunophenotypes other than early pre-B.

Concepts of Remission, Curability, and Lineage Involvement in Relationship to the Problems of Minimal Residual Leukemia

ALL is a malignant disorder resulting from the clonal proliferation of lymphoid precursors with arrested maturation. Chemocurable leukemias appear to be derived from the compartment of hemopoietic progenitors already committed to a single lineage. Intrinsic and/or acquired genetic chemoresistance represents a restriction to this chemocurability.[17,18]

Molecular techniques can assist in identifying the clonality of this disease and the lineage of lymphoblasts. This takes advantage of the normal rearrangement that occurs between the variable (V), diverse (D), joining (J), and constant (C) regions of the immunoglobulins (Ig) and the T-cell receptor genes (TCR).[19-22] These modern methods for identification of minimal residual disease (MRD) afford the hope that it may become possible to identify both children who have received sufficient treatment and, conversely, those at subsequent risk of relapse.

TABLE 1. Immunologic and Genetic Subcategories of Childhood ALL[a]

Lineage	% of Total	Immono-phenotype	Associated Cytogenetic Abnormalities	Fusion Genes/Deregulation Genes	Characteristic Features and Estimated Risk	Treatments	Reports
B-precursor:	83%	HLA DR+, CD 19/22/24/79a+, Tdt+	Hyperdiploid Hypodiploid Near tetraploid		Ploidy >50 = favorable	standard ALL	Pui[14,114] Pui[15]
Early preB -proB -common		cIgμ− -CD10+	Diploid t(4;11)(q21,q23) t(9;22)	AF4-MLL ABL-BCR	Infants, unfavorable CNS+, BM relapse 5%: older age, CNS+ Unfavorable	HR-Arm/BMT HR-Arm/BMT	Pui[116] Fletcher[115]
pre-B		cIgμ+	t(1;19)(q23,p13.3) t(11;19)	PBX1-E2A MLL-ELL			Raimondi[117]
T-cell	12%	CD 2/3/5/7+ CD(1)/(4)/(8)+ Tdt+ Acid phosphatase+	t(11,14)(p13;11q) (del 1p) Near tetraploid	RBTN2-TCRA	high WBC, CNS+, mediastinal mass Unfavorable	HR	Pui[15]
B-cell (often: FAB L3)	4%	HLA DR+, CD19/20/22/24/45+ sIgμ+, Tdt−	t(8;14) t(8,22) t(2,8), t(11,14) t(14,18) 6q−, 14q+	MYC-ICG MYC-IGL IGK-MYC	extramedullary, lymphomatous masses, CNS+	lymphoma type	Reiter[6] Nowell[118]
0-cell		No lymphoid, no myeloid markers					
Biphenotypic	1%	Lymphoid and myeloid markers	11q23	MLL	infants, high WBC Unfavorable	HR	Raimondi[119] Pui[120]

[a] Abbreviations: Tdt, terminal deoxynucleotide transferase; cIgμ, cytoplasmatic immunglobulin μ-chain expression without surface localization; sIGμ, surface immunoglobulin μ-chain; CD, cluster determination.

High-Risk Groups

Less than 20% of children are currently selected for other than "standard" therapy.[9] Such high-risk patients are defined by criteria such as poor initial prednisone response,[23,24] certain genetic abnormalities like the Philadelphia chromosome t(9;22) as well as *bcr-abl* positivity, t(4;11), T-immunophenotype or pre-pre B-ALL, myeloid markers, or more than 100,000 white blood cells/μl. The likelihood of subsequent relapse also can be predicted by the proportion of morphologically obvious marrow blast cells remaining after the first 14 days as well as by the speed of response to treatment, that is, complete remission not achieved at day 42.[25] However, changes in the type of induction treatment protocols have further influenced the previously unfavorable outlook of certain ALL-subgroups, for example, the use of lymphoma-type treatment protocols for B-cell ALL.[26]

PEDIATRIC CLINICAL STUDIES IN ALL: alloBMT AND ASCR VERSUS CHEMOTHERAPY

Because conventional chemotherapy strategies have been so successfully used in the treatment of childhood ALL, high-dose chemoradiation therapy followed by stem cell rescue has been used primarily for children with high-risk disease in first remission or for the treatment of children beyond first remission (TABLE 2).

Allogeneic BMT in ALL

Allogeneic bone marrow transplants may cure patients with acute leukemia never achieving remission with chemotherapy: About 30% of adults with ALL and 20% to 40% of children and adults with AML never achieve remission, even with intensive chemotherapy. Most die of resistant leukemia, often within 6 months or less. Biggs[27] evaluated 126 patients with resistant ALL or AML receiving alloBMT from HLA-identical siblings. The impact of this procedure was demonstrated by induction of remissions in 113 of 115 (98%) evaluable patients and a leukemia-free survival of 21% at 3 years.

TABLE 2. Indications for BMT in Children with ALL with Respect to the Source of Stem Cells

ALL[a]	Autologous Bone Marrow	HLA-Identical Sibling	Family Member Mismatched	Unrelated (mis)matched
Nonresponder	−	+	+	+
CR1[b]	−	−	−	−
CR2 very early relapse	−	+	+	+
CR2 early relapse	−	+	+	+
CR2 late relapse	+	+	(+/−)	−
>CR2	−	+	+	+

[a] CR, complete remission.
[b] Except patients with unfavorable risk features.

Testi[27a] reported in 1992 on 31 intensively pretreated children with ALL refractory to initial therapy or in first bone marrow relapse who were treated with a combination of intermediate-dose AraC and idarubicin. Twenty-four patients (77%) achieved complete remission, 8 patients relapsed early, and 2 were removed from the study. Fourteen (45% of the original 31 patients) underwent alloBMT, and 7 of them (22%) were in continuous complete remission at a median follow-up of 18 months. Thus, this study showed that it is possible to achieve complete remission in ALL children who failed initial intensive regimen.

The high-risk group is eligible for allogeneic BMT in first remission provided a family-matched donor is available: The MRC UKALL X trial was designed to facilitate a nonrandomized comparison between BMT and chemotherapy in children deemed to be at high risk.[8,9] Only 41 of 198 children presenting with a leukocyte count of more than $100 \times 10(9)/l$ at diagnosis had a human-leukocyte antigen (HLA)-compatible sibling donor of whom only 34 proceeded to BMT in first remission. All patients received induction and early intensification therapy. The nontransplant group received cranial irradiation at 24 Gy, late intensification, and two years of continuous treatment. Comparison of the 144 children who were in remission at 17 weeks and received chemotherapy with the transplant group showed no significant difference in event-free survival at five years (69% for BMT and 52% for chemotherapy). There were more treatment-related deaths in the marrow transplant group (6 vs. 4) but more relapses in the chemotherapy group (59 vs. 4). This multicenter trial thus demonstrated, first, that the number of children with HLA-identical sibling donors usually will be too small to make a significant impact on overall survival even in multicenter trials and kindles thus the question to which extent alternative donors should be used. Second, it was shown that BMT was associated with a much lower relapse rate than that with the UK ALL protocol, but results were counterbalanced in this series by the associated BMT acute and long-term toxicities. Nevertheless, for a selected group of patients with an otherwise poor prognosis, BMT-related risks appear justified in view of the chance of long-term cure.

Such a poor-prognosis group was described by Fletcher.[28] Approximately 5% of pediatric ALL contain a translocation (9;22)(q34;q11), which results in rearrangement of the *bcr* and *abl* genes. The prognostic implications of t(9;22)(q34;q11) were assessed in 434 children receiving intensive treatment for ALL. While the 4-year event-free survival was 81% in 419 children lacking t(9;22), it was 0% in the 15 children with t(9;22) ($p < 0.001$). Thus, all children with t(9;22)-positive ALL clearly need more intensive treatment approaches, including allogeneic BMT in first remission.

The BM relapse group is eligible for allogeneic BMT with a family-matched donor in second remission: Although most children who sustain a relapse of their leukemia will achieve a second remission with reinduction of conventional dose chemotherapy, the treatment of choice for the subsequent therapy still is the subject of some controversy.

The majority of studies of children treated with chemotherapy alone for ALL in second remission report event-free survival rates of 10–25%,[29–31] sometimes reaching 35%.[32,33] In contrast, the results of allogeneic BMT are often reported with event-free survival rates of 40 to 50%,[30,34–39] although others are not as favorable.[40–43] However, only early BM relapses (up to 6 months after end of initial therapy) are sure candidates,[10,37] whereas the data on late BM or combined relapses still is contradictory because some reports indicate that late relapses may benefit equally from an intensive conventional-dose relapse treatment.[44] Furthermore, it appears that the majority of extramedullary relapses are well treated with conventional chemotherapy alone.[45]

Experiences of the BFM Group

The BFM (Berlin-Freiburg-Münster) Group reported three consecutive multicenter trials for relapsing ALL, ALL-REZ-BFM 83, 85, and 87, which enrolled 326 children with BMT relapse of non-B ALL. Employing an intensive polychemotherapy regimen stratified according to type of relapse, extramedullary involvement appeared to be predictive of superior outcome in all trials as well as in the whole group with a 7-year event-free survival rate of 42% in combined versus 15% in isolated BM relapse ($p = 0.015$). Children with combined BM relapse occurring later than 6 months after completion of front-line therapy, that is, late BM relapses, reached event-free survival estimates of 60%. Thus, results of conventional polychemotherapy with BFM relapse protocols appeared equivalent to those achieved by BMT in children with late combined BM relapse.[44]

The role of alloBMT for childhood ALL in second remission after intensive primary and relapse therapy according to the BFM and CoALL protocols was also addressed by the German Cooperative Studies.[37] Fifty-one children were grafted in second remission from HLA-identical sibling donors (except for two patients who were grafted with a marrow with one antigen-mismatch). Front-line treatment was similar for 38 patients on BFM and 13 on CoALL protocols. At relapse patients were treated in consecutive relapse studies: 12 in ALL-REZ-BFM 83, 17 in ALL-REZ-BFM 85, 20 in ALL-REZ-BFM 87, and two in ALL-REZ-BFM 90. Conditioning regimens were different, consisting of TBI/CY ($n = 27$), TBI/VP16 ($n = 23$), and TBI/CY/AraC ($n = 1$).[b] Kaplan-Meier analysis yielded an event-free survival of 0.52 +/− 0.08 with 29 patients in continuous complete remission at a median follow up of 30 months. EFS for patients with late relapses was 0.47 +/− 0.12 and for patients with early relapses 0.56 +/− 0.1. There was a nonsignificant trend towards superior results for patients grafted after conditioning with TBI/VP16 and a significant advantage of alloBMT over chemotherapy for early relapses ($p < 0.01$).

The value of various treatments for isolated extramedullary relapse in children with ALL was evaluated by Borgmann.[45] Between 1983 and 1993, 165 patients up to 19 years were registered in the BFM relapse trials. While 134 patients received chemotherapy only, 17 were grafted from HLA-identical sibling donors, whereas 14 children underwent autologous BMT. Event-free survival at 5 years was 0.47% for patients receiving chemotherapy (0.76% for late relapses and 0.33% for very early and early relapses). In patients who had undergone BMT, event-free survival was 0.36% without a significant difference in the ASCR and alloBMT groups (56% for late relapses, 38% for early, and 17% for very early relapses), but small patient numbers have to be considered. It is further noteworthy that in the BMT group 71% had early or very early relapses and 45% of patients had T-cell immunophenotype (significantly more than in the chemotherapy arm), presenting thus a selection of patients with unfavorable risk factors. In the chemotherapy arm independent significant risk factors were early time of relapse and T-cell phenotype. For the latter patients the outcome was of only 14% with conventional chemotherapy. Thus, it became clear that it is very important to consider even in the case of isolated

---

[b] Abbreviations: CV, cyclophosphamide; VP16, etoposide; TBI, total body irradiation; AraC, cytarabine.

extramedullary relapse, the phenotype, the time of relapse, and the number of relapses (29% of patients had a 2nd or subsequent relapse in the BMT arm whereas only 18.5% in the chemotherapy arm) to recognize patients with a need for more complex treatments.

IBMTR and USA Experience

An interesting retrospective analysis was performed by Barrett[30] comparing the results of two cohorts of children with relapsing ALL: One cohort of 376 children was treated with marrow transplants from HLA-identical siblings, as reported to the International Bone Marrow Transplant Registry (IBMTR); the other cohort of 540 children received chemotherapy according to the Pediatric Oncology Group (POG) protocols. A preliminary analysis identified variables associated with treatment failure in both groups and was the basis for selecting cohorts by matching these variables. A possible bias associated with differences in the interval between remission and treatment was controlled by choosing matched pairs in which the duration of the second remission in the chemotherapy recipient was at least as long as the time between the second remission and transplantation in the transplant recipient. A total of 255 matched pairs was studied. The mean $(+/-$ SE) probability of relapse at five years was significantly lower among the transplant recipients than among the chemotherapy recipients (45 $+/-$ 4% vs. 80 $+/-$ 3%, $p < 0.001$). At five years the probability of leukemia-free survival was higher after transplantation than after chemotherapy (40% $+/-$ 3% versus 17% $+/-$ 3%, $p < 0.001$). The relative benefit of transplantation as compared with chemotherapy was similar in children with prognostic factors indicating a high or low risk of relapse (the duration of the first remission, age, leukocyte count at the time of the diagnosis, and phenotype of the leukemic cells). The apparent superiority of transplantation was also demonstrated for children who had first remission of less than 36 months as well as for those patients whose initial remission was greater than 36 months. However, if the duration of first remission was greater than 36 months, the survival after alloBMT was also better. Given the limitations of this type of analysis, this study strongly suggested that BMT from HLA-identical siblings resulted in improved leukemia-free survival rates with a lower probability of relapse in comparison with POG conventional-dose relapse protocols. In this analysis alloBMT seemed to erase the prognostic factors indicating a high or low risk of relapse.[30]

Similar results were previously reported by Sanders[35] for 57 children with ALL in second complete remission who received HLA-identical sibling transplants after conditioning with cyclophosphamide and TBI. Children with initial remissions of less than 36 months achieved a leukemia-free survival rate of 40%.

ALTERNATIVE STEM CELL SOURCES

Although allogeneic BMT has been reported to provide disease-free survival rates from 40 to 70% in children with recurrent ALL, its application is severely limited by the lack of sibling donors. Only about 30% of children will have a suitable, matched sibling donor. Thus, there is a need for alternative sources of stem cells, including autologous purged bone marrow, partially mismatched related donors or (mis-) matched unrelated donors.

AUTOLOGOUS STEM CELL REINFUSION IN ALL

The use of ASCR allows the application of escalated dose intensity to a larger number of patients aiming, thus, at a better eradication of leukemic clones. A potential problem associated with transplanting autologous marrow, the reinfusion of residual leukemic cells in the harvested marrow, can be addressed through purging. The most widely used purging techniques involve either immunologic or pharmacological techniques. The use of purged autologous marrow transplantation for children with ALL has been widely explored, with event-free survival rates of approximately 20 to 30% generally reported.[38,46-48] However, the efficacy of purging has been addressed in a variety of studies[49,50] but is not yet convincing and still awaits confirmation in randomized, prospective trials. Some reports claim that ASCR does contribute to the cure of relapsed children.[10,47,51,52] Nevertheless, the benefit of ASCR remains controversial because its major disadvantage, that it lacks the capacity to induce a graft versus leukemia effect (GVL), has not yet been compensated for, although there are models to induce GVL, for example, by the introduction of cyclosporin A, in the post-transplant period of ASCR. Because there is a lack of evidence of GVL at the most commonly involved extramedullary sites, CNS and the testicles, the lack of GVL effect might be less disturbing when ASCR is used for early isolated extramedullary relapses, as suggested by some studies to introduce further dose escalation to improve prognosis.[52-54]

Single Center Study Reports

The crucial factor of time to relapse was stressed by the report of Sallan[47] who reported a 2-year disease-free survival of 29% for 44 children with ALL beyond first complete remission receiving a purged autologous transplant. After an initial remission duration of less than 24 months, however, the disease-free survival rate was only 10% compared to a probability of 51% for patients with first-remission durations greater than 24 months.

The possible bias of patients selectively transferred to a center for ASCR was addressed by Lonnerholm.[55] Twenty-five children with high-risk ALL and treated with purged, autologous BMT achieved a very favorable outcome of 65% long-term disease-free survival with 8 out of 25 (72%) in continuous complete remission at a median follow-up of 50 months (range 5–71 months).

The Dana-Farber Cancer Institute has had an active ASCR program for children with recurrent ALL[52] since 1980. Sixty-six children underwent ASCR after TBI-containing conditioning regimens followed by infusion of autologous marrow purged with two monoclonal antibodies directed against CD19 and CD10 and complement. The disease-free survival rate was 47% for patients with a first remission of at least 2 years, compared with a rate of 10% for those with a shorter first remission. Considering only the 51 children with ALL in second or subsequent remission of at least 24 months, the event-free survival (+/− SE) at 3 years after ASCR was 53% +/− 7%. In multivariate analysis, the most significant predictors of event-free survival were duration of longest pre-ASCR remission and remission duration immediately before ASCR.

Multicenter Study Reports

The reports of 552 patients with ALL who underwent ASCR at 17 centers were analyzed by Billet.[52] Only four series were limited to pediatric patients. Although conditioning regimens varied greatly, more than 80% of patients received at least TBI/CY. Purging was used in 483 (87%) patients, and failure of engraftment was reported in only three patients. The rates of disease-free survival in these series clustered between 25 and 35%. The most common cause of treatment failure after ASCR was relapse, which occurred in 40 to 85% of patients. Early deaths from toxicity occurred in 5 to 21% of patients.

Data of 75 children who underwent ASCR were obtained from the BMT Registry of the AIEOP (Italian Association of Pediatric Hemato/Oncology). Fifty-six children were transplanted after marrow +/− other site(s) relapse and 19 after an isolated extramedullary relapse with various transplant preparative regimens. Forty-four (58.6%) of 75 patients relapsed again following ASCR with a 5-year DFS of 27.8%. In this retrospective study, ASCR for patients with ALL in second complete remission following marrow relapse did not offer an encouraging result (13% probability of disease-free survival at 5 years), whereas ASCR following an (early) isolated extramedullary relapse resulted in nearly 70% DFS. Although ASCR was suggested as an option for this latter group of patients by the authors, the results demonstrate a better outcome in a better risk group and no superiority of ASCR over conventional treatment.[56]

The role of ASCR for early, isolated extramedullary relapse of childhood ALL was also addressed by Rosetti.[54] Twelve children underwent ASCR following high-dose cytarabine (HD AraC) plus 14.4 Gy hyperfractionated TBI for early isolated extramedullary relapse of ALL, while in first BM remission. No patient received intrathecal prophylaxis following ASCR. The toxicity of HD AraC plus hyfr-TBI was acceptable and well controlled with supportive therapy. Because 8 out of 12 children were in continuous complete remission, it was suggested that this particular approach may be very efficient in eradicating leukemia from extramedullary sites of ALL relapse. In view of the three early BM relapses after ASCR, however, a more intensive preparative regimen may be considered, that is, additional melphalan as suggested by the French group, to avoid early systemic relapse.

The BFM group also investigated the prognostic impact of fractionated (f) TBI (12 Gy)/VP16 and purged ASCR in 22 children with high-risk, relapsed ALL,[57] that is, relapse within 24 months of diagnosis; any relapse of T- or B-ALL; and multiple relapses, in second ($n = 13$), third, or subsequent complete remission ($n = 9$). Before transplantation all patients had been treated according to intensive German BFM front-line or BFM relapse protocols. The Kaplan-Meier estimation showed a probability of event-free survival of 18% and a probability of relapse of 80%. Thus, results were not satisfactory, demonstrating the need to develop better therapy concepts in the ASCR setting or to consider the use of allogeneic stem cells even including alternative donors.

The failure of purged ASCR in high-risk ALL in first complete remission as a therapeutic alternative to cranial irradiation and maintenance chemotherapy was highlighted by Gilmore.[58] Twenty-seven bone marrows taken in remission were purged using monoclonal antibodies (anti-CD7 for T lineage and anti-CD10 and/or anti-CD19 for B-lineage leukemias) plus rabbit complement. Engraftment was seen in 26 of the patients. The actuarial disease-free survival was 32% at 7 years (median follow-up 3.4 years). The main cause of treatment failure was disease recurrence with an overall actuarial risk of 67%—76% for T-ALL (five of nine), 62% for common ALL (five of 10), two of five pre-B, and none of three patients with B-ALL.

Similar outcomes after ASCR and alloBMT were reported for 74 children with relapsed ALL when comparable subsets of patients were selected.[60] Specifically, patients who were not in complete remission, those with T-cell leukemia, t(9;22) or those with only extramedullary relapse, were excluded from both groups. The event-free survival at 3 years was 47% for the 57 ASCR patients and was 53% for the 17 patients receiving an alloBMT. However, antileukemia potential was not comparable since the relapse rate was higher in the ASCR group, whereas the incidence of nonleukemic deaths was higher in the alloBMT group.

Borgmann compared treatment results for children who underwent ASCR with those who had conventional chemotherapy.[59] All patients were on BFM or CoALL multicenter front-line or relapse trials. Two groups of patients were selected for matched-pair analysis based on variables associated with treatment outcome and duration of second remission. The event-free survival probability at 9 years was 32% for patients receiving chemotherapy versus 26% for patients who underwent ASCR. Because ASCR was used as the final treatment step, it is possible that its relative ineffectiveness was due to the lack of continuation therapy after transplant. Thus, it was stressed that ASCR should be complemented by subsequent immuno-therapy, molecular biotherapy, chemotherapy, or a combination of those.

CONCLUSIONS

The consecutive BFM studies have supplied important data based on conventional-dose chemotherapy trials to define patients at risk. These trials demonstrated[10] that patients with an early bone marrow relapse of non-T-ALL, that is, occurring less than 6 months after the end of maintenance therapy, and very early bone marrow relapses, that is, occurring within 18 months after diagnosis, or bone marrow relapses in T-cell ALL have little chance to survive (<5% with conventional dose chemotherapy).[33] In contrast, in the hands of the BFM group, isolated extramedullary relapses have a high probability of 77% to remain disease-free after conventional chemotherapy alone.[59]

Whenever a more intensive approach appears to be indicated, allogeneic SCT appears as the better strategy, whereas autotransplants are clearly lacking the GVL effect. The additional immune-mediated effect of GVL with the potential to eradicate residual leukemic cells is even present after intensive preparative regimens. This became apparent in different studies finding that the probability of relapse was higher, and the time to relapse shorter in the ASCR group in comparison with alloBMT.[35,60] However, most studies report a higher toxic death rate associated with alloBMT. This important difference in causes of treatment failure as well as quality of life and functional status post transplant have to be considered in treatment decisions. In comparable subsets of relapse patients, that is, excluding patients with high-risk disease, ASCR and alloBMT appear equivalent, demonstrating a positive impact of ASCR related to the highly intensive myeloablative radio- and/or chemotherapy schedules being used with the advantage of a rather short treatment period. New therapy approaches including immune therapy to induce GVL effect in the autologous setting are underway, using either cyclosporin A[61] or the cytokine interleukin 2 or both. Studies on the immunostimulatory effect of interleukin-2[2,62–64] kindled interest to further stimulate the potential advantage of the GVL effect in order to improve the quality of second remission after both alloBMT and ASCR. It should be mentioned that the duration of first remission (<24 months versus ≥24 months) on event-free survival after BMT apparently has a significant impact in the autologous setting[47,60] as well as in conventional dose relapse protocols.[29,30,65] This is less clear when alloBMT is performed, that is, some confirm this prognosis factor[30,66,67] while others don't.[35–37] In addition, related mismatched donors or (mis)-

matched unrelated donors could be considered as an option for high-risk patients beyond first remission.

Once a patients gets enrolled in a SCT program according to given guidelines, adequate treatment to extramedullary sites, in particular the CNS and testes, still is an important issue[45] to prevent post-SCT relapse in these sites. Irradiation doses and sites will depend on the initial disease involvement and/or presentation at time of relapse, the inherent leukemia risk, previous prophylaxis/treatment of extramedullary sites as well as the preparative regimen before SCT.[37,54,68]

Unfavorable risk group as defined by BFM criteria at diagnosis for elective alloBMT in first remission: late/non-response, t(9;22), t(4;11), poor prednisone response (plus one or more of the following), risk factor >1.7, myeloid marker(s), pre T/T-ALL.

High risk groups at relapse for elective alloBMT: (1) very early bone marrow and combined relapses, that is, occurring within 18 months after diagnosis; (2) early bone marrow and combined relapse, occurring less than 6 months after the end of maintenance therapy; and (3) isolated bone marrow and combined relapses in T-cell ALL.

Indications

Choice of the stem cell source in relation to remission status is summarized in TABLE 2.

- CR1: High-risk patients do benefit from allogeneic BMT from an HLA-identical sibling.
- CR2: Two-thirds of children do benefit from allogeneic BMT. Late bone marrow relapses beyond 30 months may benefit from autologous BMT, which is, however, not clearly superior over intensive conventional dose relapse chemotherapy. However, when allogeneic BMT is considered, any allogeneic source appears justified in very early and early BM relapses, whereas for late BM relapses only an HLA-identical donor should be considered. Early isolated extramedullary relapses may benefit from ASCR.
- >CR2: One-third of children do benefit from allogeneic BMT. If an HLA-identical sibling is not available, alternative allogeneic donor sources should be considered. The value of autologous BMT is questionable unless additional immunotherapy is given.
- no CR: single cases benefit from allogeneic BMT (HLA-identical sibling but also alternative allogeneic donor sources).

ACUTE MYELOGENEOUS LEUKEMIA

History and progress of treatment for acute myelogenous leukemia (AML) in children has evolved over the past decade. Although it represents only 15 to 20% of all childhood acute leukemias, >30% of deaths from leukemia are still a consequence of AML, and only 30 to 50% of children with newly diagnosed AML are expected to achieve a long-term remission with current strategies.[5,69–72]

Before the 1970s, nearly every child with AML died. Modern intensive chemotherapy, bone marrow transplantation, and advanced supportive care of critically ill patients have improved the outlook for children with this fatal disease. Induction of remission in AML needs intensive multiagent chemotherapy, causing prolonged bone marrow aplasia. The introduction of AraC in combination with anthracyclines for induction therapy was a major breakthrough in the treatment of AML. The original "7 and 3" schedule has been adopted by most pediatric AML trials, which have been extensively reviewed by Vormoor.[70] Greater advances in treatment are expected as headway is made in understanding the complex biology of this heterogeneous disease.

FAB Classification and Associated Cytogenetic Abnormalities

The incidence of the various FAB classification subtypes, their immunophenotype, associated cytogenetic abnormalities, and fusion genes as well as associated outcome are summarized in TABLE 3.

Some recent reports further highlight the association of certain cytogenetic abnormalities with FAB subtypes and show an important influence on prognosis. AML with t(8;16) or its variant t(8;V) is found in a high proportion of infants and children, often with a bleeding tendency and disseminated intravascular coagulopathy. Only one-third of these *de novo* patients remain in the first complete remission following multiagent chemotherapy and BMT. This disorder is classified morphocytochemically, as an M5, M4, or M4/M5 variant. The high specificity of t(8:16) or its variants to the unique M4/M5 type leukemia and the role of a gene on 8p11 in this specific transformation is supported by various reports.[73]

The translocation t(1;22)(p13;q13) is a nonrandom marker specifically associated with AML in young children,[74] and its occurrence appears restricted to the FAB subtype M7. Its overall incidence in children with acute megakaryocytic leukemia (AMKL) was 27.8% and was 66.7% in infants with AMKL according to the AML-BFM group. When compared with an age-matched group of children with AMKL lacking this translocation, patients with the t(1;22) had a lower median value of the peripheral white blood cell (WBC) count, a higher median hemoglobin level, and a hypocellular BM with marked fibrosis. Despite intensive chemotherapy, AMKL in children appears to be associated with a poor prognosis. Five such patients received aggressive chemotherapy. However, two died within 15 months of diagnosis without entering remission, whereas one out of three patients who entered remission died 7 months after diagnosis, most likely from intramedullar hemorrhage. Only two out of five children with the t(1;22) who received ASCR are alive and in complete remission 23 and 40 months after diagnosis, respectively. ASCR or alloBMT was suggested for this subset of patients.

A retrospective analysis of 24 cases of childhood AMKL was performed by St. Jude Children's Research Hospital, Memphis, Tennessee.[76] Leukemic cell morphology and cytochemistry were consistent with the M7 classification in 17 cases, and all cases tested expressed megakaryocytic surface antigens. There were five Down's syndrome patients and one patient with chronic myeloid leukemia (Ph+) in blastic crisis. Twelve patients had significant hepatosplenomegaly. AMKL patients were significantly younger than other AML patients ($p = 0.0001$) and had poorer responses to therapy ($p = 0.03$, univariate analysis only). Ten of 24 failed induction, and only five are disease-free at 6 months to 4.5+ years. AMKL, thus, usually affects young children, frequently producing marked organomegaly. It comprises approximately 10% of pediatric AML cases and responds poorly to intensive AML therapies.

Acquired pure monosomy7[77] is associated with various myeloproliferative disorders (MPD), myelodysplasias (MDS), and AML in children and with a poor prognosis.[76] Several myeloid cell lineages are involved in the proliferation, which partly

TABLE 3. FAB Classification and Childhood AML[a]

FAB Class	Incidence % of AML[b] Mean/Range	Histochemistry	Immunophenotype	Associated Cytogenetic Abnormalities	Fusion Genes/Deregulation Genes	Characteristic Features and Estimated Risk	Treatment Intensity/Trials[b]	Reports
M0, undifferentiated	4%	MP−	MPO+, CD33, CD13	−7		Favorable	CHT	
M1, myeloblastic undifferentiated	14% (10–20)	MP+	CD33, CD13, CD11, HLA DR+	+8, −5, −7		Favorable	CHT	Baranger[76]
M2, myeloblastic undifferentiated	25% (20–27)	MP+	CD33, CD13, CD11	t(8;21)	EOT, AML-1	Favorable	CHT	
M3, promyelocytic	8% (3–17)	MP+	CD33, CD13, CD11	t(15;17)	RAR-alfa/PML	Bleeding favorable	Retinoid-sensitive	Huang[121] Warrell[122]
M4, myelomonocytic	23% (16–27)	MP+ NSE+	CD33, CD13, CD11 HLA DR+	inv 16 t(8;16)	MLL, HRX, ALL-1		CHT or CHT+SCT	Stark[73]
M5, monoblastic	18% (14–26)	NSE+	CD33, CD13, CD11 HLA DR+	t(1;11) t(9;11), t(8;16)		Unfavorable	CHT+SCT	Odom[123]
M6, erythroblastic	4% (1–5)	PAS+ MP+	glycophorine+spectrin +HLA DR+			Unfavorable	CHT+SCT	
M7, megakaryoblastic	5% (2–8)	PPO+	glycoprotein IIb, IIIa (CD41) or IIIa (CD61)	t(1;22)		Unfavorable Young children Organomegaly	CHT+SCT	Lyon[74] Ribeiro[75]

[a] MP = myeloperoxidase; NSE = nonspecific esterase; PPO = platelet peroxidase; CMT = conventional chemotherapy; SCT = stem cell transplantation.
[b] UK trial: Stiller, 1994, BFM 83/87; Creutzig, 1993, AEIOP; Amadori, 1993, CCSG 2861; Wells, 1994, POG 8498; Ravindranath, 1991.

explains the difficulties of cytological classification and suggests that a pluripotent stem cell is at the origin of the disease. In a series of 14 children with malignant blood disorders with pure monosomy 7 (8 MPD, 2 refractory anemia with excess of blasts, and 4 AML) the outcomes of MPD and RAEB were characterized by a high risk of rapid blastic transformation and resistance to polychemotherapy. BMT seems to be a good treatment option since a child with AML was reported to survive in complete remission 2 years after ASCR.

Description of Risk Criteria According to BFM Criteria

The AML-BFM Study Group reported in 1990 on the impact of therapy intensification and risk groups in study AML-BFM 83 as compared with study AML-BFM 78.[78] The induction treatment in the second study was intensified by addition of an 8-day ADE (cytarabine, daunorubicin, etoposide) resulting in a reduction of the relapse rate, but not the rate of induction failures. The probability of a 6-year EFS increased from 38% in study AML-BFM 78 to 49% in study AML-BFM 83 (p = 0.08). The improvement of the 6-year event-free interval was significant in the second study (61% versus 47%, $p < 0.05$) but was restricted to the FAB types M1 through M4 (EFI: 67% versus 45%, $p < 0.01$). Thus, two different risk groups (low and high) were identified by combinations of predominantly pretherapeutic parameters. The low-risk group, comprising 37% of the patients achieving complete remission, included the FAB types with granulocytic differentiation and specific additional features: FAB M1 with Auer rods, FAB M2 with white blood cell count of less than $20,000/\mu l$, FAB M3 all patients, and FAB M4 with eosinophilia. The 6-year Kaplan-Meier estimation of the event-free interval was 91% compared with 42% in the high-risk group.

These prognostic factors predicting relapse-free survival were confirmed by a further retrospective analysis of studies AML-BFM 83 to 87.[79] Approximately 70% of children are high-risk patients with an expectancy of only 37% event-free survival. This group of children thus has a need for intensified treatment approaches, that is, introduction of HAM (HD AraC with mitoxantrone), which has been proven to be very efficient in adult refractory AML.

The Role of SCT in AML

BMT for children with alloAML developed from adoption of allogeneic BMT in adults.[80,81] Matched, related BMT in first remission achieved long-term disease-free survival in well over half the young patients. However, as only 25 to 30% of patients have a matched sibling donor, other options for BMT have been explored, including ASCR and unrelated donor BMT.[82–85] There are many issues that require further study regarding autologous transplant, including the efficacy and benefits of marrow purging.[82] Unrelated donor transplants offer encouraging results in suitable patients, but bring with them the increased risk of complications.[83,84]

BMT also was successful in some children in second remission or who have refractory disease.[81,86–88] A recent development in BMT of children with AML was the use of nonradiation-containing conditioning regimens to avoid potential long-term sequelae of TBI.[89–91] Nonradiation regimens appear to be as efficacious as radiation-containing regimens for these children.[89,90]

The superiority of BMT over chemotherapy has recently been challenged, as improved supportive care and extremely intensive therapy have resulted in approximately 75% of children achieving a remission and 30 to 50% of patients surviving.[80] Recent studies have focused on answering the question regarding the role of BMT in children in first remission. Strategies to improve the outcome of BMT and comparison of the sequelae of BMT to those on high-dose chemotherapy are issues that need to be addressed.

Pediatric Clinical Trials in AML: Chemotherapy versus Transplantation

Conventional chemotherapy protocols (CCT): Treatment results of three consecutive German childhood AML studies have been reported by the AML-BFM Group.[72] Between December 1978 and April 1991, 543 children under the age of 17 years entered the three consecutive multicenter studies, AML-BFM-78, -83, and -87. The treatment strategy of BFM 78 consisted of an 8-week induction/ consolidation regimen employing seven different drugs together with cranial irradiation, followed by continuous maintenance for two years. The main alteration in the second study, BFM-83, was the addition of an intensive 8-day ADE induction course (AraC, daunorubicin, VP16). In the BFM-87 trial, two courses of HD-AraC and VT16 were given after consolidation. Complete remission rates were 80% in trials I and II, and 78% in trial III. The event-free survival rates at 4.5 years were 35% in study I, 49% in study II, and 45% in study III. Only 10 children underwent BMT in first complete remission during the given observation period, but interestingly 7 out of 10 are still in first complete remission after a maximum follow-up time of 3.5 years. However, the addition of HD-AraC together with VP16 given after induction/consolidation treatment did not further reduce the incidence of relapses in childhood AML.

Cranial irradiation in childhood AML could have an important role according to the AML-BFM 87 study.[79] In 100 patients with a low risk of CNS recurrences (no initial CNS involvement and WBC at diagnosis $< 70.000/\mu l$), a significantly ($p = 0.007$) higher probability of the relapse-free interval at 5 years (pRFI: 78%) was found in irradiated patients compared with nonirradiated patients (pRFI: 41%). The authors concluded that cranial irradiation should be an integral part of the treatment of all AML patients not undergoing BMT, since residual blasts in the CNS appear to escape systemic chemotherapy leading thus to recurrence of disease not only in the CNS but also in the bone marrow.

The Children's Cancer Study Group (CCGS) conducted a prospective trial in which 589 patients with AML were randomized to one of two induction approaches involving five active chemotherapeutic agents, one with intensive timing.[93] All patients achieving complete remission received a total of four induction courses. They were then allocated to either alloBMT, if a compatible family donor was present, or randomized to aggressive nonmyeloablative therapy or to myeloablative therapy followed by ASCR. Superior results were documented for patients receiving intensive timing irrespective of the postremission therapy: 211 patients on the intensive chemotherapy arm reached an event-free survival of 42% at 3 years versus 195 patients on the standard treatment arm with an event-free survival of 27% ($p = 0.0002$). This study clearly demonstrated that without controlling for the type of induction regimen, results of various BMT studies comparing different preparative regimens will be difficult to interpret.

Single-center studies of SCT in AML: Hermann[94] reported results in favor of ASCR because it spared severe morbidity and/or mortality as observed in the alloBMT patients (two deaths from severe acute GVHD, two from encephalopathy and cardiomyopathy). Transplant candidates were children with AML and high risk of relapse defined by an initial WBC greater than $20 \times 10(9)/1$ and FAB M 5 to M 7. The disease-free survival for the 29 patients grafted in remission was 65%, without a statistical difference in disease-free survival (DFS) between patients receiving either alloBMT or ASCR.

Because Shaw observed a high relapse rate in the ASCR group, the group decided to escalate the dose of busulfan (BU) from 16 mg/kg to 600 mg/m^2 using a single dose of BU.[95] After six cycles of induction chemotherapy 43 consecutive children with AML had received either alloBMT ($n = 11$) or unpurged ASCR ($n = 23$) following a preparative regimen with BU/CY. Disease-free survival rates were 90% for alloBMT patients (median follow up 42 months) and 61% for the ASCR group. In the alloBMT group one out of 11 relapsed and one died of toxicity, whereas there were no toxic deaths among the ASCR patients but 8 out of 23 relapsed.

An intensive regimen in childhood AML, incorporating high-dose melphalan (180 mg/m^2) and ASCR in first complete remission was investigated by Tiedemann.[96] As published by Shaw,[95] patients received an intensive induction regimen consisting of six cycles. The 5-year event-free survival rate was 68%. However, 21 out of 32 patients had FAB types M1, M2, M3, and M4 eosinophilic and would not have been transplant candidates in the hands of other groups (see BFM criteria).

Multicenter studies of SCT in first-remission AML: The AIEOP ("Associazione Italiana di Ematologia Oncologia Pediatricia")/GITMO Group reported all children with AML grafted in first complete remission in 11 Italian centers between 1980 and 1990.[97] The preparative regimens in these 59 children receiving an alloBMT from an HLA-identical sibling ($n = 57$) or from an identical twin ($n = 2$) included TBI/CY ($n = 50$) or melphalan ($n = 2$) or BU/CY ($n = 7$). GVHD prophylaxis consisted of cyclosporin A ($n = 48$), methotrexate ($n = 7$) or cyclosporin and methotrexate ($n = 2$). Actuarial relapse-free survival was 58% at 66 to 137 months. Risk of nonrelapse deaths was 33% in the period from 1980 until 1987, but only 4% in the period from 1988 until 1990 ($p = 0.02$). The group thus suggested that alloBMT in first complete remission should be considered as consolidation therapy if a matched sibling is available.

The French experience with alloBMT in children with AML in first complete remission was recently analyzed using the data base of the French BMT registry (Societe Francaise de Greffe de Moelle). Twenty-three children were conditioned with busulfan and 120 mg/kg CY (BU/CY-120 group), whereas 19 children received a higher CY dose (200 mg/kg) (BU/CY-200 group). During the same time period, 32 patients were prepared with TBI/CY. The probability of relapse was 54%, 13%, and 10% for the BU/CY120, BU/CY200, and TBI groups, respectively ($p < 0.05$). In multivariate analysis, a conditioning regimen with BU/CY120 was significantly associated with a higher risk of relapse ($p = 0.02$; relative risk, 3.62). The probability of transplant-related mortality was 0% for BU/CY-120, 5% for BU/CY-200, and 10% for TBI regimens. The event-free survival rates were 46%, 82% and 80%, respectively, for the three groups, with median follow-up durations of 28 months, 31 months, and 48 months. In the multivariate analysis, two factors adversely affected event-free survival: a conditioning regimen with BU/CY-120 ($p = 0.07$) and a long interval from diagnosis to BMT ($>$ or $= 120$ days, $p = 0.08$). Busulfan/CY-120 was shown to be a well-tolerated preparation, but one that results in a high risk of relapse for children with AML in first complete remission. Most importantly, this high risk of relapse was not observed when the dose of CY was increased to 200 mg/kg.[92]

ASCR versus alloBMT and/or CCT: The AIEOP Group also conducted a prospective study to assess the comparative values of alloBMT and ASCR with sequential postremission chemotherapy (SPC) in children with AML in first remission.[98] One hundred sixty-one assessable patients younger than 15 years of age with newly diagnosed AML were treated uniformly with two courses of daunorubicin and standard-dose AraC. After initial consolidation with a course of daunorubicin, cytarabine, and thioguanine (DAT), patients in complete remission were randomized to receive either ASCR or SPC, except for those with an HLA-matched sibling who were assigned to undergo alloBMT. SPC consisted of three additional courses of DAT, followed by three pairs of drugs administered sequentially for a total of six cycles. Overall, 127 of 161 patients attained complete remission (79%). For the 127 complete responders, the 5-year probability of disease-free survival was 31%, with a cumulative risk of relapse of 64%. The 5-year disease-free survival was 51% for the alloBMT group ($n = 24$), significantly higher ($p = 0.03$) than that observed for the other cohorts: 21% for ASCR ($n = 35$), 27% for SPC ($n = 37$) and 34% for a group of 31 nonrandomized patients. Bone marrow relapse was the most frequent cause of postremission failure in all therapeutic subgroups, including the BMT cohort, in which no deaths attributable to the toxicity of the procedure were recorded. The results of this study showed that alloBMT was more effective than ASCR or SPC in preventing leukemia relapse and extended disease-free survival duration in children with AML in first remission.

A similar study design was chosen by the European Organization for Research and Treatment of Cancer (EORTC) and the Gruppo Italiano Malatti Ematologiche Maligne del'Adulto (GIMEMA): alloBMT in case of a matched, related sibling while the others were randomized to receive either ASCR or a second course of intensified chemotherapy (ICHT) consisting of HD-AraC and daunorubicin. Induction treatment was initiated with AraC and daunorubicin and those in complete remission had further consolidation treatment with an intermediate dose of AraC and amsacrine. The median age was 33 years (range, 11 to 59), and only very few patients were younger than 15 years. A total of 623 patients achieved complete remission, but only 343 completed the assigned treatment arm: 144 had a MDS and underwent alloBMT, whereas 95 received ASCR and 104 were elected for ICHT. The preparative regimen was equal in both transplant arms: TBI/CY. The DFS rates were 55% for alloBMT, 48% for ASCR and 30% ICHT. Again, the relapse rate was highest in the intensive therapy arm and lowest with alloBMT. However, mortality rate was highest with alloBMT and lowest with ICHT. This trial demonstrated that alloBMT resulted in better disease-free survival. However, BMT might also be appropriate soon after relapse or second complete remission.[99]

A prospective study comparing alloBMT with aggressive post-remission chemotherapy including HD-AraC in children with AML in first remission was undertaken by the French Society of Pediatric Hematology and Immunology.[100] One hundred seventy-one children were enrolled in the LAME89/91 protocol. Induction chemotherapy included cytarabin and mitoxantrone. After achieving complete remission, patients either received two additional consolidation courses or, in case of an HLA-identical sibling donor, underwent alloBMT. Complete remission was achieved in 87%. Thirty-two were eligible for alloBMT. The disease-free survival rates at 4 years were 72% in the BMT group and 48% in the chemotherapy group ($p = 0.02$). Thus, it appeared that alloBMT was capable of significantly reducing the risk of relapse in those children.

The POG group performed a randomized study comparing the value of ASCR with intensive consolidation chemotherapy as treatments for children with AML in first remission. After two courses of induction treatment (DAT, HD-AraC), 552 out of 649 enrolled patients entered remission (85%). A total of 232 patients were not eligible for randomization. Of the other patients, 117 were assigned to the intensive chemotherapy arm and 115 underwent ASCR. The event-free survival rate at 3 years was 34%. When the two arms were compared by intention-to-treat analyses, no difference in prognosis was detected (38% for the ASCR arm and 36% for the intensive chemotherapy arm). There was a lower relapse rate (31% vs. 58%, $p < 0.001$) but a higher rate of treatment-related mortality (15% vs. 2.7%, $p = 0.005$) in the group treated with ASCR than in the intensive chemotherapy group. The event-free survival at 3 years for nonrandomized patients in the intensive chemotherapy arm was 39%, whereas it was 52% for a contemporaneous group of patients receiving a matched, related donor (sibling) alloBMT. The study concluded that ASCR and intensive chemotherapy equally prolong event-free survival in AML patients, but alloBMT results warrant an additional comparative study.[101]

Another recent study evaluated the feasibility and results of BMT after induction and intensification in 159 patients with *de novo* AML.[81] Depending on the availability of an HLA-identical sibling, the intention of treatment after CT was alloBMT or ASCR. There was a trend for better leukemia-free survival at 4 years for patients assigned to the ASCR group (50% +/− 6%) compared with the alloBMT group (31% +/− 7%) ($p = 0.08$). The favorable results in patients who received ASCR (no toxic deaths and 37% +/− 7% probability of relapse at 4 years) contrast with the poor outcome òf allografted patients in this series (11 patients with transplant-related mortality).

SCT in AML beyond first remission: The role of ASCR was evaluated by Petersen in AML patients in untreated first-relapse (REL1) or in second remission (REM2).[102] Preparative regimens included various doses of BU/CY or BU/CY/TBI. Children and adults were included. Disease-free survivals at 2 years were 45% and 32% in REL1 and REM2, respectively. No conclusions could be drawn on the benefits of adding TBI to busulfan/CY. These results suggest ASCR may produce long-term leukemia-free survival. There was no apparent advantage of REM2 patients over REL1 patients for whom autologous marrow had been stored in first remission. However, storage of these marrows would be impractical without the coordinated support of treating physician and transplant center.

The Seattle group reported 126 patients transplanted during untreated first relapse including pediatric and adult patients.[88] Again various regimens and doses were used including TBI/CY or BU/CY +/− AraC and BCNU. The disease-free survival rate of all included patients was 23% at 5 years. The major message of this report was that patients receiving TBI at 15.75 Gy and CY had lower incidence of relapse and that GVHD prophylaxis with MTX alone produced a better survival, although the rate of aGVHD was higher. However, TBI/CY at average dose was related with a higher relapse rate in comparison with the other, more intensive regimens. Thus, this report confirms a dose–response relationship in AML, although escalated doses carry higher transplant-associated risks.

Conclusions

In AML the standard-risk patient, as defined by criteria such as FAB M1/M2-Auer rods positive, all FAB M3 and FAB M4 with atypical eosinophiles appears not as a candidate in first remission when modern intensive polychemotherapy treatment protocols are used. Patients presenting other criteria or more than 5% blasts in the bone marrow at day 15 or certain cytogenetic abnormalities are at high risk in first remission and should be considered for alloBMTas well if a family-matched donor is available. Autologous SCR in first remission AML remains,

TABLE 4. BMT in Children with AML with Respect to the Source of Stem Cells

	Autologous SCR	HLA-Identical Sibling	Family Member Mismatched	Unrelated, Matched
AML				
Non-responder	−	+	+	+
CR1-HR[a]	+/−	+	−	−
CR2	+	+	+	+
>CR2	+	+	+	+

[a] CR, complete remission.

however, a controversial issue. In contrast, in second remission allogeneic as well as ASCR are superior over conventional chemotherapy.

High-risk criteria in de novo *AML*: more than 5% of blasts in the bone marrow at day 15, M4 without eosinophiles, M5, M6, certain cytogenetic abnormalities: t(1;11), t(9;11), t(1;22), monosomy 7.

Conclusion: Indications for BMT in children with AML with respect to the source of stem cells are summarized in TABLE 4.

- CR1: BMT appears indicated in high-risk patients. Many studies describe no difference in ASCR and alloBMT.
- CR2: ASCR will save one-third of the children, but alloBMT appears preferable.
- >CR2: single cases may benefit from BMT.

FURTHER IMPORTANT ASPECTS

Infants

Leukemia is rare in infancy with an equal predominance of lymphoblastic and myeloblastic cases. Acute lymphoblastic leukemia in infants under one year is characterized by a high leukocyte count, organomegaly, and early B-cell phenotype, sometimes with evidence of monocytoid differentiation and cytogenetic abnormalities. Southern blot analysis revealed Ig heavy chain rearrangement in almost all infants with ALL.[103] A higher incidence of germline IgG and TCR loci was also reported in infants with ALL in comparison with older children.[104] It seems probable that different genetic and environmental factors are involved in the genesis of infant leukemia. This is reflected in its poor prognosis, which is reported to be within a range of 20 to 30% with conventional chemotherapy.[73,105–107] Although the toddler (aged 1–2) tends to develop typical childhood ALL that is responsive to treatment, he still remains vulnerable to late effects of therapy, particularly radiation. In AML the distribution of subtypes differs in the younger and older child. There is a higher incidence of the FAB types M4 and M5 (M6, M7) compared with older children. Ig-κ, TCR-β, and TCR-γ chain genes showed a germline configuration in infants with AML.[108]

Bone marrow transplantation in infants presents special challenges, particularly because of the need to use conditioning regimens containing cytotoxic agents during a period of rapid somatic growth.[8,107–110]

The experience of two consecutive MRC UKALL trials (VIII and X) in infants with ALL underlines their poor prognosis. Only 14 out of 48 infants were alive when treated with conventional-dose chemotherapy alone. In a subsequent trial (UKALL X pilot), 40 additional infants have been enrolled in a program with high-dose chemotherapy followed by ASCR or alloBMT. The results showed no improvement over previous trials, largely due to the number of remission deaths. Event-free survival at 5 years was 40% for children over 26 weeks but under 10% for younger children. These results showed the susceptibility of infants to the toxicity of intensive chemotherapy and do not support the use of short-term high-dose chemotherapy alone in the management of infant leukemia under the age of 4 months.[110]

Johnson[109] updated the largest study in children under 2 years of age treated by marrow transplantation for AML stressing the necessity of developing conditioning regimens that avoid total body radiation on the basis of late effects experienced by older children.

Mismatched Related and Unrelated Bone Marrow Donor Transplants

The donor of choice—an HLA matched sibling—is only available about 30% of the time for children who might benefit from alloBMT. To provide this potentially curative therapy to patients without a matched related donor, marrow transplants using less well-matched related donors or unrelated donors have been performed. With the development of National Marrow Donor Programs throughout the world, the probability of identifying an unrelated donor has increased. Almost 400 transplants from unrelated donors in children have been reported to date.[83–85,112,113] Potential problems with the use of donors who are not HLA-matched include increased risk of nonengraftment, GVHD, and infection. When compared to transplants performed from identical siblings, the probability of severe, acute GVHD is increased in all of them. That kindled the interest for T-cell depletion of the graft[85] with promising results in terms of GVHD rates. Future studies are imperative to lower the risk of this potentially promising approach offering the chance of cure to children without matched family or unrelated donors.

OVERALL CONCLUSION: WHICH CHILDREN BENEFIT FROM STEM CELL TRANSPLANT?

There are several childhood diseases, for example, inborn errors of metabolism or some advanced non-neoplastic hematologic disorders, in which allogeneic BMT is indicated because the disease is fatal and there is no other life-saving alternative. In addition, risks of morbidity are relatively small in these children.

However, the most frequent reason for alloBMT, that is, acute leukemia, also remains the most controversial. A major problem still is the rarity of randomized, comparative studies that eliminate selective bias in assigning patients to alternative treatment arms. SCT is a highly selective process, because most patients are assigned to this procedure only if they have achieved and maintained remission for at least 3 months. Clinical complete remission is a prerequisite for autologous stem cell sampling, whereas alloBMT is hampered by the need of a compatible donor.

About 30% of adults with acute lymphoblastic leukemia (ALL) and 20% to 40% of children and adults with AML never achieve remission, even with intensive chemotherapy. Most die of resistant leukemia, often within 6 months or less. However, alloBMT may cure such patients with a 3-year probability of disease-free survival of 21% and 23% for ALL and AML patients, respectively.[27]

It appears justified to suggest alloBMT in CML, ALL in second complete remission and AML in first complete remission, depending on the type of induction chemotherapy but surely in second complete remission. However, there is more controversy regarding the ALL high-risk group in first complete remission and ALL in second complete remission (late relapses). However, it should be a common goal to treat such patients in large, well-designed trials and the transplant decision should not be taken on an individual basis. Thus, open questions may be resolved in the forthcoming years.

SUMMARY

Hematopoetic stem cell transplantation (SCT) often represents a unique opportunity for curing children with leukemia. Nevertheless, selecting the patient who could really benefit from this procedure remains a controversial issue. The current consensus is as follows: About 20% of children with ALL can be defined as high-risk patients by criteria such as t(9;22), t(4;11), no complete remission at day 42, poor prednisone response, and T-immunophenotype or pre-pre B-ALL, myeloid markers or more than 100,000 white blood cells/μl. This high-risk group is eligible for alloBMT in first remission, provided a family-matched donor is available. At relapse the majority of patients will benefit from alloBMT, and alternative donor sources can be considered in high-risk patients. Only early alloBMT relapses (up to 6 months after end of initial therapy) are sure candidates, whereas late relapses, especially extramedullary sites, may equally benefit from an intensive conventional relapse treatment. However, any alloBMT relapse beyond second remission should be transplanted with allogeneic stem cells (bone marrow or peripheral stem cells). In particular, family mismatched donors or matched unrelated donors may be acceptable in high-risk cases beyond first remission. In contrast, ASCR in ALL seems not to be superior to conventional therapy. In AML the standard-risk patient, defined by criteria such as FAB M1/M2-Auer rods positive, all FAB M3, and FAB M4, is not a candidate for SCT in first remission. Patients presenting other criteria or more than 5% of blasts in the bone marrow at day 15 are at high risk in first remission and should be considered for alloBMT if a family matched donor is available. ASCR in first remission AML remains a controversial issue. In contrast, in second remission alloBMT as well as ASCR are superior to conventional chemotherapy.

REFERENCES

1. HOROWITZ, M. M. 1993. Results of bone marrow transplants from human leukocyte antigen-identical sibling donors for treatment of childhood leukemias. A report from the International Bone Marrow Transplant Registry. Am. J. Pediatr. Hematol. Oncol. **15**(1): 56–64.
2. MELONI, G. 1992. Autologous bone marrow transplantation followed by interleukin-2 in children with advanced leukemia: A pilot study. Leukemia **6**(8): 780–785.
3. PINKEL, D. 1993. Bone marrow transplantation in children. J. Pediatr. **122**(3): 331–341.
4. HOELZER, D. 1993. Acute lymphoblastic leukemia—progress in children, less in adults [editorial; comment]. N. Engl. J. Med. **329**(18): 1343–1344.
5. PUI, C. H. 1995. Treatment of childhood leukemias. Curr. Opin. Oncol. **7**(1): 36–44.
6. REITER, A. 1994. Chemotherapy in 998 unselected childhood acute lymphoblastic leukemia patients. Results and conclusions of the multi-centre trial ALL BFM 86. Blood **84:** 3122–3133.

7. RIVERA, G. K. 1993. Treatment of acute lymphoblastic leukemia: 30 years of experience at St. Jude Children's research hospital. N. Engl. J. Med. **329:** 1289–1295.

8. CHESSELLS, J. M. 1995. Intensification of treatment and survival in all children with lymphoblastic leukaemia: results of UK Medical Research Council trial UKALL X. Medical Research Council Working Party on Childhood Leukaemia. Lancet **345**(8943): 143–148.

9. CHESSELLS, J. M. 1992. Bone marrow transplantation for high-risk childhood lymphoblastic leukaemia in first remission: Experience in MRC UKALL X. Lancet **340**(8819): 565–568.

10. EBELL, W. 1992. Chemotherapy versus bone marrow transplantation in childhood acute lymphoblastic leukaemia. BFM Study Group. Eur. J. Pediatr. **151**(Suppl 1): S50–S54.

11. RIEHM, H. 1990. Results and significance of 6 randomised trials in four consecutive ALL-BFM studies. Haematol. Blood Trans. **33:** 349–450.

12. CORTES, J. E. 1995. Acute lymphoblastic leukemia. A comprehensive review with emphasis on biology and therapy. Cancer **76**(12): 2393–2417.

13. MILLER, D. R. 1990. Acute lymphoblastic leukemia in children: An update of clinical, biological, and therapeutic aspects. Crit. Rev. Oncol. Hematol. **10**(2): 131–164.

14. PUI, C. H. H. 1989. Prognostic importance of structural abnormalities in children with hyperdiploid (>50 chromosomes) acute lymphoblastic leukemia. Blood **73:** 1963–1967.

15. PUI, C. H. 1990. Near triploid and near tetra-ploid acute lymphoblastic leukemia of childhood. Blood **76:** 590–596.

16. PUI, C. H. 1990. Clinical presentation karyotypic characterisation and treatment outcome of childhood acute lymphoblastic leukemia with a near haploid or hyploid <45 line. Blood **75:** 1170–1177.

17. JASMIN, C. 1991. Concepts of remission, curability and lineage involvement in relationship to the problems of minimal residual leukaemia. Baillieres Clin. Haematol. **4**(3): 599–607.

18. CONSOLINI, R. 1993. Evaluation of minimal residual disease in high risk childhood acute lymphoblastic leukemia using an immunological approach during complete remission. AIEOP Cooperative Group for Immunology of Acute Leukemias. Haematologica **78**(5): 297–305.

19. KROLEWSKY, J. J. 1989. Molecular genetic approaches in the diagnosis and classification of lymphoid malignancies. Hematol. Pathol. **3:** 45–61.

20. FELIX, C. A. 1987. T-cell receptor alpha, beta and gamma genes in T-cell and pre-B cell acute lymphoblastic leukemia. J. Clin. Invest. **80:** 540–556.

21. ADRIAANSEN, H. J. 1991. Immunoglobulin and T-cell gene receptor gene rearrangements in acute non-lymphocytic leukemias: Analysis of 54 cases and a review of the literature. Leukemia **9:** 744–751.

22. SCHUTT, S. 1990. Clinical applications of the study of immunoglobulin and T-cell receptor gene rearrangements in acute leukemia and non-Hodgkin's lymphoma in children. Klin. Padiatr. **202**(4): 218–223.

23. SCHRAPPE, M. 1994. [Concept and interim result of the ALL-BFM 90 therapy study in treatment of acute lymphoblastic leukemia in children and adolescents: The significance of initial therapy response in blood and bone marrow]. Klin. Padiatr. **206**(4): 208–221.

24. GAJJAR, A. 1995. Persistence of circulating blasts after 1 week of multiagent chemotherapy confers a poor prognosis in childhood acute lymphoblastic leukemia. Blood **86**(4): 1292–1295.

25. EDEN, O. B. 1991. Results of Medical Research Council Childhood Leukemia Trial UKALL VIII. Br. J. Haematol. **78:** 187–196.

26. REITER, A. 1992. Favourable outcome of B-cell acute lymphoblastic leukemia in childhood: A report of three consecutive studies of the BFM group. Blood **80:** 2471–2478.

27. BIGGS, J. C. 1992. Bone marrow transplants may cure patients with acute leukemia never achieving remission with chemotherapy. Blood **80**(4): 1090–1093.

27a. TESTI, A. M. 1992. Treatment of primary refractory or relapsed acute lymphoblastic leukemia (ALL) in children. Am. Oncol. **3**(9): 765–767.

28. FLETCHER, J. A. 1991. Translocation (9;22) is associated with extremely poor prognosis in intensively treated children with acute lymphoblastic leukemia. Blood **77**(3): 435–439.

29. HENZE, G. 1991. Six-year experience with a comprehensive approach to the treatment of recurrent childhood acute lymphoblastic leukemia (ALL-REZ BFM 85). A relapse study of the BFM group. Blood **78**(5): 1166–1172.

30. BARRETT, A. J. 1994. Bone marrow transplants from HLA-identical siblings as compared with chemotherapy for children with acute lymphoblastic leukemia in a second remission [see comments]. N. Engl. J. Med. **331**(19): 1253–1258.

31. MINIERO, R. 1995. Relapse after first cessation of therapy in childhood acute lymphoblastic leukemia: A 10-year follow-up study. Italian Association of Pediatric Hematology-Oncology (AIEOP). Med. Pediatr. Oncol. **24**(2): 71–76.

32. SADOWITZ, P. D. 1993. Treatment of late bone marrow relapse in children with acute lymphoblastic leukemia: A Pediatric Oncology Group Study. Blood **81**: 602–609.

33. HENZE, G. 1994. Chemotherapy for relapsed childhood acute lymphoblastic leukemia: Results of the BFM study group. Haematol. Blood Transf. **36**: 374–379.

34. JOHNSON, F. L. 1990. Role of bone marrow transplantation in childhood lymphoblastic leukemia. Hematol. Oncol. Clin. N. Am. **4**(5): 997–1008.

35. SANDERS, J. E. 1987. Marrow transplantation for children with acute lymphoblastic leukemia in second remission. Blood **70**: 324–326.

36. KERSEY, J. H. 1987. Comparison of autologous and allogeneic bone marrow transplantation for treatment of high-risk refractory acute lymphoblastic leukemia. N. Engl. J. Med. **317**: 461–467.

37. DOPFER, R. 1991. Allogeneic bone marrow transplantation for childhood acute lymphoblastic leukemia in second remission after intensive primary and relapse therapy according to the BFM- and CoALL-protocols: Results of the German Cooperative Study. Blood **78**(10): 2780–2784.

38. GORIN, N. C. 1993. Stem cell transplantation in acute leukemia. Ann. N. Y. Acad. Sci. **770**: 262–287.

39. MOUSSALEM, M. 1995. Allogeneic bone marrow transplantation for childhood acute lymphoblastic leukemia in second remission: Factors predictive of survival, relapse and graft-versus-host disease. Bone Marrow Transplant. **15**(6): 943–947.

40. PINKEL, D. 1992. Therapy of acute lymphoid leukemia in children. Leukemia **6**(Suppl 2): 127–131.

41. BUTTURINI, A. 1987. Which treatment for childhood acute lymphoblastic leukemia in second remission? Lancet **i**: 429–432.

42. RIVERA, G. K. 1986. Intensive retreatment of childhood acute lymphoblastic leukemia in first bone marrow relapse. N. Engl. J. Med. **315**: 273–278.

43. DINSMORE, R. 1983. Allogeneic bone marrow transplantation for patients with acute lymphoblastic leukemia. Blood **62**: 381–388.

44. BÜHRER, C. 1993. Superior prognosis in combined compared to isolated bone marrow relapses in salvage therapy of childhood acute lymphoblastic leukemia. Med. Pediatr. Oncol. **21**(7): 470–476.

45. BORGMANN, A. 1995. Autologous bone-marrow transplants compared with chemotherapy for children with acute lymphoblastic leukaemia in a second remission: A matched-pair analysis. The Berlin-Frankfurt-Munster Study Group. Lancet **346**(8979): 873–876.

46. KERSEY, J. H. 1987. Comparison of autologous and allogeneic bone marrow transplantation for treatment with high-risk refractory acute lymphoblastic leukemia. N. Engl. J. Med. **317**: 461–467.

47. SALLAN, S. E. 1989. Autologous bone marrow transplantation for acute lymphoblastic leukemia. J. Clin. Oncol. **7**: 1594–1601.

48. CAHN, J. Y. 1991. The TAM regimen prior to allogeneic and autologous bone marrow transplantation for high-risk acute lymphoblastic leukemias: A cooperative study of 62 patients. Bone Marrow Transplant. **7**: 1–4.

49. BILLETT, A. L. 1993. Autologous bone marrow transplantation in childhood acute lymphoid leukemia with use of purging. Am. J. Pediatr. Hematol. Oncol. **15**(2): 162–168.

50. DOUAY, L. 1995. Amifostine improves the antileukemic therapeutic index of mafosfamide: Implications for bone marrow purging. Blood 86(7): 2849–2855.
51. SCHROEDER, H. 1991. High dose melphalan and total body irradiation with autologous marrow rescue in childhood acute lymphoblastic leukaemia after relapse. Bone Marrow Transpl. 7(1): 11–15.
52. BILLETT, A. L. 1993. Autologous bone marrow transplantation after a long first remission for children with recurrent acute lymphoblastic leukemia. Blood 81(6): 1651–1657.
53. UDERZO, C. 1995. High-dose vincristine, fractionated total-body irradiation and cyclophosphamide as conditioning regimen in allogeneic and autologous bone marrow transplantation for childhood acute lymphoblastic leukaemia in second remission: A 7-year Italian multicentre study. Br. J. Haematol. 89(4): 790–797.
54. ROSSETTI, F. 1993. ABMT for early isolated extramedullary relapse of childhood ALL. Bone Marrow Transpl. 12(5): 549.
55. LONNERHOLM, G. 1992. Autologous bone marrow transplantation in children with acute lymphoblastic leukemia. Acta Paediatr. 81(12): 1017–1022.
56. COLLESELLI, P. 1994. Autologous bone marrow transplantation for childhood acute lymphoblastic leukemia in remission: First choice for isolated extramedullary relapse? Bone Marrow Transpl. 14(5): 821–825.
57. SCHMID, H. 1993. Fractionated total body irradiation and high-dose VP-16 with purged autologous bone marrow rescue for children with high risk relapsed acute lymphoblastic leukemia. Bone Marrow Transpl. 12(6): 597–602.
58. GILMORE, M. J. 1991. Failure of purged autologous bone marrow transplantation in high risk acute lymphoblastic leukaemia in first complete remission. Bone Marrow Transpl. 8(1): 19–26.
59. BORGMANN, A. 1995. Isolated extramedullary relapse in children with acute lymphoblastic leukemia: A comparison between treatment results of chemotherapy and bone marrow transplantation. BFM Relapse Study Group. Bone Marrow Transpl. 15(4): 515–521.
60. PARSONS, S. K. 1996. Relapsed acute lymphoblastic leukemia: Similar outcomes for autologous and allogeneic marrow transplantation in selected children. Bone Marrow Transpl. 17: 763–768.
61. GARIN, L. 1996. Strong increase in the percentage of the CD8bright+ CD28− T-cells and delayed engraftment associated with cyclosporine-induced autologous GVHD. Eur. J. Haematol. 56(3): 119–123.
62. BAUMGARTEN, E. 1994. Low-dose natural interleukin-2 and recombinant interferon-gamma following autologous bone marrow grafts in pediatric patients with high-risk acute leukemia. Leukemia 8(5): 850–855.
63. SOIFFER, R. J. 1993. Monoclonal antibody-purged autologous bone marrow transplantation in adults with acute lymphoblastic leukemia at high risk of relapse. Bone Marrow Transplant. 12(3): 243–251.
64. ROBINSON, N. 1996. Phase I trial of interleukin-2 after unmodified HLA-matched sibling bone marrow transplantation for children with acute leukemia. Blood 87(4): 1249–1254.
65. UDERZO, C. 1990. Treatment of isolated testicular relapse in childhood acute lymphoblastic leukemia: An Italian multicenter study. Associazione Italiana Ematologia ed Oncol. Pediatr. J. Clin. Oncol. 8(4): 672–677.
66. UDERZO, C. 1995. Treatment of childhood acute lymphoblastic leukemia in second remission with allogeneic bone marrow transplantation and chemotherapy: Ten-year experience of the Italian Bone Marrow Transplantation Group and the Italian Pediatric Hematology Oncology Association. J. Clin. Oncol. 13(2): 352–358.
67. WEISDORF, D. 1987. Allogeneic bone marrow transplantation for acute lymphoblastic leukemia in remission: Prolonged survival associated with acute graft versus host disease. J. Clin. Oncol. 5: 1348–1355.
68. RINGDEN, O. 1993. Methotrexate, cyclosporine, or both to prevent graft-versus-host disease after HLA-identical sibling bone marrow transplants for early leukemia. Blood 81(4): 1094–1100.

69. HURWITZ, C. A. 1995. Treatment of patients with acute myelogenous leukemia: review of clinical trials of the past decade. J. Pediatr. Hematol. Oncol. **17**(3): 185–197.
70. VORMOOR, J. 1996. Therapy of childhood acute myelogenous leukemias. Ann. Hematol. **73**(1): 11–24.
71. LIE, S. O. 1996. A population-based study of 272 children with acute myeloid leukaemia treated on two consecutive protocols with different intensity: best outcome in girls, infants, and children with Down's syndrome Nordic Society of Paediatric Haematology and Oncology (NOPHO). Br. J. Haematol. **94**(1): 82–88.
72. RITTER, J. 1992. Treatment results of three consecutive German childhood AML trials: BFM-78, -83, and -87. AML-BFM-Group. Leukemia **6**(Suppl 2): 59–62.
73. STARK, B. 1995. A distinct subtype of M4/M5 acute myeloblastic leukemia (AML) associated with t(8:16)(p11:p13), in a patient with the variant t(8:19)(p11:q13)—case report and review of the literature. Leuk. Res. **19**(6): 367–379.
74. LION, T. 1992. The translocation t(1;22)(p13;q13) is a nonrandom marker specifically associated with acute megakaryocytic leukemia in young children. Blood **79**(12): 3325–3330.
75. RIBEIRO, R. C. 1993. Acute megakaryoblastic leukemia in children and adolescents: A retrospective analysis of 24 cases. Leuk. Lymph. **10**(4-5): 299–306.
76. BARANGER, L. 1990. Monosomy-7 in childhood hemopoietic disorders. Leukemia **4**(5): 345–349.
77. KALRA, R. 1995. Monosomy 7 and activating RAS mutations accompany malignant transformation in patients with congenital neutropenia. Blood **86**(12): 4579–4586.
78. CREUTZIG, U. 1990. Identification of two risk groups in childhood acute myelogenous leukemia after therapy intensification in study AML-BFM-83 as compared with study AML-BFM-78. AML-BFM Study Group. Blood **75**(10): 1932–1940.
79. CREUTZIG, U. 1992. Comparison of chemotherapy alone with allogeneic bone marrow transplantation in first full remission in children with acute myeloid leukemia in the AML-BFM-83 and AML-BFM-87 studies—matched pair analysis. Klin. Padiatr. **204**(4): 246–252.
80. DINNDORF, P. 1995. Bone marrow transplantation for children with acute myelogenous leukemia. J. Pediatr. Hematol. Oncol. **17**(3): 211–224.
81. SIERRA, J. 1996. Feasibility and results of bone marrow transplantation after remission induction and intensification chemotherapy in de novo acute myeloid leukemia. Catalan Group for Bone Marrow Transplantation. J. Clin. Oncol. **14**(4): 1353–1363.
82. SELVAGGI, K. J. 1994. Improved outcome for high-risk acute myeloid leukemia patients using autologous bone marrow transplantation and monoclonal antibody purged bone marrow. Blood **83**(6): 1698–1705.
83. LADENSTEIN, R. 1996. BMT from unrelated or family donors other than HLA identical siblings. GVHD prophylaxis, incidence and treatment. Bone Marrow Transpl. **18**(2): 86–91.
84. BALDUZZI, A. 1995. Unrelated donor marrow transplantation in children. Blood **86**(8): 3247–3256.
85. CORNISH, J. 1996. Unrelated donor bone marrow transplant in childhood ALL. The role of T-cell depletion. Bone Marrow Transpl. **18**(2): 31–35.
86. TAKAHASHI, S. 1994. Recombinant human glycosylated granulocyte stimulating factor (rhG.CSF) combined regimen for allogeneic bone marrow transplantation in refractory acute myeloid leukemia. Bone Marrow Transpl. **13**: 239–245.
87. BROWN, R. A. 1995. High-dose etoposide, cyclophosphamide, and total body irradiation with allogeneic bone marrow transplantation for patients with acute myeloid leukemia in untreated first relapse: A study by the North American Marrow Transplant Group. Blood **85**(5): 1391–1395.
88. CLIFT, R. G. 1992. Allogeneic marrow transplantation during untreated first relapse of acute myeloid leukemia. J. Clin. Oncol. **19**(11): 1723–1729.
89. CHAO, N. J. 1993. Busulfan/etoposide—initial experience with a new preparatory regimen for autologous bone marrow transplantation in patients with acute nonlymphoblastic leukemia. Blood **81**(2): 319–323.

90. CRILLEY, P. 1990. Bone marrow transplantation following busulfan and cyclophosphamide for acute myelogenous leukemia. Bone Marrow Transpl. **5**(3): 187–191.
91. SANZ, M. A. 1993. Busulfan plus cyclophosphamide followed by autologous blood stem-cell transplantation for patients with acute myeloblastic leukemia in first complete remission: A report from a single institution [see comments] J. Clin. Oncol. **11**(9): 1661–1667.
92. MICHEL, G. 1994. Allogeneic bone marrow transplantation for children with acute myeloblastic leukemia in first complete remission: impact of conditioning regimen without total-body irradiation—a report from the Societe Francaise de Greffe de Moelle. J. Clin. Oncol. **12**(6): 1217–1222.
93. WOODS, W. G. 1996. Timed-sequential induction therapy improves postremission outcome in acute myeloid leukemia: a report from the Children's Cancer Group. Blood **87**(12): 4979–4989.
94. HERMANN, J. 1992. Results of allogenic and autologous bone marrow transplantation in children with acute myeloid leukemia. Kinderarztl-Prax. **60**(2): 35–39.
95. SHAW, P. J. 1994. Childhood acute myeloid leukemia: Outcome in a single centre using chemotherapy and consolidation with busulphan/cyclophosphamide for bone marrow transplantation. J. Clin. Oncol. **12**(10): 2138–2145.
96. TIEDEMANN, K. 1993. Results of intensive therapy in childhood acute myeloid leukemia, incorporating high-dose melphalan and autologous bone marrow transplantation in first complete remission. Blood **82**(12): 3730–3738.
97. DINI, G. 1994. Allogeneic bone marrow transplantation in children with acute myelogenous leukemia in first remission. Associazione Italiana di Ematologia e Oncologia Pediatrica (AIEOP) and the Gruppo Italiano per il Trapianto di Midollo Osseo (GITMO). Bone Marrow Transpl. **13**(6): 771–776.
98. AMADORI, S. 1993. Prospective comparative study of bone marrow transplantation and postremission chemotherapy for childhood acute myelogenous leukemia. The Associazione Italiana Ematologia ed Oncologia Pediatrica Cooperative Group. J. Clin. Oncol. **11**(6): 1046–1054.
99. ZITTOUN, R. A. 1995. Autologous or allogeneic bone marrow transplantation compared with intensive chemotherapy in acute myelogenous leukemia. European Organization for Research and Treatment of Cancer (EORTC) and the Gruppo Italiano Malattie Ematologiche Maligne dell-Adulto (GIMEMA) Leukemia Cooperative Groups. N. Engl. J. Med. **332**(4): 217–223.
100. MICHEL, G. 1996. Allogeneic bone marrow transplantation vs aggressive post-remission chemotherapy for children with acute myeloid leukemia in first complete remission. A prospective study from the French Society of Pediatric Hematology and Immunology (SHIP). Bone Marrow Transpl. **17**(2): 191–196.
101. RAVINDRANATH, Y. 1996. Autologous bone marrow transplantation versus intensive consolidation chemotherapy for acute myeloid leukemia in childhood. Pediatric Oncology Group. N. Engl. J. Med. **334**(22): 1428–1434.
102. PETERSEN, F. B. 1993. Autologous marrow transplantation for patients with acute myeloid leukemia in untreated first relapse or in second complete remission. J. Clin. Oncol. **11**(7): 1353–1360.
103. LUDWIG, W. D. 1989. Phenotypic and genotypic heterogeneity in infant acute leukemia. I. Acute lymphoblastic leukemia. Leukemia **3**(6): 431–439.
104. FELIX, C. A. 1987. Immunoglobulin and T cell receptor gene configuration in acute lymphoblastic leukemia of infancy. Blood **70**(2): 536–541.
105. SATHER, H. 1986. Age at diagnosis in childhood acute lymphoblastic leukemia. Med. Pediatr. Oncol. **14**: 166–172.
106. CHRIST, W. M. 1986. Clinical and biological features predict a poor prognosis in acute lymphoid leukemias in infants: A pediatric oncology group study. Blood **67**: 135–140.
107. PUI, C. H. 1995. Biology and treatment of infant leukemias. Leukemia **9**: 762–769.
108. KOLLER, U. 1989. Phenotypic and genotypic heterogeneity in infant acute leukemia. II. Acute nonlymphoblastic leukemia. Leukemia **3**(10): 708–714.
109. JOHNSON, F. L. 1992. Allogeneic marrow transplantation in the treatment of infants with cancer. Br. J. Cancer Suppl. **18**: S76–S79.

110. CHESSELLS, J. M. 1994. Acute lymphoblastic leukaemia in infancy: Experience in MRC UKALL trials. Report from the Medical Research Council Working Party on Childhood Leukaemia. Leukemia 8(8): 1275–1279.
111. EMMINGER, W. 1992. Treatment of infant leukemia with busulfan, cyclophosphamide +/− etoposide and bone marrow transplantation. Bone Marrow Transpl. 9(5): 313–318.
112. CASPER, J. 1995. Unrelated bone marrow donor transplants for children with leukemia or myelodysplasia. Blood 85(9): 2354–2363.
113. GROVAS, A. 1994. Unrelated donor bone marrow transplants in children. Cell Transplant. 3(5): 413–420.
114. PUI, C. H., D. L. WILLIAMS & S. C. RAIMONDI, et al. 1987. Hypoploidy is associated with a poor prognosis in childhood acute lymphoblastic leukemia. Blood 70: 247–253.
115. FLETCHER, J. A., N. TU, R. TANTRAVAHI & S. E. SALLAN. 1992. Extremely poor prognosis of pediatric acute lymphoblastic leukemia with translocation (9;22): updated experience. Leuk. Lymph. 8(1-2): 75–79.
116. PUI, C. H. 1991. Clinical characteristics and treatment outcome of childhood acute lymphoblastic leukemia with the t(4, 11;q21;q23): a collaborative study of 40 cases. Blood 71: 440–447.
117. RAIMONDI, S. C. 1990. Cytogenetics of pre-B-cell acute lymphoblastic leukemia with emphasis on prognostic implications of the t(1;19). J. Clin. Oncol. 8: 1380–1388.
118. NOWELL, P. C. 1990. Chromosome translocations and oncogenes in human lymphoid tumors. Am. J. Clin. Pathol. 94: 229–237.
119. RAIMONDI, S. C. 1989. Childhood lymphoblastic leukemia with chromosomal breakpoints at 11q23. Blood 73: 1627–1634.
120. PUI, C. H. 1991. Characterization of childhood acute leukemia with multiple myeloid and lymphoic markers at diagnosis and at relapse. Blood 78: 1327–1337.
121. HUANG, M. E. 1988. Use of all-*trans* retinoic acid in the treatment of acute promyelocytic leukemia. Blood 72: 567–572.
122. WARREL, R. P. 1993. Acute promyelocytic leukemia. N. Engl. J. Med. 329: 177–189.
123. ODOM, L. F., B. C. LAMPKIN, R., TANNOUS, J. D. BUCKLEY & G. D. HAMMOND. 1990. Acute monoblastic leukemia: A unique subtype—a review from the Children's Cancer Study Group. Leuk. Res. 14(1): 1–10.

Minimal Residual Disease in Leukemia in Children[a]

BERNHARD KORNHUBER, UWE EBENER,
ECKHARD NIEGEMANN, AND SIBYLLE WEHNER

Clinic of Pediatrics
Department of Pediatric Hematology and Oncology
University Hospital
J. W. Goethe University
Frankfurt/Main, Germany

In multicenter studies much data concerning treatment and diagnosis of acute leukemia in children are generated prospectively. These findings are assessed in common with the treatment results. The objective is to detect facts of prognostic relevance at the time of diagnosis and to decide what importance they will have on the therapy to be applied.[1,2] Minimal residual disease (MRD) (or minimal residual cancer or minimal residual leukemia) has to be defined and assessed in this connection. It is as important to consider relevant facts in time for them to inform therapeutic strategies as it is not to overestimate otherwise interesting findings that will not be relevant for therapy.

DEFINITION OF MINIMAL RESIDUAL DISEASE

Because microscopic cytological examination is insufficiently sensitive, the remission of acute leukemia is defined by proof of less than 5% blast cells in the bone marrow. More sensitive techniques show definitely that, in spite of attained remission, malignant cells are still present. At the time of diagnosis the absolute number of leukemic cells is approximately 10^{12} and approximately 10^8 to 10^9 at the time of remission.[3] This residual population of blast cells can be detected by means of immunological and molecular genetic methods. So the meaning of minimal residual leukemia begins to be clearer.[4]

Molecular Genetic Methods

The precondition for detecting MRD starts with the exact immunological and molecular genetic classification of the blast cells of each patient at the time of diagnosis. By this examination it is possible to control the most important markers during the course of the disease very accurately.

Often, but not always, leukemias are detectable by specific chromosomal abnormalities. The genes concerned either take part directly in cell proliferation or in cell differentiation or control them. There are two groups of genes implied: (1) Oncogenes are homologous to retroviral genes. Cellular genes (proto-oncogenes)

[a] This work was supported by Hilfe für Krebskranke Kinder Frankfurt e.V.

become activated oncogenes either by structural alteration (mutation or deletion) or by amplification of the gene. (2) In contrast to this dominant "gain of function," tumor-suppressor genes are involved in loss or inactivation of genes whose proteins suppress cancer, so-called "loss of function" (recessive).

These well-known genetic markers could be detected in a specific and highly sensitive way by polymerase chain reaction (PCR) and reverse transcriptase-PCR (RT-PCR).[5,6] Proof and identification of chromosomal translocations may be brought about by using PCR either on the DNA level or after transcription in cDNA by means of reverse transcriptase on the RNA level. In order to guarantee a high degree of sensitivity, so-called nested PCR is employed as a rule in investigating MRD. By this technique we understand that a second amplification is performed with oligonucleotide primers specific for the sequences within the sector recognized by the first primers, following the first PCR or RT-PCR.

In chronic myeloid leukemias (CML) the appearance of the genetic marker BCR-ABL has been especially well investigated.[7-9] By modification of the specimen preparations through isolation of nucleic acids, among others, it is possible to raise the sensitivity to 10^{-8}. But in that case, the BCR-ABL rearrangement can also be detected in specimens of healthy donors (in 22 out of 73 adults and 1 out of 22 children) without its being a contamination.[10]

An additional powerful tool for analysis of translocations and deletions and mapping of genes and identification of human chromosomes is the fluorescence *in situ* hybridization technique (FISH). Different chromosome paints and specific DNA probes are now available for the application of this method (e.g., detecting the translocation in Philadelphia chromosome).[11,12]

Immunological Techniques

Experiences in immunological methods have been known for more than 15 years. Antibodies directed towards the leukemia-associated antigens are used; normally, these are monoclonal antibodies that can be detected applying a fluorescent dye or an enzymatic reaction. Double and triple labeling using a second or third monoclonal antibody has become available.[6,13-15] This possibility is especially significant when physiological antigens are to be detected, and the percentage of leukemic blast cells in the bone marrow is below 20%. The immunological characterization can be applied to cells in suspension or to cells in smears. The first, flow cytometry, requires a greater number of cells; the latter is a time-consuming procedure, but gives the morphologist an opportunity to assess the morphology and the immunological labeling at the same time and thus to identify the antigens he is looking for under the microscope in atypical cells.

Sensitivity of Methods

PCR is more sensitive compared to the immunological techniques. Very careful work in separate laboratories is necessary in order to avoid nonspecific positive findings caused by contamination. The different sensitivities of the mentioned techniques can be quantified (see TABLE 1).

Doubtless, molecular genetic investigations are the most sensitive but also the most advanced and are thus the less appropriate techniques for making judgments about their prognostic relevance.[19,20]

TABLE 1. Sensitivity of Various Methods in Detecting MRD

Methods	Sensitivity
Cytomorphology	10^{-2}
FISH	10^{-2}
Immunophenotyping (f.i. FACS, APAAP)	10^{-4}
PCR/RT-PCR	10^{-5}–10^{-6}
	table 1[16,17,18]

WHY DO WE WANT TO RECOGNIZE MINIMAL RESIDUAL DISEASE?

After a defined period of collecting parameters of applied techniques mentioned before, the results must be evaluated. Even today it is not yet possible to draw relevant conclusions from the results of important multicenter treatment studies regarding their correlation with PCR results.

Flow cytometry and immunocytology are reliable techniques that, concerning their sensitivity, only slightly surpass the cytological investigations of panoptical preparations stained according to the Pappenheim-Wright method. The immunological findings, however, are a valuable addendum for cytological findings when leukemic blast cells are scarce. Therefore, it should be required that the blast cell frequency of less than 5% in bone marrow preparations should be assured by immunocytological findings, provided that the marker profile is known and did not change in relapse. A remainder of leukemic cells detected by immunological techniques represent a prognostic relevant finding that must be considered when deciding on therapy.[6,13–15]

If several repeated examinations during the course of a disease confirm the molecular genetic results at the time of diagnosis, PCR and RT-PCR positivity seem to be of prognostic relevance. This is mentioned in a series of publications by a number of pediatric oncology centers. Nevertheless, neither prospective nor retrospective findings of big, multicenter studies have been published yet. The BFM group started a study in 1990 providing molecular genetic controls of minimal residual leukemia.[21] This was a prospective study. It was agreed in advance that the sender would not be informed of the results of the molecular genetic examinations in order to avoid arbitrary changes in treatment in cases of positive results as long as the therapeutic relevance of these results is not proven. A study design requiring examination of several bone marrow samples under conditions of withholding the result from physicians and parents did not meet with enough compliance. The study is not yet finished.

Most of the samples became PCR negative after 4 to 8 weeks of ALL treatment. Only a few patients continued to show positive PCR results after 8 weeks. There seems to be a difference between the prognostic relevance of BCR-ABL and other markers, for example, the T-cell receptor (TcR). The data of the BFM studies have not yet been published.[21]

The quota of relapses in the group of PCR-positive patients in remission is higher than for patients in the negative group. The most important finding so far is that patients whose bone marrow became PCR negative and later became positive developed a cytologically detectable relapse within 6 months. Similar findings were reported in other studies. A greater number of patients is reported by Brisco *et al.*: 26 patients out of 38, who after having reached first remission still showed MRD according to PCR analysis, were subject to relapse, whereas only 6 patients out of 50 in remission without MRD positivity were subject to relapse.[10]

THE IMPORTANCE OF MRD IN AUTOLOGOUS BONE MARROW TRANSPLANTS AND PERIPHERAL BLOOD STEM CELL PREPARATIONS

Peripheral blood stem cell (PBSC) apheresis and autologous bone marrow transplants (ABMT) are principally used in high-risk patients, especially in relapsed patients. Normally, these patients are in a conventional remission at the time of the transplantation, but there is still a remaining population of blast cells to be perceived by PCR. These residual leukemic blast cells are also detectable in the peripheral blood stem cell harvests and in autologous bone marrow. The number of blast cells in the transplants is considerably more than the number of the blast cells in the patient before the application of the myeloablative therapy. Therefore it is important to raise the question whether or not after a successful myeloablative therapy these leukemic blast cells in the harvest are relevant to a possible relapse and therefore should require purging. If purging should prove necessary more questions will arise:

1. Will the number of healthy cells in the transplant be affected during the purging procedure?
2. Will the number of the leukemic blast cells in the transplant be considerably reduced by purging?
3. What kind of purging technique should be used? Chemotherapeutical or immunological purging? If the latter, should the immunological purging be positive or negative?

Point 1:

- At this time, purging techniques mostly also lead to a reduction in the number of cells to be transplanted.
- The number of leukemic blast cells is reduced by purging, but often the complete elimination of leukemic blast cells is not accomplished.[22]
- For purging, chemotherapy and immunological techniques can be applied. Among the immunological techniques, positive purging is possible; that means $CD34^+$ cell selection using suitably coated columns when transplanting stem cells and simultaneously eliminating all the other cells (positive purging) or an attempt to eliminate leukemic blast cells by columns that retain cells carrying the leukemic markers (negative purging).

These techniques require the use of immunoaffinity columns that are coated with the corresponding monoclonal antibodies and retain the wanted cells. In order to carry out immunological purging in closed systems, suitably prepared columns, pumps, and bags are necessary. However, this apparatus is still being tested.

Immunological purging often is brought about by suitable immunobeads. Magnetic microbeads coated with antibodies make it possible to eliminate the bound cells with a magnet. By using all possibilities and with the necessary experience, this technique can be applied with great success.

Immunological purging, however, cannot be employed in every case because immunological beads are not available for each leukemic subtype. In such cases, chemotherapeutic purging should be applied. Under certain circumstances, a combination of both techniques will be considered.

According to the present investigations, the attempt to eliminate MRD from autologous bone marrow transplants or from stem cell aphereses seems to be reasonable. Personally, we consider positive purging to be the better technique when autologous peripheral blood stem cells are concerned. The final answers to the still-open questions will be found in extensive prospective studies, the results of which will take years to prove the effectiveness of the different purging procedures.

At present it is not clear whether relapses after autologous BMT or PBSC transfusion are due to residual leukemia in the patient or in the autograft.

CONCLUSION

On the basis of the literature and our experience, we propose that the new techniques for investigating MRD should be employed. Though the evaluation of long-term, prospective multicenter studies is not yet completed, many publications indicate that these studies will provide confirmation that the detection of MRD is of prognostic relevance and will be taken into account in future studies in order to modify treatment. These techniques are also of benefit in autografts, because an unfavorable MRD finding has to be eliminated by purging. Indications also suggest that multiple purging of an autograft is a reliable method of eliminating leukemic blast cells.

SUMMARY

Many chromosomal translocations involved in leukemia have been defined at the molecular level in recent years. In addition to advancing the understanding of pathological mechanisms underlying the transformation process, the cloning and sequencing of the genes altered by the translocations have provided new tools for diagnosis and monitoring of patients. In particular, the polymerase chain reaction (PCR) method yields sensitive and accurate diagnostic and prognostic information.

Minimal residual disease (MRD) is not clearly defined. In ALL we define MRD as fewer than 5% blast cells in the bone marrow by conventional cytology and proof of leukemic cells with more sensitive methods. The techniques for detecting MRD are imaging for detection of single leukemic cells in the blood, bone marrow, or other tissues by means of immunocytology or PCR/RT-PCR. Highly sensitive PCR, immunocytology, FACS analysis, or conventional cytology are important tools to use in the process of deciding on appropriate therapy. Detection limits at present are 10^{-2} for cytology and FISH, up to 10^{-4} for immunological procedures, and 10^{-5} to 10^{-6} for PCR. But multiple methods also imply the possibility of mistakes (e.g., PCR). The question must be raised what method should be decisive in assessing MRD for evaluating autologous peripheral blood stem cells (PBSC) or autologous bone marrow transplants? Prospective studies will have to answer the question whether MRD should be treated or not and whether purging of bone marrow or PBSC is useful or damaging. When applied, should a positive or a negative immunopurging or a chemotherapeutic purging be used? MRD refers to the organism of the patient as well as to the peripheral blood stem cells and autologous bone marrow that had been taken before myeloablative therapy and kept for retransfusion.

REFERENCES

1. MAURER, J., *et al.* 1991. Detection of chimeric BCR-ABL genes in acute lymphoblastic leukaemia by the polymerase chain reaction. Lancet **337:** 1055–1058.
2. SHAI, I., *et al.* 1992. Detection and clinical relevance of genetic abnormalities in pediatric acute lymphoblastic leukemia: A comparison between cytogenetic and polymerase chain reaction analyses. Leukemia **7:** 671–678.
3. COPELAN, E. A. & E. M. McGUIRE. 1995. The biology and treatment of acute lymphoblastic leukemia in adults. Blood **85:** 1151–1168.
4. DREXLER, H. G., *et al.* 1995. Detection of chromosomal translocations in leukemia–lymphoma cells by polymerase chain reaction. Leuk. Lymph. **19:** 359–380.
5. BEISHUIZEN, A., *et al.* 1995. Molecular biology of acute lymphoblastic leukemia: Implications for Detection of Minimal Residual Disease. Acute Leukemias V: 460–474.
6. ZOUBEK, A., *et al.* 1995. Minimal metastatische—und minimal residuelle Erkrankung bei Patienten mit Ewing-Tumoren. Klin. Pädiatr. **207:** 242–247.
7. POTTER, M. N., *et al.* 1993. Molecular evidence of minimal residual disease after treatment for leukaemia and lymphoma: An updated meeting report and review. Leukemia **7:** 1302–1314.
8. RADICH, J. P., *et al.* 1995. Polymerase chain reaction detection of the BCR-ABL fusion transcript after allogeneic marrow transplantation for chronic myeloid leukemia: Results and implications in 346 patients. Blood **85:** 2632–2638.
9. DIEKMANN, L., *et al.* 1994. Presence or re-appearance of BCR-ABL-positive cells years after allogeneic bone marrow transplantation for chronic-phase chronic myelogenous leukemia in patients in hematological remission. Acta Haematol. **92:** 169–175.
10. BIERNAUX, C., *et al.* 1995. Detection of major *bcr-abl* gene expression at a very low level in blood cells of some healthy individuals. Blood **86:** 3118–3122.
11. LICHTER, P., *et al.* 1988. Detection of chromosome aberrations in metaphase and interphase tumor cells by in situ suppression hybridization using chromosome-specific library probes. Hum. Genet. **80:** 235–246.
12. ARNOLDUS, E. P. J., *et al.* 1990. Detection of the Philadelphia chromosome in interphase nuclei. Cytogenetic. Cell Genet. **54:** 108–111.
13. EBENER, U., *et al.* 1995. Rapid immunodiagnosis of childhood leukemia using microwave-stimulated APAAP-complex system. Anticancer Res. **15:** 1257–1261.
14. LAVABRE-BERTRAND, T., *et al.* 1994. Leukemia-associated changes identified by quantitative flow cytometry: I. CD10 expression. Cytometry **18:** 209–217.
15. KNAPP, W., *et al.* 1994. Flow cytometric analysis of cell-surface and intracellular antigens in leukemia diagnosis. Cytometry **18:** 187–198.
16. KEATING, A. 1991. Is marrow purging necessary or clinically useful? Bone Marrow Transplant **7:** 61–65.
17. GOLDMAN, J. M. & T. HUGHES. 1991. Detection and significance of minimal residual disease in patients with leukaemia and lymphoma. Bone Marrow Transplant **7:** 66–68.
18. ATTA, J., *et al.* 1995. High titers of residual leukemia after chemotherapy as a rationale for in vitro purging in patients with BCR/ABL-positive acute lymphoblastic leukemia. Exp. Hematol. **23:** 76.
19. BARTRAM, C. R. 1993. Molecular genetic techniques for detection of minimal residual disease in acute lymphoblastic leukemia: Possibilities and limitations. Rec. Results Cancer Res. **131:** 149–155.
20. LION, T. 1994. Clinical implications of qualitative and quantitative polymerase chain reaction analysis in the monitoring of patients with chronic myelogenous leukemia. Bone Marrow Transplant **14:** 505–509.
21. BARTRAM, C. R. 1996. Personal communication.
22. MARTIN, H., *et al.* 1995. Purging of peripheral blood stem cells yields BCR-ABL–negative autografts in patients with BCR-ABL–positive acute lymphoblastic leukemia. Exp. Hematol. **23:** 1612–1618.

Correlation between Morphologic and Nonmorphologic Prognostic Markers of Neuroblastoma

VIJAY V. JOSHI AND GREGORY J. TSONGALIS

Department of Pathology
Connecticut Children's Medical Center
and
Hartford Hospital
Hartford, Connecticut 06102

In the first four years of life, neuroblastoma (NB) is the commonest extracranial solid tumor, accounting for 41% of such tumors.[1] NB is a heterogeneous tumor with respect to its genetic, biologic, clinical, and morphologic features. The clinical behavior of NB varies from highly aggressive with fatal outcome to regression following minimal or no therapy. Maturation to ganglioneuroma has also been observed.[2] Despite advances in therapy, the overall prognosis of NB still remains poor as compared with other pediatric malignancies. Attempts are being made by an international committee to bring a reasonable degree of uniformity to diagnostic approach, staging, and assessment of response to therapy of NB.[3] Knowledge of various prognostic factors and characterization of NBs into low-, intermediate-, and high-risk subgroups on the basis of various prognostic markers would be helpful in designing appropriate therapy so that overall prognosis can be improved. Knowledge of prognostic factors may also be helpful in understanding the pathogenesis of NB.

The prognostic markers of NB can be divided into four categories: (1) biochemical (serum neuron–specific enolase, ferritin, lactate dehydrogenase, etc.); (2) genetic (1 p deletion, N-myc copy number, DNA index, trk-A gene expression, etc.); (3) clinicobiological (age, stage, site, etc.); and (4) morphologic (Shimada classification, histologic grading, etc.). For the purpose of this article, the emphasis of which is on morphologic markers, we have included biochemical, genetic, and clinicobiological markers under the broad category of nonmorphologic markers. It should be noted that some of the nonmorphologic markers such as N-myc copy number, expression of P-glycoprotein, trk-A gene, or CD_{44} cell surface glycoprotein can be determined by morphologic methods such as fluorescence *in situ* hybridization (FISH) technique or specialized immunocytochemical staining procedures. The term *morphologic markers* is used in the restricted sense to include only those histologic features detected in routine hematoxylin- and eosin-stained sections or sections stained with readily available immunoperoxidase stains for well-established cell constituents (e.g., S-100).

The purpose of this article is: (a) to review briefly the various prognostic markers of NB; (b) to describe the presence or absence of correlation between the morphologic and nonmorphologic markers; and (c) to briefly discuss the clinical relevance of the knowledge of correlation.

MORPHOLOGIC PROGNOSTIC MARKERS

Occurrence of spontaneous maturation of NB led to the speculation that the degree of maturation exhibited by a given case of NB may be related to its eventual outcome. The following initial observations lent support to this speculation. Thus in 1914, Wahl[4] demonstrated the association of a high degree of differentiation with good prognosis. In 1959, Stout[5] described a higher incidence of metastasis in neuroblastic tumors showing completely differentiated ganglioneuromas in one portion and undifferentiated or poorly differentiated NB in a contiguous portion (these tumors would be designated as nodular ganglioneuroblastomas by current criteria). Horn et al.[6] in their series of 44 cases found the more differentiated NBs to have better prognosis.

Beckwith and Martin[7] were among the first investigators to demonstrate systematically the value of degree of maturation as a morphologic prognostic marker of NB. The tumors were divided into four well-defined histologic grades based upon proportion of differentiating tumor cells. The histologic grades 1 to 4 ($>$50%, 5–50%, $<$5%, and 0% differentiating tumor cells in grades 1, 2, 3, and 4, respectively) reflected an ascending scale of aggressive behavior. These authors described the following morphologic criteria for differentiating neuroblasts, which have been used as the baseline by subsequent investigators: nuclear and cytoplasmic enlargement, clear borders of eosinophilic cytoplasm, and the presence of neuritic processes. The method of dividing NB into different grades described by Beckwith and Martin[7] has been used by subsequent investigators. In the same series of 50 cases of NB, the authors observed that lymphoid infiltrates were present in the relatively differentiated tumors and in tumors of patients who survived.[8] Lawler and Aherne confirmed the correlation between lymphoid infiltration and longer survival.[9] Makinen[10] divided NBs into two groups on the basis of the presence or absence of features of differentiation. There was statistically significant longer survival in patients with NBs showing evidence of differentiation. Hughes et al.[11] also demonstrated good prognosis in patients with NBs showing a high level of differentiation. Sandstedt et al.[12] divided NBs into three grades on the basis of amount of fibrillary matrix and demonstrated that survival was better in grade 1 vs. 2 and in grade 2 vs. 3 NBs.

In more recent years, three major histologic classifications related to prognosis that are based on a large series of cases with adequate histologic sampling and detailed statistical analysis have been described.[13–15] The salient features of the previous and more recent prognostic histologic classifications are given in TABLE 1. Of these, the Shimada classification[13] is based on the following histologic features: (1) amount of Schwannian stroma; (2) nuclear morphology (mitosis and karryorrhexis index, MKI); and (3) degree of differentiation. The criteria for age-linked prognostic subgroups designated as favorable histology (FH) and unfavorable histology (UH) in 295 cases based on these features are given in TABLE 2. In a series of 211 cases of NB, Joshi et al. described a histologic grading system related to prognosis based on mitotic rate (MR) and calcification (TABLE 3).[14] The original histologic grading system was modified by inclusion of MKI instead of MR (TABLE 4).[15] The survival differences between FH and UH subgroups and between the original or modified histologic grades were statistically significant.

In addition to the series in which systematic prognostic classification of NBs into histologic grades or FH/UH categories is described (TABLE 1), the more recent reports summarized in TABLE 5 demonstrate one or more individual morphologic features to be significantly related to survival.[16–18]

It should be noted that NBs frequently show considerable heterogeneity of morphologic features in different areas of the same tumor, while chemotherapy

TABLE 1. Morphologic Prognostic Markers of Neuroblastoma

Author(s)/Year	Prognostic Feature	No. of Cases	No. of Grade[a]/Group[b] and Their Criteria	Comment
Beckwith & Martin (1968)	Degree of differentiation along gangliocytic line.	50	Grades I, II, III, & IV (>50, 5–50, <5 & 0% differentiating cells, respectively).	Lymphocytic infiltration also of prognostic value. Statistical analysis not given.
Lauder & Aherne (1972)	Degree of lymphocytic infiltration.	23	Grades 1, 2, 3, 4, & 5 (No, occasional, moderately dense, dense with and without follicles, respectively).	—
Makinen (1972)	Degree of differentiation along gangliocytic line.	54	Groups 1, 2a, and 2b (no differentiation, some differentiation without ganglion cells, and some differentiation with ganglion cells, respectively).	Necrosis, calcification, vascular invasion also of prognostic value.
Hughes et al. (1974)	Degree of differentiation along ganglioicytic line.	90	Grades I, II, & III (undifferentiated cells and ganglion cells, undifferentiated cells and differentiating cells, only undifferentiated cells, respectively).	Statistical analysis not given. No relationship between lymphocytic infiltration and survival.
Sandstedt et al. (1983)	Amount of neuropil.	70	Grades I, II, & III (dominant, moderate, and no neuropil, respectively).	Degree of differentiation but not rosettes found to be of prognostic value.
Shimada et al. (1984)	Amount of Schwannian stroma, nuclear morphology (mitosis karyorrhexis index—MKI), degree of differentiation.	295	Favorable and unfavorable histology (see TABLE 2 for details).	Linkage of age with morphologic features is essential for categorization into favorable and unfavorable histology.
Joshi et al. (1992)	Mitotic rate and calcification.	221	Grades 1, 2, & 3 (see TABLE 3).	Ganglion cells and tumor giant cells also found to be of prognostic significance. Grades with or without linkage to age found to be of prognostic value.
Joshi et al. (1996)	Mitosis karyorrhexis index, calcification.	223	Grades 1, 2, & 3 (see TABLE 4).	Grades with or without linkage found to be of prognostic significance.

[a] High grades are associated with *poor* prognosis.
[b] Higher group is associated with *better* prognosis.

TABLE 2. Prognostic Categorization of Neuroblastic Tumors According to Shimada Classification

Category	Favorable Histology (Survival)	Unfavorable Histology (Survival)
NB, stroma rich[a] at any age	Well differentiated[b] (100%) and intermixed (92%) subtypes	Nodular subtype (18%)
NB, stroma poor at age <18 months	Any subtype[c] with MKI <200 (84%)	Any subtype MKI >200 (4.5%)
NB, stroma poor at age 18 to 60 months	Differentiating subtype with MKI <100 (84%)	Differentiating subtype with MKI >100 or undifferentiated subtype with any MKI (4.5%)
NB, stroma poor at age >60 month	None	All subtypes (4.5%)

[a] NB, stroma-rich category is designated as ganglioneuroblastoma in the modified conventional terminology.

[b] This subtype is designated as ganglioneuroblastoma of borderline subtype in modified conventional terminology.

[c] Two subtypes, undifferentiated (<5%) differentiating neuroblasts and differentiating (≥5% differentiating neuroblasts) are recognized. Abbreviation: MKI, mitosis karyorrhexis index.

produces secondary changes such as necrosis, hemorrhage, calcification, and fibrosis and as well as induces maturation. Therefore, adequacy of histologic sample, consisting of multiple sections from the tumor resected prior to therapy, is of crucial importance in the assessment of prognostic significance of various histologic features of NB. Furthermore the patients included in the study of morphologic prognostic markers should receive standardized therapy and have adequate follow-up after therapy. Most or all of these criteria were met in the series described by Shimada et al.[13] and Joshi et al.[14,15]

NONMORPHOLOGIC PROGNOSTIC MARKERS

The clinical, biochemical, and genetic prognostic markers of NB are summarized in TABLE 6.[19–36] Of the various markers described in TABLE 6, N-myc and DNA index (DI) are the most widely used for stratification for therapeutic protocols.

TABLE 3. Original Histologic Grading of Neuroblastomas[a]

Grade	Criteria	Survival[b]
1	Calcification *and* low mitotic rate[c]	89.83%
2	Calcification *or* low mitotic rate	77.63%
3	Absence of calcification *and* of low mitotic rate[d]	33.53%

[a] Shimada classification is used for prognostic categorization of ganglioneuroblastoma.

[b] Survival at 5 years; the survival differences are statistically significant ($p < 0.001$).

[c] Low mitotic rate = ≤10/10 high-power fields.

[d] Absence of low mitotic rate = presence of high mitotic rate (>10/10 high-power fields).

TABLE 4. Modified Histologic Grading System of Neuroblastoma

Grade	Criteria	Survival[a]
1	Calcification *and* low MKI[b]	87.15%
2	Calcification *or* low MKI	72.23%
3	Absence of both[c]	18.19%

[a] Survival at 5 years; survival differences are statistically significant ($p < 0.001$)
[b] Low MKI (mitosis karyorrhexis index): $\leq$4% mitotic karyorrhectic cells (MKCs) ($\leq$200 MKCs/5000 tumor cells).
[c] Absence of low MKI = presence of high MKI, i.e., >4% (MKCs).

Other, more recently demonstrated markers such as trk-A expression, multidrug resistance–associated protein (MRP), and telomerase activity are promising from the conceptual and practical points of view, but need confirmation of the initial observations.

CORRELATION BETWEEN MORPHOLOGIC AND NONMORPHOLOGIC PROGNOSTIC MARKERS

Correlation between the original histologic grading system of NB[14] and nonmorphologic markers including DNA index, N-myc gene copy number per haploid genome, and serum lactate dehydrogenase (LDH) has been demonstrated in an analysis of 275 cases of NB.[37] A statistically significant association of low histologic grades (1 and 2) was found with DI of >1 (hyperdiploidy), single copy of N-myc gene per haploid genome, and serum LDH of <1500 IU/L (p value for each <0.001) (TABLE 7). In a recent preliminary analysis of 181 cases of NB (TABLE 8), similar statistically significant associations were found between modified low histologic grades (1 and 2), single copy of N-myc gene per haploid genome, and serum LDH of <1500 IU/L; and between modified grade 3 and >1 N-myc copy number and

TABLE 5. Individual Morphologic Features Related to Prognosis in Neuroblastomas

Authors/Year	No. of Cases	Morphologic Feature(s) Related to Prognosis	Comment
Shimada *et al.* (1985)	36	S-100 positivity (associated with better outcome).	Only cases ofundifferentiated neuroblastoma were studied. Age was linked to morphologic prognostic groups.
Chatten *et al.* (1988)	420	High mitotic rate ($\geq$10/10 hpf),[a] necrosis, foam cells: poor prognosis; multinucleation, calcification, ganglion cells: good prognosis.	Only stage III and IV patients studied. In stage IV patient—post-therapy samples were included.
Meitar *et al.* (1996)	50	Vascular index (VI) (total no. of blood vessels mm^2 of tumor) $\leq$4: good prognosis; >4: poor prognosis.	Correlation of VI with N-myc copy number and histology also noted.

[a] hpf, high power fields.

TABLE 6. Nonmorphologic Prognostic Markers of Neuroblastoma

Marker/Reference	Relationship to Prognosis	Mechanism	Comment
Clinical			
Age[14,19,20]	Better prognosis in children <1 year.[14,19] or <2 yr.[20] of age.	? Greater tendency to maturation in infants and young children as a result of normal "embryological" processes.	Age-related therapeutic protocols have been used by various cooperative oncology groups.
Stage[14,20]	Poor prognosis in stages III & IV (C & D) tumors as compared with stage I, II, & IV-S (A, B, & D-S) tumors.	? Tumors having advanced stage at diagnosis are biologically more aggressive since their inception.	Age and stage have been combined to define prognostic subgroups for effective therapy. Stage D-S may be a heterogenous group consisting of an admixture of tumors with good and poor prognosis.
Biochemical			
Serum neuron specific enolase (NSE)[21]	>100 ng/ml associated with poor prognosis.	High levels of NSE may reflect larger tumor burden and metastatic disease since neuroblastoma cells contain NSE	There is correlation between high NSE levels and advanced stages of neuroblastoma.
Serum ferritin[20,22]	>150 ng/ml associated with poor outcome.	Tumor cells are the likely source of serum ferritin. Higher levels may reflect large tumor burden.	Strong association between NSE, age, stage, and ferritin has been observed.
Serum lactate dehydrogenase (LDH)[23]	$\geq$1500 IU/L associated with poor outcome.	Same as for NSE.	LDH is a readily available and inexpensive assay.
Urinary catecholamines[20]	VMA : HVA ratio of <1 associated with poor outcome.	HVA is a metabolite of dopamine. Tumor excreting higher quantities of HVA relative to VMA are more primitive (and aggressive) tumors lacking dopamine-B-hydroxylase and, therefore, capable of synthesizing only dopamine (and not adrenaline and noradrenaline).	Catecholamines are expressed in microgram/mg of creatinine and corrected for age.
Genetic			
Deletion of 1 p or 14 q[24,25]	Deletion associated with poor outcome.	Loss of heterozygosity (LOH) suggests presence of tumor suppressor genes in this region that are lost or inactivated.	1 p deletion is seen in approximately 30 to 40% neuroblastomas.
N-myc oncogene copy number[26–28]	N-myc amplification associated with poor outcome.	N-myc gene product forms a heterodimer with helix-loop-helix protein called "Max." This heterodimer activates transcription of genes involved in cell proliferation. Amplification of N-myc (>3 copies) leads to high proliferative activity (and, therefore, aggressive behavior) of tumor cells.	There is association between stage III and IV and amplification of N-myc, which can also be determined by fluorescence *in situ* hybridization.[28]

DNA index (DI)[29]	Diploidy (DI = 1) associated with poor outcome.	Tumors with DI of 1 have higher percentage of S-phase cells. Diploid tumors respond poorly to therapy.	DI loses its prognostic significance in children >2 yrs. of age.
P-glycoprotein (PGP)[30]	PGP expression associated with poor outcome.	PGP is the product of multidrug resistance gene (MDR1). PGP functions as an energy-dependent drug efflux pump of broad specificity. Thus, high PGP expression renders drug resistance.	Complex multilayered immunohistochemical method used for demonstration of PGP expression. The method's reproducibility has not been tested.
Multidrug resistance–associated protein (MRP)[31]	High level of MRP expression associated with poor outcome.	MRP gene codes for a membrane bound glycoprotein which mediates multidrug resistance.	Prognostic value of MRP is retained after adjustment for effects of N-myc amplification, trk-A expression, age, and stage by multivariate analysis.
trk-A gene expression[32,33]	High level of trk-A gene expression highly predictive of favorable outcome.	trk-A gene encodes for nerve growth factor associated with neural differentiation (apoptosis may also be affected).	trk-A gene expression can also be determined by immunocytochemical stain.
Ha-ras gene expression[33]	Coexpression with trk-A gene associated with favorable outcome.	Coexpression with trk-A gene suggests linkage to similar nerve growth factor–initiated signal transduction pathway.	Ha-ras gene expression demonstrated by immunocytochemical stain.
Telomerase activity[34,35]	High telomerase activity associated with poor outcome.	High telomerase activity associated with prolonged survival of neuroblastoma cells with loss of apoptotic mechanism.	Correlation between high telomerase activity and N-myc amplification has been demonstrated.
CD_{44} cell surface expression[36]	High level of CD_{44} expression associated with favorable outcome.	CD_{44} is involved in cell-cell or cell–extracellular matrix interactions (which may influence metastatic potential).	Correlation between high CD_{44} expression and young age, early stage, lack of N-myc amplification, and favorable histology has been demonstrated. CD_{44} expression can be demonstrated by immunocytochemical stain.

TABLE 7. Correlation between Original HG[a] and Other Prognostic Factors in Neuroblastoma

Prognostic Marker	Grades 1 & 2	Grade 3	Total	p value
DI[b]				
>1	78	0	78	
1	15	3	18	<0.001
N-myc copy no.[c]				
1	73	4	77	
>1	10	7	17	<0.001
Serum LDH[d]				
<1500	186	11	197	
≥1500	12	15	27	<0.001

[a] HG: Histologic grades were available in 275 cases.
[b] DI: DNA index was available in 96 of the 275 cases in which HG was available.
[c] N-myc copy number was available in 94 cases.
[d] LDH: Lactate dehydrogenase in IU/L was available in 224 cases.

serum LDH of ≥1500.[15,38] Oppedal *et al.*,[39] in a small series of 18 cases, have demonstrated correlation between histologic grades determined by the criteria of Beckwith and Martin,[7] N-myc copy number, and DI.

Shimada *et al.*[40] compared the histologic phenotypes in NBs with and without N-myc amplification. In 45 of 51 (88%) NBs with N-myc amplification, high MKI (i.e., >4% mitotic karryorrhectic cells) and undifferentiated histology were seen, whereas the aforementioned histologic phenotype was seen in only 21 of 162 (13%) NBs without N-myc amplification. Correlation between normal serum ferritin level (≤150 ng/ml) and favorable histology (FH) has been reported by Evans *et al.*[20] in a series of 55 cases. The most recent report on correlation between morphologic and nonmorphologic markers is by Meitar *et al.*[18] These authors demonstrated statistically significant correlation between vascular index (VI) (number of blood vessels per mm^2 of tumor) and N-myc copy number. Thus 50% of tumors with VI of >4 had N-myc amplification, whereas only 16% with VI of ≤4 had N-myc amplification.

In all the aforementioned series, the number of cases in which data on all the

TABLE 8. Correlation between Modified HG[a] and Other Prognostic Factors in Neuroblastoma

Marker	Grades 1 & 2	Grade 3	Total
NM[b] = 1	46	6	52
NM > 1	2	4	6
			$p = 0.001$
LDH[c] < 1500	136	31	167
LDH > 1500	4	10	14
			$p < 0.001$

[a] HG: Histologic grades were available in 181 cases.
[b] NM: N-myc copy number per haploid genome was available in 58 of the 181 cases in which HGs were available.
[c] LDH: Serum lactate dehydrogenase expressed in IU\L was available in 181 cases.

prognostic markers were available was small. Therefore, a multivariate analysis could not be done. However, after univariate analysis in the series reported by Joshi et al.,[37] the original histologic grades retained their prognostic value after adjusting for the effects of DI, N-myc copy number, and serum LDH ($p = 0.036$, 0.011, and 0.001, respectively). In particular, patients with low LDH and those lacking N-myc amplification, which ordinarily are favorable features, did poorly if their histologic grade was 3 (8 of 11 and 4 of 4 deaths, respectively). The analysis of DI was somewhat inconclusive because there were only 4 deaths among 96 patients in whom DI was available.

In contrast to the aforementioned reports on the association of favorable histology or low histologic grades with nonmorphologic markers indicative of good prognosis, Cohn et al.,[41] in a study of six localized NBs (stage A or B) with N-myc amplification, found lack of correlation between N-myc copy number and histologic grade or prognostic histologic subgroups by Shimada classification. Four of these six patients survived despite N-myc amplification, and the NBs in these survivors had FH by Shimada classification or low histologic grades.

CORRELATION WITHIN THE GROUP OF MORPHOLOGIC PROGNOSTIC MARKERS

Besides the correlation between morphologic and nonmorphologic markers reviewed above, correlation is also found between various morphologic prognostic systems themselves. Joshi et al.[14,15] demonstrated correlation between age-linked original and modified histologic grades with the FH and UH categories of Shimada classification, which is inseparably linked with age (84% and 76% concordance, respectively, for original and modified histologic grades with FH/UH subgroups of Shimada classification).

CORRELATION WITHIN THE GROUP OF NONMORPHOLOGIC PROGNOSTIC MARKERS

Correlations have also been demonstrated between various nonmorphologic markers. Nonmorphologic markers indicative of poor prognosis tend to be associated with each other and vice versa. Thus, association is found between (a) N-myc amplification and DI of 1 (diploidy);[42] (b) lack of trk-A expression and presence of N-myc amplification (correlation between trk-A expression and prognostic histologic subgroups based on Shimada classification was also seen);[43] (c) expression of MRP gene expression and N-myc amplification;[44] (d) high serum NSE and ferritin levels;[20] (e) CD_{44} expression and lack of N-myc amplification;[36] (f) allelic loss of 1 p and N-myc amplification;[24] and (g) high stages (C and D or 3 and 4) and markers of poor prognosis such as N-myc amplification.[27]

Multivariate analysis to determine which of these prognostic markers remain significant after adjusting for the other associated markers was not done except in the series reported by Norris et al.[31] and by Nakagawara et al.[32] Norris et al.[31] demonstrated that high levels of MRP expression retained its prognostic value for poor survival even after adjusting for effects of N-myc amplification and other prognostic markers (age, trk-A expression, and stage). With the same analysis, N-myc amplification did not retain its prognostic value after adjusting for effects of MRP expression and other prognostic markers. Nakagawara et al.[32] found that although on univariate analysis trk-A gene expression was highly significant ($p <$ 0.001), multivariate analysis after adjusting for age, stage, N-myc copy number, and tumor site did not retain the statistical significance of trk-A gene expression.

COMMENT

Three questions may be raised regarding the study of correlation between morphologic and nonmorphologic prognostic markers of NB: (a) what are the mechanisms at the cellular level that would explain the presence of correlation? (b) what is the relative impact of an individual prognostic marker in comparison with others? and (c) what is the clinical or therapeutic relevance of the presence of correlation?

The possible or known mechanisms of various nonmorphologic markers are summarized in TABLE 6. The morphologic markers that correlate with nonmorphologic markers may merely represent histologic expression(s) of the mechanism(s) of the latter. For example, the N-myc gene product forms a heterodimer with a helix-loop-helix protein called "Max."[45] This heterodimer activates transcription of genes involved in cell proliferation.[45,46] Amplification and overexpression of N-myc would lead to abundance of the heterodimeric complex with subsequent cell proliferation.[47] The histologic expression of cell proliferation is high mitotic rate, which is the major criterion for both original and modified histologic grade 3 (high MR and high MKI, respectively). Hence, the correlation between N-myc amplification and histologic grade 3 of NB. Similarly DI of 1, which is associated with a significantly higher proportion of S-phase cells in NB, correlates with original histologic grade 3, which has a high mitotic rate. Elevation of serum LDH seen in many different types of malignant diseases is considered to be a reflection of a large tumor burden. The large size of tumor burden may be related to the high proliferative activity of the tumors. Thus, the association of high serum LDH values with high original or modified histologic grade 3 can be explained on the basis of high proliferative activity of these NBs.

The statistical significance with respect to prognosis of an individual prognostic factor after adjustment by univariate analysis for the effect of another individual prognostic factor has been done for histologic grades as described above. It appears from such analyses that histologic grades retain their prognostic significance after adjusting for other individual prognostic factors such as N-myc copy number and serum LDH.[37,38] In the small series of six cases of localized NB with N-myc amplification, reported by Cohn et al.,[41] it was demonstrated that the histology was of a greater prognostic value. From the data in the aforementioned series[37,38,41] and from the practical point of view of cost and ready availability, it appears to us that histologic evaluation of NBs should form an important component of prognostic evaluation.

Reproducibility of histologic assessment is a major concern of the pediatric oncologists, molecular biologists, and also pathologists working in this field. A study of one of the major prognostic histologic features, MKI, has shown that histologic assessment of NBs is reproducible with good concordance between two different pathologists.[48] At present, a concordance study for various histologic features related to prognosis such as degree of differentiation, MKI, MR, and calcification is being carried out by an International Neuroblastoma Pathology Subcommittee of the International Neuroblastoma Risk Group Committee. Another limiting factor for morphologic prognostic factor may be sample size.[14] However, an open biopsy of an adequate size performed prior to chemotherapy may suffice. The prognostic value of histologic features in relation to that of nonmorphologic markers also needs to be assessed by multivariate analysis.

The practical significance of studying correlation between different prognostic factors is to identify subsets of NBs, each with certain prognostic markers indicative of a certain grade of malignancy. For these subsets, the prognostic markers that are statistically most significant, are readily and reproducibly determined, and are inexpensive should be chosen. Evans *et al.*[20] and Brodeur[49] have attempted to identify such subsets. By using age and serum ferritin level, Evans *et al.* described three subsets with good (normal ferritin, age <2 yrs.), intermediate (normal ferritin, age ≥2 yrs.), and poor (abnormal ferritin) prognosis having 2-year survival of 93%, 58%, and 19%, respectively.[20] Brodeur[49] has divided NBs into three "genetic" types 1, 2, and 3 that have good (95%), intermediate (28–50%), and poor (<5%) 3-year survival, respectively. He included N-myc copy number (1 in type 1, amplified in type 3), trk-A expression (high in type 1, low or absent in type 3), DI (>1 in type 1, 1 in type 3), 1 p and 14 q deletion (absent in type 1, present in type 3), and age and stage in defining these types. There is overlap of some of these features in type 2 NBs. Neither Evans *et al.* nor Brodeur included histologic assessment in defining these prognostic subsets. Histologic assessment is a statistically strong, readily reproducible, and inexpensively determinable prognostic factor. Therefore, we would like to suggest that histologic features should be included in the criteria for prognostic subsets of NBs. Stratification based on clusters of prognostic markers can be used in designing risk-specific therapy. Thus, study of correlation between various prognostic markers of NB is of direct practical and clinical importance.

SUMMARY

Morphologic (Shimada classification—SC, original and modified histologic grades—OHG and MHG) and nonmorphologic (serum LDH, 1 p del, DNA index, N-myc copy number, telomerase activity, and expression of MRP, MDR1, and TRK) prognostic markers for NB have been reviewed. The functional role of these nonmorphologic markers in the development and progression of this disease include abnormal cell proliferation, resistance to chemotherapeutic agents, and induction of apoptosis. A statistically significant association between high OHG/MHG (grade 3), DNA index of 1 (diploidy), >1 copy of N-myc per haploid genome and serum LDH of ≥1500 IU/l ($p < 0.001$ for each) has been described. In SC, undifferentiated histology and high MKI are associated with N-myc amplification. However, a lack of correlation between morphology and N-myc amplification has been found in localized NB. Confirmation of these observations must now be obtained on larger numbers of prospectively studied cases. Data on correlation for various prognostic markers could provide guidelines for identification of subsets of NB having strongly significant, readily determinable, reproducible, and relatively inexpensive prognostic markers that could ultimately be used to design an algorithm for risk-specific therapy.

ACKNOWLEDGMENTS

The authors thank Dr. R. P. Castleberry, Dr. A. B. Cantor, Dr. P. V. Rao, and Dr. J. J. Shuster for their contributions to the pathologic studies of Pediatric Oncology Group neuroblastoma cases reviewed in this article.

REFERENCES

1. Ross, J. A., *et al.* 1996. Childhood cancer in the United States. A geographical analysis of cases from the Pediatric Co-operative Clinical Trials Group. Cancer **77:** 201–207.
2. Fox, F., *et al.* 1959. Maturation of sympathicoblastoma into ganglioneuroma. Cancer **12:** 108–116.
3. Brodeur, G. M., *et al.* 1993. Revisions of the international criteria for neuroblastoma diagnosis, staging and response to treatment. J. Clin. Oncol. **11:** 1466–1477.
4. Wahl, N. R. 1914. Neuroblastoma: With study of a case illustrating the three types that arise from the sympathetic nervous system. J. Med. Res. **30:** 205–260.
5. Stout, A. P. 1949. Tumors of the peripheral nervous system. *In* Atlas of Tumor Pathology, Sect. II, Fascicle 6: 37–45. Armed Forces Institute of Pathology. Washington, DC.
6. Horn, R. C., C. E. Koop & W. B. Kiesenwester. 1956. Neuroblastoma in childhood. Lab. Invest. **5:** 106–119.
7. Beckwith, J. B. & R. F. Martin. 1968. Observations of histopathology of neuroblastomas. J. Pediatr. Surg. **3:** 106–110.
8. Martin, R. F. & J. B. Beckwith. 1968. Lymphoid infiltrates in neuroblastomas: Their occurrence and prognostic significance. J. Pediatr. Surg. **3:** 161–164.
9. Lawler, I. & W. Aherne. 1972. The significance of lymphocytic infiltration in neuroblastoma. Br. J. Cancer **26:** 321–330.
10. Makinen, J. 1972. Microscopic patterns as a guide to prognosis of neuroblastoma in childhood. Cancer **29:** 1637–1646.
11. Hughes, M., H. B. Marsden & M. K. Palmer. 1974. Histologic patterns of neuroblastoma related to prognosis and clinical staging. Cancer **34:** 1706–1711.
12. Sandstedt, B., B. Jereb & G. Eklund. 1983. Prognostic factors in neuroblastomas. Acta Path. Microbiol. Immunol. Sect. A **91:** 365–371.
13. Shimada, H., *et al.* 1984. Histopathologic prognostic factors in neuroblastic tumors: Definition of subtypes of ganglioneuroblastoma and an age-linked classification of neuroblastomas. J. Nat. Cancer Inst. **73:** 405–416.
14. Joshi, V. V., *et al.* Age-linked prognostic categorization based on a new histologic grading system of neuroblastomas: A clinicopathologic study of 211 cases from the Pediatric Oncology Group. Cancer **69:** 2197–2211.
15. Joshi, V. V., *et al.* 1996. Modified histologic grading of neuroblastoma by replacement of mitotic rate with mitosis karyorrhexis index: A clinicopathologic study of 223 cases from the Pediatric Oncology Group. Cancer **77:** 1582–1588.
16. Shimada, H., *et al.* 1985. Prognostic subgroups for undifferentiated neuroblastoma: Immunohistochemical study with anti-S-100 protein antibody. Hum. Pathol. **16:** 471–476.
17. Chatten, J., *et al.* 1988. Prognostic value of histopathology in advanced neuroblastoma: A report from Children's Cancer Study Group. Hum. Pathol. **19:** 1187–1198.
18. Meitar, D., *et al.* 1996. Tumor angiogenesis correlates with metastatic disease, N-myc amplification and poor outcome in human neuroblastoma. J. Clin. Oncol. **14:** 405–414.
19. Breslow, N. & B. McCann. 1971. Statistical estimation of prognosis for children with neuroblastoma. Cancer Res. **31:** 2098–2103.
20. Evans, A. E., *et al.* 1987. Prognostic factors in neuroblastoma. Cancer **59:** 1853–1859.
21. Zeltzer, P. M., *et al.* 1986. Serum neuron specific enolase in children with neuroblastoma: Relationship to stage and disease course. Cancer **57:** 1230–1234.
22. Hann, H. L., *et al.* 1985. Prognostic importance of serum ferritin in patients with stage III and IV neuroblastoma: The Children's Cancer Study Group experience. Cancer Res. **45:** 2843–2848.
23. Shuster, J. J., *et al.* 1992. Serum lactate dehydrogenase in childhood neuroblastoma: A Pediatric Oncology Group recursive partitioning study. Am. J. Clin. Oncol. **15:** 295–303.
24. Caron, H., *et al.* 1996. Allelic loss of chromosome 1 p as a predictor of unfavorable outcome in patients with neuroblastoma. N. Engl. J. Med. **334:** 225–230.
25. Fong, C. T., *et al.* 1992. Loss of heterozygosity for chromosomes 1 or 14 defines subsets of advanced neuroblastomas. Cancer Res. **52:** 1780–1785.

26. BRODEUR, G. M., *et al.* 1984. Amplification of N-myc in untreated human neuroblastomas correlates with advanced disease stage. Science **224:** 1121–1124.

27. SEEGER, R. C., *et al.* 1985. Association of multiple copies of the N-myc oncogene with rapid progression of neuroblastomas. N. Engl. J. Med. **313:** 1111–1116.

28. SHAPIRO, H. N., *et al.* 1993. Detection of N-myc gene amplification by fluorescence in situ hybridization: Diagnostic utility for neuroblastoma. Am. J. Pathol. **142:** 1339–1346.

29. LOOK, A. T., *et al.* 1984. Cellular DNA content as a predictor of response to chemotherapy in infants with unresectable neuroblastoma. N. Engl. J. Med. **311:** 231–235.

30. CHAN, H. S. L., *et al.* 1991. Glycoprotein expression as a predictor of the outcome of therapy for neuroblastoma. N. Engl. J. Med. **325:** 1608–1614.

31. NORRIS, M. D., *et al.* 1996. Expression of the gene for multi-drug-resistance–associated protein and outcome in patients with neuroblastoma. N. Engl. J. Med. **334:** 231–238.

32. NAKAGAWARA, A., *et al.* 1993. Association between high levels of expression of the TRK gene and favorable outcome in human neuroblastoma. N. Engl. J. Med. **328:** 847–854.

33. TANAKA, T., *et al.* 1995. trk A gene expression in neuroblastoma: The clinical significance of an immunohistochemical study. Cancer **76:** 1086–1095.

34. HIYAMA, E., *et al.* 1995. Correlating telomerase activity levels with human neuroblastoma outcomes. Nature Med. **1:** 249–255.

35. BRODEUR, G. M. 1995. Do the ends justify the means? Nature Med. **1:** 203–205.

36. COMBARET, V., *et al.* 1996. Clinical relevance of CD_{44} cell surface expression and N-myc gene amplification in a multivariate analysis of 121 pediatric neuroblastomas. J. Clin. Oncol. **14:** 25–34.

37. JOSHI, V. V., *et al.* 1993. Correlation between morphologic and other prognostic markers of neuroblastoma: A study of histologic grade, DNA index, N-myc gene copy number and lactic dehydrogenase in patients in the Pediatric Oncology Group. Cancer **71:** 3173–3181.

38. JOSHI, V. V., *et al.* 1996. Unpublished data.

39. OPPEDAL, B. R., *et al.* 1989. N-myc amplification in neuroblastomas: Histopathological, DNA ploidy, and clinical variables. J. Clin. Pathol. **42:** 1148–1150.

40. SHIMADA, H., *et al.* 1995. Identification of subsets of neuroblastomas by combined histopathologic and N-myc analysis. J. Nat. Cancer Inst. **87:** 1470–1476.

41. COHN, S. L., *et al.* 1995. Lack of correlation of N-myc amplification with prognosis in localized neuroblastomas: A Pediatric Oncology Group Study. Cancer Res. **55:** 721–726.

42. LOOK, A. T., *et al.* 1991. Clinical relevance of tumor cell ploidy and N-myc gene amplification in childhood neuroblastoma. A Pediatric Oncology Group study. J. Clin. Oncol. **9:** 581–591.

43. BORRELLO, M. G., *et al.* 1993. trk and ret proto-oncogene expression in human neuroblastoma specimens. High frequency of trk expression in nonadvanced stages. Int. J. Cancer **56:** 540–545.

44. BORDEN, S. B., *et al.* 1994. Expression of the multi-drug resistance associated protein (MRP) gene correlates with amplification and overexpression of N-myc oncogene in childhood neuroblastoma. Cancer Res. **54:** 5036–5040.

45. BLACKWOOD, E. M. & R. N. EISENMAN. 1991. A helix-loop-helix zipper protein that forms a sequence-specific DNA-binding complex with myc. Science **25:** 1211–1217.

46. MAKELA, T. P., *et al.* 1992. Alternative forms of Max as enhancers or suppressors of myc-ras co-transformation. Science **256:** 373–377.

47. WENZEL, A., *et al.* 1991. The N-myc oncoprotein is associated in vivo with the phosphoprotein Max (p20/22) in human neuroblastoma cells. EMBO J. **10:** 3703–3712.

48. JOSHI, V. V., *et al.* 1991. Evaluation of Shimada classification in advanced neuroblastoma with a special reference to the mitosis karyorrhexis index. A report from the Children's Cancer Group. Mod. Pathol. **4:** 139–147.

49. BRODEUR, G. M. 1994. Molecular pathology of human neuroblastomas. Semin. Diagn. Pathol. **11:** 118–125.

Chemotherapy of Acute Myelocytic Leukemia in Children[a]

SVERRE O. LIE,[b] GUDMUNDUR K. JONMUNDSSON,[c]
LOTTA MELLANDER,[d] MARTTI A. SIIMES,[e]
MINNA YSSING,[f] AND GÖRAN GUSTAFSSON[g]

[b]Department of Pediatrics
University Hospital, Rikshospitalet
N-0027 Oslo, Norway

[c]The Children's Department
The University Hospital, Landspitalinn
101 Reykjavik, Iceland

[d]Department of Pediatrics
East Hospital
S-41685 Göteborg, Sweden

[e]Childrens Hospital
University of Helsinki
SF-00290 Helsinki, Finland

[f]Pediatric Department G.G.K. 4064
University Hospital, Rigshospitalet
DK-2100 Copenhagen, Denmark

[g]Department of Pediatrics
Karolinska Hospital
S-10401 Stockholm, Sweden

Acute myelocytic leukemia (AML) in both children and adults is a more complex and resistant disease than acute lymphocytic leukemia. Progress has been slower and therapy more complicated, but with intensive myelosuppressive induction and further postremission therapy one-third of such patients may now achieve long-term survival and probably be cured.[1-5]

In the Nordic countries (Denmark, Finland, Iceland, Norway, and Sweden) hospital services are, in principle, free to all citizens. Private hospitals are very few, and "lost to follow-up" patients a rarity. These countries offer, therefore, a unique possibility for performing population-based studies and to trace patients for follow-up. The Nordic Society of Paediatric Haematology and Oncology (NOPHO) was established in 1981, and we now have documented information on every case of any childhood acute leukemia in a population of 23 million inhabitants. Examples will be drawn from our experience during the last decade.[6]

The major focus of this chapter will be on therapy, but a brief review of epidemiology and biology will be given as a background.

---

[a] Written by the authors on behalf of the Nordic Society of Paediatric Haematology and Oncology (NOPHO).

TABLE 1. Number and Incidence (Inc) of Acute Leukemias in the Nordic Countries from July 1984 through December 1992

	Total	% AML	$Inc_{TOT}{}^{a}$	$Inc_{AML}{}^{a}$
Denmark	358	18	4.6	0.8
Finland	390	12	4.8	0.6
Iceland	22	23	4.1	0.9
Norway	289	18	4.2	0.8
Sweden	606	16	4.7	0.8
Nordic countries	1665	16	4.6	0.7

[a] Per 100,000 children < 16 years per year.

EPIDEMIOLOGY

TABLE 1 shows the incidence of all acute leukemias in the Nordic countries: 4.6 new cases are diagnosed every year per 10^5 children younger than 16 years of age. AML accounts for 16% of these, which means that slightly less than one child per 100,000 per year develop this disease. In black children the relative incidence of AML is significantly higher due to the lower incidence of ALL.[1]

The age distribution of our patients is seen in FIGURE 1. The distinct peak incidence in the lower age group is higher than has been reported in other material. Part of this can be explained by the high frequency of Down's syndrome in our series. Children with trisomy 21 actually account for 13% of all new cases.[7,8]

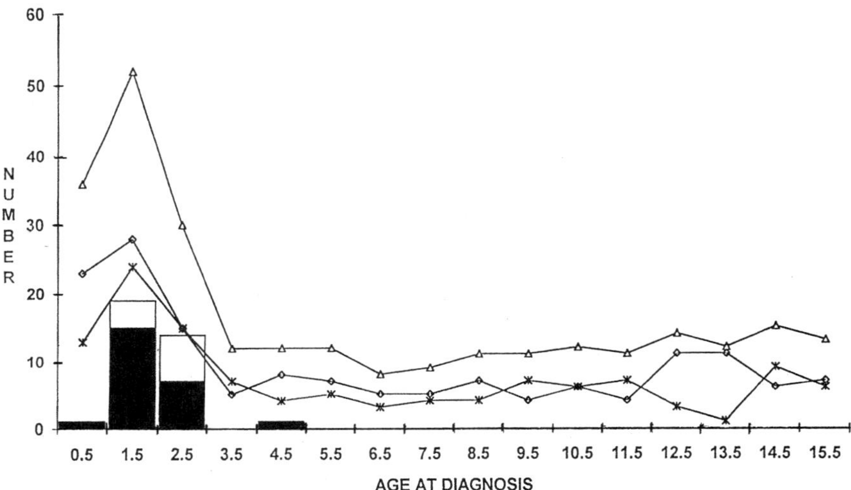

FIGURE 1. Age distribution of children with acute myelocytic leukemia. ■ Down's syndrome females; □ Down's syndrome males; —△— Total; —◇—females; and —*— males.

BIOLOGY

The French–American–British (FAB) classification from 1985 distinguished seven subgroups (M1-M7) on the basis of conventional morphology, cytochemical, and immunological methods.[9] M0 has been added by the same working group as a very undifferentiated leukemia.[10]

In children with AML, the types M1 and M2 account for 30–40% and M4 for about 25%. The megakaryoblastic leukemia (M7) is strongly associated with Down's syndrome.

Specific chromosomal abnormalities have been found in the various FAB subclasses. Translocation between chromosomes 8 and 21 is found almost exclusively in M1 and M2. All cases with M3 should carry the translocation t(15:17). Furthermore, M5 is associated with t(9;11). Abnormalities of chromosome 16 are seen predominantly in patients with M4. A recent review has been published by the Children's Cancer Group.[11] The chromosomal translocations often affect transcription factors that are involved in regulation of myeloid differentiation. In recent years great interest has been focused on the chromosomal changes involving 11q23 which is commonly seen in infants. The gene has recently been characterized and given various names (MLL, HRX, and ALL-1). It is related to a homebox gene in *Drosophila* which, when mutated, gives rise to bizarre malformations in chest and abdomen. Also, the chromosomal products of t(8;21) and t(15;17) have recently been defined. It is one of the most exciting developments in the biology of leukemias that the promyelocytic (M3) leukemia responds to all-*trans* retinoic acid (ATRA). Accordingly, it may be treated separately from all the other forms of AML.[12,13] The α-receptor for ATRA is located at the breakpoint on chromosome 17.

THERAPY

Before the 1970s, nearly every child with acute myelocytic leukemia died. After effective agents, including cytosine arabinoside (ara-C) and anthracyclines, became available in the 70s, a temporary remission could be induced in some patients, but most of them relapsed. It is the increased intensity of later protocols that has made it possible to cure at least some of the children. The first primary objective is to achieve a complete remission.

Remission Induction

TABLE 2 presents the remission induction results of some recent studies. Failure to enter remission may be due either to resistant disease or to death in aplasia. It is clear from the table that lack of intensity may cost death from resistant disease while increasing intensity beyond a certain level makes death from aplasia unacceptably high. In spite of several randomized studies, the most effective induction regimen that exists still consists of a 5- to 7-day course of ara-C (by continuous infusion or twice daily) with three days of anthracycline with or without other agents such as VP 16. With such regimens, remission rates of 70–85% in children with AML can be expected.

A topic of great debate is the effect of induction therapy, not only on remission rates but also on long-term survival. In their recently published paper, Woods and coworkers[14] have shown that very intensive timed sequential induction therapy

TABLE 2. Results of Chemotherapy in Childhood AML

	CCG-213[17]	POG-8498[18]	POG (1996)[19]	CCG-2891 intens[14]	CCG-2891 stand[14]	CCG-2861[20]	AIEOP[21]	BFM-83[16]	BFM-87[16]	MCR-10[22]	NOPHO-84[8]	NOPHO-88[8]
Years entered	1986–89	1984–88	1988–93	1989–93	1989–93	1986–89	1987–90	1982–86	1986–91	1988–93	1984–88	1988–92
Number evaluable	591	285	552	295	294	142	161	173	210	270	105	118
Death in aplasia (%)	6	7	11	14	26	13	7	7	5	9	8	12
Resistant disease (%)	15	8	4	11	4	11	14	13	17	9	14	3
Complete remission (%)	77	85	85	75%	70%	76	79	80	78	91	78	85
pDFS	0,39	0,45				0,4	0,31	0,61	0,52	0,56	0,43	0,56

improves the outcome in AML. When their protocol (DCTER, which is an intensive protocol of five drugs in four days) is given seven days apart regardless of blood counts, the results are much better than the standard time interval when the second DCTER is given after bone marrow recovery. The price to pay is that 11% die of toxicity during induction compared to 4% with the standard timing.

Post Remission Therapy

TABLE 1 also presents the disease-free survival of the studies mentioned above. It is always difficult to compare studies, but it is evident that many different protocols may lead to the same final result. Space does not allow a detailed discussion of the various protocols. However, most of the studies use many cytostatics in a cyclic fashion and in high-intensity dosages. The most simple post-remission therapy is in NOPHO-84, where the children are given four courses of high-dose ara-C only.[15] It is quite clear from the table that more is not necessarily better.

The ultimate intensity therapy is bone marrow transplantation (BMT). With increasingly intensive chemotherapy the role of BMT is more difficult to define. TABLE 3 presents some studies in which BMT has been compared in a randomized fashion with post-remission chemotherapy only. Those studies in which BMT has proven advantageous are early studies in which chemotherapy had a lower intensity than in most recent studies.

The subset promyelocytic leukemia is unfortunately rare in children, but should be treated separately. Probably the best therapy is to induce remission by ATRA, and then consolidate with chemotherapy. Several studies are currently addressing this question.

Because there still is a high hematopoetic relapse rate in children with AML, the true incidence of central nervous leukemia in AML remains unknown. However, most of the current protocols use high-dose therapy that would also have an effect on the central nervous system. One of the best results in AML has been presented by the German group. In a challenging paper from 1993, they propose that cranial irradiation may also reduce the risk of bone marrow relapse in AML.[10]

As survival has increased, the possibility of looking for risk factors have also been possible. Again, the German group has clearly identified two risk groups in their patients and are stratifying the therapy accordingly. With the exception of patients with M3 leukemia and children with Down's syndrome, no other group

TABLE 3. Bone Marrow Transplantation in Childhood AML

| | | alloBMT | | Auto BMT | Chemotherapy Only |
	No. in CR	No	DFS (%)	DFS (%)	DFS (%)
CCG 251[17]	381	85	50	—	36
CCG 213[20]	439	92	46	—	38
CCG 2861[23]	108	16	55	51	—
MRC AML-10[24]	766	174	58	54	52
AIEOP/LAM 87[21]	127	22	51	21	27
	1821	389 (21%)			

NOTE: CR, complete remission; CCG, Children's Cancer Group.

has so far stratified their patient in risk groups, although many have reported on high white cell burden as a risk factor. In the Nordic studies we could not identify white cells as a risk factor. However, gender did carry a prognostic significance with girls doing better than boys.[8]

FUTURE DIRECTIONS

There are two reasons for treatment failures: therapy-related mortality and relapse. Part of the improvement seen during the last decade is certainly related to better supportive care, but 20–25% of the children still fail induction therapy and 5–10% die during intensive post-remission therapy. Reducing intensity will certainly increase the problem of resistant disease. The use of growth factors is still controversial in AML.

New and more effective drugs are not on the horizon. Better variants of the presently used drugs are continuously being developed, such as the various analogues of anthracyclines. However, we do not even know what is the best combination of the presently available drugs. Correct doses and timing of ara-C is still a topic of debate. And what about the role of mitoxantrone, VP16, and amsacrin? It is hard to see that we will ever find a protocol so good that it will be universally accepted. To lose a child as a result of therapy-related complications is among the most difficult events to handle in pediatric oncology. Only centers with experience and all possible supportive care facilities can take on the responsibility of caring for these children.

REFERENCES

1. BOULAD, F. & N. A. KERNAN. 1993. Treatment of childhood acute nonlymphoblastic leukemia: A review. Cancer Invest. **11:** 534–553.
2. LIE, S. O. 1989. Acute myelogenous leukaemia in children. Eur. J. Pediatr. **148:** 382–388.
3. LIE, S. O. 1995. Treatment of acute myeloid leukaemia in children. Bailliere's Clin. Paediatr. **3:** 757–778.
4. VORMOOR, J., *et al.* 1996. Therapy of childhood acute myelogenous leukemias. Ann. Hematol. **73:** 11–24.
5. HURWITZ, C. A., *et al.* 1995. Treatment of patients with acute myelogenous leukemia: Review of clinical trials of the past decade. J. Pediatr. Hematol. Oncol. **17:** 185–197.
6. LIE, S. O. & G. GUSTAFSSON. 1992. Progress in the treatment of childhood leukaemias. Ann. Med. **24:** 319–323.
7. SLØRDAHL, S. H., *et al.* 1994. Leukemic blasts with markers of four cell lineages in Down's syndrome ("megakaryoblastic leukaemia"). Med. Pediatr. Oncol. **21:** 254–258.
8. LIE, S. O., *et al.* 1996. A population based study of 272 children with acute myeloid leukaemia treated on two consecutive protocols with different intensity: Best outcome in girls, infants, and children with Down's syndrome. Br. J. Hematol. **94:** 82–88.
9. BENNETT, J. M. 1985. Proposed revised criteria for the classification of acute myeloid leukaemia. Ann. Med. **103:** 620–629.
10. BENNETT, J. M. 1991. Proposals for the recognition of minimally differentiated acute myeloid leukaemia (AML-MO). Br. J. Haematol. **78:** 325–329.
11. BARNARD, D. R., *et al.* 1996. Morphologic, immunologic, and cytogenetic classification of acute myeloid leukemia and myelodysplastic syndrome in childhood: A report from the Children's Cancer Group. Leukemia **10:** 5–12.
12. WARRELL, R. P., Jr., *et al.* 1993. Acute promyelocytic leukaemia. N. Engl. J. Med. **329:** 177–189.

13. OLSSON, I., *et al.* 1996. Cell Differentiation in acute myeloid leukemia. Eur. J. Haematol. **57:** 1–16.
14. WOODS, W. G., *et al.* 1996. Timed-sequential induction therapy improves postremission outcome in acute myeloid leukemia: A report from the Children's Cancer Group. Blood **87:** 4979–4989.
15. LIE, S. O., *et al.* 1990. High-dose ara-C as a single-agent consolidation therapy in childhood acute myelogenous leukaemia. Haematol. Blood Transf. **33:** 215–221.
16. CREUTZIG, U., *et al.* 1993. Does cranial irradiation reduce the risk for bone marrow relapse in acute myelogenous leukaemia? Unexpected results of the childhood acute myelogenous leukaemia study BFM-87. J. Clin. Oncol. **11:** 279–286.
17. NESBIT, M. E., *et al.* 1994. Chemotherapy for induction of remission of childhood acute myeloid leukaemia followed by marrow transplantation or multiagent chemotherapy: A report from the Children's Cancer Group. J. Clin. Oncol. **12:** 127–135.
18. RAVINDRANATH, Y., *et al.* 1991. High-dose cytarabine for intensification of early therapy of childhood acute myeloid leukaemia: A Pediatric Oncology Group Study. J. Clin. Oncol. **9:** 572–580.
19. RAVINDRANATH, Y., *et al.* 1996. Autologous bone transplantation versus intensive consolidation chemotherapy for acute myeloid leukemia in childhood. N. Engl. J. Med. **334:** 1428–1434.
20. WELLS, R. J., *et al.* 1994. Treatment of newly diagnosed children and adolescents with acute myeloid leukaemia: A Children's Cancer Group Study. J. Clin. Oncol. **12:** 2367–2377.
21. AMADORI, S., *et al.* 1993. Prospective comparative study of bone marrow transplantation and postremission chemotherapy for childhood acute myelogenous leukaemia. J. Clin. Oncol. **11:** 1046–1054.
22. STEVENS, R. F., *et al.* 1994. Improved outcome in paediatric acute myeloid leukaemia: Results of the MRC AML 10 trial. Med. Pediatr. Oncol. **23:** 172.
23. WOODS, W. G., *et al.* 1994. Timing intensive induction therapy improves post-remission outcome in acute myeloid leukaemia (AML) irrespective of the use of bone marrow transplantation (BMT). Blood **84:** 914.
24. BURNETT, A. K., *et al.* 1994. The role of BMT in addition to intensive chemotherapy in AML in first CR: Results of the MRC AML-10 trial. Blood **84:** 994.

Non-Hodgkin's Lymphomas: Epidemiology and Treatment

IAN T. MAGRATH

Lymphoma Biology Section
Pediatric Branch
National Cancer Institute–NIH
Lymphoma Biology Section
10 Center Drive, MSC 1928, Building 10/13N240
Bethesda, Maryland 20892-1928

INTRODUCTION

The non-Hodgkin's lymphomas (NHLs) in children fall into two main groups, each of which can be divided into tumors of B- and T-cell lineage—those of precursor cell origin and those of more mature phenotype (TABLE 1).[1] All NHLs of precursor origin, that is, the lymphoblastic lymphomas, have morphology identical to acute lymphoblastic leukemia (ALL), express terminal deoxyribonucleotide transferase (TdT), and encompass the same immunophenotypes. Unlike ALL, however, which, at least in the industrialized nations, consists predominantly (approximately 85%) of precursor B-cell neoplasms, lymphoblastic lymphomas are predominantly (85–90%) of T-cell origin.

Precursor Lymphoid Neoplasms

Precursor B- and precursor T-cell lymphoblastic lymphomas account for approximately one-third of NHL in children in the United States. They are morphologically and immunophenotypically indistinguishable from their leukemic counterparts, and T and B subtypes have quite distinct presentations. Very few pre-T-cell lymphomas present with limited disease, and the majority (more than 75% of pre-T lymphoblastic lymphoma) arise in the thorax as anterior superior mediastinal masses, reflecting their thymocyte origin. Pleural effusions, pericardial effusions, and lymphadenopathy, most often in the neck, supraclavicula fossae, or axillae are also frequently present; whereas intraabdominal disease is uncommon, being confined to small nodules or lymph nodes in most cases. In contrast, most pre-B lymphomas present with limited disease, such as localized lymphadenopathy or isolated bone involvement.

Mature B-Cell Lymphoid Neoplasms

Mature B-cell neoplasms in children, which account for more than 50% of childhood lymphomas in the United States, predominantly fall into the category of small noncleaved cell lymphomas of Burkitt's and non-Burkitt's subtypes, which are referred to as Burkitt's lymphomas and Burkitt-like lymphomas in the Revised European American Lymphoma classification. These tumors, which nearly always express monoclonal surface immunoglobulin (Ig), usually IgM, but lack terminal deoxyribonucleotide transferase (TdT), merge histologically with large B-cell

TABLE 1. Major Groupings of Childhood Non-Hodgkin's Lymphoma

Immunophenotype	Histological Category	Frequently Involved Sites
Precursor B-cell	Lymphoblastic lymphoma	Isolated bone or lymph node
B-cell	Small noncleaved cell lymphoma, large B-cell lymphomas	Abdomen disease, jaw (Africa)
Precursor T-cell	Lymphoblastic lymphoma	Thorax (esp. thymus)
T-cell	Anaplastic, large-cell lymphoma, other peripheral T-cell lymphomas	Lymph nodes, skin

lymphomas. The latter account for a small fraction (approximately 10%) of non-Hodgkin's lymphomas in children. Burkitt-like lymphoma is intermediate, morphologically, between Burkitt's and large B-cell lymphoma and probably does not represent a homogeneous pathological entity. It may be a composite group of tumors comprising the more pleomorphic variants of both Burkitt's lymphoma and large B-cell lymphomas.[2] Consequently, the composition of this rather ill-defined histological category is likely to differ in children and adults. Indeed, a fraction of these tumors in adults contain 14;18 chromosomal translocations, which are present in some 25 to 30% of large B-cell lymphomas in adults but are rare in childhood NHL.[2] Many Burkitt-like and even some large B-cell lymphomas, although certainly a much higher proportion in children than in adults, contain the same chromosomal translocations as Burkitt's lymphoma (see below), although rather limited cytogenetic information is available. Nonetheless, such information as does exist suggests that a fraction of large B-cell lymphomas in children are closely related to Burkitt's lymphoma. Others appear to differ—for example, the large B-cell lymphomas that arise in the mediastinum and the rare anaplastic large-cell lymphomas of B lineage. A few large B-cell lymphomas, particularly those in the adolescent child, may be biologically similar to the most frequent types of large B-cell lymphoma occurring in adults, which commonly contain translocations involving the *bcl*-6 gene or, somewhat less often, *bcl*-2. The relative proportions of these subtypes, however, are not well defined because of the lack of cytogenetic and molecular studies.

The clinical presentations of large B-cell lymphomas, doubtless reflecting their biological heterogeneity, are variable. Lymph node involvement, overall, is more common than in the small, noncleaved cell lymphomas; but many, like the small, noncleaved cell lymphomas, present as intraabdominal lymphomas. Whether the clinical differences reflect differences in the normal counterpart cells, namely, of peripheral lymphoid versus mucosal-associated lymphoid tissue (MALT) is not known, but is worthy of consideration. Evidence that Burkitt's lymphoma is an aggressive lymphoma of MALT origin is provided not only by the high frequency of bowel and mesenteric involvement, but also by the frequent involvement of the jaw in African children (50–70% of patients at presentation), probably arising in MALT surrounding the developing molar teeth, and the breast in lactating women.[3] Breast tissue is known to be infiltrated by B blasts of MALT origin during lactation that are involved with the production of IgA antibodies for transfer to the infant.[3–5]

Mature T-Cell Lymphoid Neoplasms

Mature T-cell lymphomas, that is, post thymic T-cell lymphomas, account for the predominant form of anaplastic large-cell lymphoma, which, in turn, is by far the most common type of peripheral T-cell lymphoma in children, accounting for approximately half of all large-cell lymphomas (i.e., some 7–10% of all childhood NHL). Non-anaplastic T-cell lymphomas of peripheral origin are rare in children.

Treatment

Treatment approaches to the NHLs in children are rather straightforward. The precursor neoplasms appear to require relatively prolonged therapy—at present, believed to be at least a year—and are most often treated with standard ALL treatment protocols of 18 months to 2 years duration regardless of immunophenotype. All the remaining NHLs can be treated with short-duration intensive chemotherapy protocols described below. However, peripheral T-cell lymphomas in children, whether anaplastic or not, have also been successfully treated with ALL-like regimens.

With the exception of Burkitt's lymphoma, little epidemiological information is available with respect to the childhood NHLs. Burkitt's lymphoma provides, however, a paradigm, from which extrapolations can be made to other NHLs; and hence the epidemiology of this disease will be discussed in some detail.

EPIDEMIOLOGY AND PATHOGENESIS OF BURKITT'S LYMPHOMA

Although a number of expatriot pathologists working in Africa in the first half of the 20th century noted that facial "sarcomas" and lymphomas occurred at high frequency in African children,[5–10] it was Denis Burkitt who literally put the disease now named after him on the map. He observed that children with jaw tumors often had histologically similar tumors at multiple other sites, particularly in the abdomen,[11] and subsequently demonstrated, by making safaris and writing to outlying hospitals, that the tumor occurred at high frequency in a broad belt across Africa, extending approximately 15° north and south of the equator with a southern prolongation on the eastern side of the continent.[12,13] Although Burkitt was initially under the impression that the tumor he was studying was a sarcoma, O'Conor and Davies, in a review of the malignant tumors of children in the Kampala Cancer Registry, which included patients from all of Uganda, recognized that malignant lymphoma accounted for some 50% of all childhood malignant tumors in the registry.[14,15] It therefore became clear that Burkitt's tumor was a lymphoma.

Soon after these observations in Africa, O'Conor, Wright (who had also worked in Uganda), and others reported histologically indistinguishable tumors in children in the United States and Europe,[16–18] observations that led to considerable discussion regarding the definition of Burkitt's lymphoma. At about the same time, Davis, who with O'Conor had recognized the high frequency of malignant lymphoma in Ugandan children, saw strong similarities in the map of the distribution of Burkitt's lymphoma to that of the distribution of yellow fever.[19] He suggested, therefore, that Burkitt's lymphoma might be caused by a vectored virus. Shortly thereafter, Epstein and his colleagues identified a new type of herpes virus (subsequently called Epstein-Barr virus, or EBV) in a cell line derived from Burkitt's lymphoma.[20] The Henles, working in Philadelphia, developed a test for antibodies to EBV and were able to demonstrate not only that African patients with Burkitt's lymphoma had high titers of antibodies against EBV, but that EBV is a ubiquitous virus that infects most normal individuals in populations in widely separated parts of the world.[21–23]

Even so, African patients with Burkitt's lymphoma have an approximately eightfold higher geometric mean titer of antibody against the EBV viral capsid antigen, VCA, than normal controls[23] and frequently possess antibodies against EBV early antigens, which are not usually detectable in normal individuals positive for anti-VCA.[24] Subsequently, the identification of EBV nuclear antigen(s) by Reedman and Klein[25] led to the recognition that viral DNA is present in all tumor cells and that the majority of African Burkitt's lymphomas are associated with EBV. Outside Africa, however, the association with EBV is much more variable, and only some 20% of North American and European tumors contain EBV DNA, although in most developing countries, 50% to 90% are EBV positive.

Because of the close similarity of the distribution of Burkitt's lymphoma in Africa with that of mosquito-vectored virus diseases, it seemed highly probable that mosquitos were, in some way, involved in the transmission of the disease. However, insect transmission is clearly not an important (if it occurs at all) route of EBV transmission, and it now seems likely that Dalldorf's suggestion that malaria, rather than a virus, might account for the geographical distribution of the tumor in Africa is correct.[26] In Africa, the incidence rate of Burkitt's lymphoma is approximately 5 to 10 cases per 100,000 children below the age of 16 years, which is considerably higher than its incidence in the United States and Europe, which is approximately three per million children per year. This difference in incidence is probably accounted for by early infection with EBV and the holoendemicity of malaria in sub-Saharan Africa.

Chromosomal Translocations of Burkitt's Lymphoma

An important clue to the understanding of the pathogenesis of Burkitt's lymphoma is the invariable presence (it should, perhaps, be considered to be a *sine qua non* for a diagnosis of Burkitt's lymphoma) of a nonrandom translocation. Approximately 85% are between chromosomes 8(q24) and 14(q32), while 15% have a variant translocation—either an 8;22(q24;q12) or a 2;8(p11;q24) translocation.[27,28] The gene residing at the common chromosomal breakpoint of all three translocations (8q24) is c-*myc*, a transcription factor involved in cellular proliferation and apoptosis.[29–31] This gene is juxtaposed, by virtue of the translocations, to immunoglobulin sequences—either heavy chain sequences on chromosome 14q32, or light chain sequences on chromosomes 2p11 (kappa), or 22q12 (lambda), hence the general term of *myc*/Ig translocations.[32] This results in transcriptional deregulation of c-*myc*, presumably at least in part due to its juxtaposition to transcriptional enhancers within the immunoglobulin loci, a second element being the structural changes within the c-*myc* regulatory region that invariably accompany the translocation.[32,33] The net consequence of the translocation appears to be that c-*myc* is regulated as if it were an immunoglobulin gene, that is, it is constitutively expressed in these immunoglobulin-synthesizing tumor cells.

Recently, we have shown that in a significant proportion of Burkitt's lymphomas with an 8;14 translocation (at least half, and probably more), the J region on one of the two chromosome 14 alleles—that are involved in the translocation—is unrearranged.[34] This has several implications. First, the translocation must usually occur before D-J joining, the first element of immunoglobulin gene rearrangement; that is, it must occur in a pro-B cell. This probability is supported by evidence from a quite different direction, namely the occasional occurrence of very immature human lymphoid neoplasms, that is, expressing TdT, or with a pre-B-cell phenotype, in which an 8;14 translocation is present.[35–38] Secondly, the immunoglobulin allele uninvolved in the translocation must successfully rearrange subsequent to the translocation, since essentially all Burkitt's lymphomas synthesize immunoglobulin. This also implies that a functional immunoglobulin molecule is required for lymphomagenesis, perhaps because the cell must escape the apoptotic fate that awaits B cells that have failed to successfully create an antibody molecule.

The occurrence of *myc*/Ig translocations in immature B cells is consistent with experimental evidence that malaria increases the proportion of immature B cells (pro-B cells) in the bone marrow of mice.[39] Moreover, in these same mice, decreases in mature B-cell populations are readily demonstrable. This finding is, in turn, entirely consistent with Wedderburn's observations that mice infected with malaria—like mice transgenic for c-*myc*[40]—are more likely to develop lymphoma when simultaneously infected with Maloney leukemia virus.[41] It seems highly likely that malarial infection of the bone marrow in African children also results in an expansion of pro-B-cell populations, consequently increasing the likelihood of the development of a c-*myc*/Ig translocation. A cell bearing such a translocation, however, is probably destined to undergo apoptosis because of the presence of a deregulated c-*myc* gene, (as do such cells in transgenic mice), unless it develops additional genetic changes that permit it to pass checkpoints in the cell cycle that detect the aberrant Myc expression. This hypothesis, if correct, predicts that cells bearing *myc*/Ig translocations might arise in relatively high numbers in normal (malarious) African children, since the likelihood of developing a mutation that will prevent apoptosis must be extremely small. It is also consistent with the mature phenotype of Burkitt's lymphoma cells, which, in addition to surface IgM[42] and characteristics consistent with a germinal follicle cell (CD10 and CD77 expression[43]), includes the presence of mutations in the hypervariable region of the immunoglobulin molecule (which is usually interpreted to indicate that the cell has been exposed to and activated by antigen, or perhaps some other mitogen[44]) and the secretion of immunoglobulin.[45] It seems likely that the cells, once they have developed lesions permitting them to bypass the cell cycle and differentiation checkpoints, go on to mature to the level of a germinal center cell.

EBV and Burkitt's Lymphoma

EBV seroconversion is known to occur much earlier in populations of low socioeconomic status, or in those who live in crowded circumstances, than in populations in which families are smaller and have more spacious accommodations.[46,47] In Africa, seroconversion occurs in the majority of children by the age of 3 years,[47–49] occurring earlier in rural areas compared to urban zones,[50] and is usually not associated with a clinical syndrome.[48] In contrast, in the industrial nations, where seroconversion is delayed until adolescence or beyond, seroconversion is associated with the development of infectious mononucleosis in a fraction of individuals.[51] It is possible, as originally suggested by de Thé,[52] that early infection with EBV predisposes a child to the development of EBV-associated Burkitt's lymphoma, perhaps because of differences in the immune system in infants compared to older individuals. In Africa, the pattern of infectious diseases encountered early in life may also alter the response to EBV infection. This has been clearly shown in the case of malaria. Although humoral responses appear unaffected, individuals with acute malaria have an increased number of circulating EBV-containing B cells due to impairment of cytotoxic T-cell control of EBV-infected B cells. This interaction between EBV and malaria is a reasonable explanation for the very high incidence of Burkitt's lymphoma in equatorial Africa.

While a definitive conclusion regarding the pathogenetic role of EBV in Burkitt's lymphoma cannot be reached solely on the basis of epidemiological observations, molecular studies provide additional insights and are the most likely source of unequivocal evidence for a pathogenetic role for the virus. It is clear that EBV could not have entered the tumor cells after tumorigenesis, since it is not only present in all tumor cells, but is monoclonal—that is, all tumor cells were derived from the same EBV-infected cell.[56] In addition, the only EBV gene that is invariably expressed in Burkitt's lymphoma is EBNA-1, a transactivating gene that regulates the replication of the viral episomes at the time of cell division and thus ensures that the viral presence is equally maintained in all progeny cells.[57-58] EBNA-1, unlike the other EBV latent genes (i.e., genes expressed in the absence of virus replication) is not processed for expression at the cell surface in the context of HLA class I antigens.[59] Thus, EBNA-1 fails to excite a cytotoxic T-cell response.[60] Recently, it was shown that small, resting lymphocytes present in the circulation of normal individuals express, like Burkitt's lymphoma, only EBNA-1, and it is now generally agreed that the viral strategy is to maintain a permanent reservoir of viral genomes in resting B cells which, because they express only EBNA-1, do not excite a T-cell cytotoxic response.[61-62] The increase in number of circulating EBV-containing cells in malaria and HIV-infected individuals,[53-55,63] however, is consistent with the probability that the number of EBV-containing cells is normally limited, in some way, by T cells, since T-cell functions are impaired by these infections. The similarity of the pattern of latent gene expression in Burkitt's lymphoma to that in circulating, resting B cells suggests that both cell types regulate EBV-latent gene expression via the same pathways. This could explain, in part, the differentiation of the tumor cell beyond the pro-B cell, in which the translocation is believed to occur; differentiation through the level of a resting B cell may ensure that the immunogenic latent proteins, EBNAs 2–6 and LMPs 1 and 2, are switched off, thus permitting the cell to escape the regulatory control of cytotoxic T cells.

Recently, EBNA-1 subtypes that differ in their amino acid sequence in the DNA binding and dimerization domain of the protein have been identified.[64-66] One of these subtypes, which we have referred to as V-leu,[66] is present in Burkitt's lymphomas, but not in normal lymphocytes, suggesting that specific mutations in EBNA-1 are required for tumorigenicity. These data, coupled to the observation that mice transgenic for EBNA-1 develop lymphomas,[67] strongly support a role for EBNA-1 in the pathogenesis of Burkitt's lymphoma. Whether acute malaria increases the likelihood of mutations arising in EBNA-1 is unknown, but worthy of study.

TREATMENT AND PROGNOSIS OF PEDIATRIC NON-HODGKIN'S LYMPHOMA

Complications Consequent on Site or Volume of Tumor

Patients with non-Hodgkin's lymphoma are prone to develop a number of the complications that frequently require urgent attention at the time of presentation (TABLE 2).[68] They include intestinal obstruction, perforation, or hemorrhage; airway obstruction from pharyngeal tumor or mediastinal tumor (patients with large mediastinal masses and airway obstruction are at very high risk for acute cardiac or respiratory failure if subjected to anesthesia); and respiratory or cardiac insufficiency from pleural effusions and pericardial effusions. Among the most common complications at presentation, particularly in small, noncleaved cell lymphomas, however,

TABLE 2. Management of the Child with Suspected Non-Hodgkin's Lymphoma

1. Identify and address problems requiring emergency management:
    - Airway obstruction or other respiratory compromise
    - Cardiac tamponade or arrythmia
    - Raised intracranial pressure, paraparesis, etc.
    - Superior or inferior caval obstruction
    - Renal obstruction or metabolic nephropathy
    - Other abnormalities: gastrointestinal obstruction, gastrointestinal bleeding, thrombocytopenia or neutropenia with fever, hypercalcemia.
2. Confirm diagnosis histologically and immunophenotypically.
3. Determine extent of disease (staging studies) and risk category.
4. Initiate specific chemotherapy as rapidly as possible after emergency problems have been controlled.

are biochemical abnormalities related to the high proliferative rate of the tumor cells. The likelihood of uric acid nephropathy before the commencement of chemotherapy or of the development of biochemical abnormalities immediately after chemotherapy, which have collectively become known as the "acute tumor lysis syndrome," correlate directly with the tumor burden, but have also been reported to be more common in African patients treated with combination therapy than those treated with single-agent cyclophosphamide.[69]

In the presence of uricosemia, the reduction of serum uric acid to normal levels can usually be accomplished within 24 to 48 hours by alkaline diuresis and high-dose allopurinol administration. Only patients with long-standing renal compromise; outlet tract obstruction; or, less commonly, massive involvement of the kidneys by tumor will require renal dialysis or, if unavailable, peritoneal dialysis. It is essential to correct the existing biochemical abnormalities before the initiation of chemotherapy and to continue to maintain high urine flow rates for a few days after the initiation of specific therapy.

Chemotherapy

The primary therapeutic modality for non-Hodgkin's lymphomas is chemotherapy. While a fraction of patients with Burkitt's lymphoma in Africa, particularly those with limited disease, have been cured with only one or two doses of cyclophosphamide,[70] most patients treated in this way will relapse and require further therapy, although many will still ultimately achieve prolonged survival. Nevertheless, there are major advantages to combination drug regimens in Burkitt's lymphoma, as in all NHLs. As recently as the mid-1970s, it was unclear whether the outcome of treatment in the pediatric NHLs was influenced by histology. In addition, while radiation had been the mainstay of lymphoma therapy for decades, the advent of combination chemotherapy brought with it two very different treatment philosophies. One was derived from the chemotherapy regimens used for ALL, in which radiation could not logically be used as a primary therapeutic modality, and the other from the remarkable successes achieved in the chemotherapy of Burkitt's lymphoma, which had recently been described in Africa where radiation was unavailable.[70–72] The application of therapeutic principles derived from the treatment of ALL seemed logical, since NHLs in children, for the most part, morphologically resemble ALL (LL is indistinguishable from L1 or L2 ALL, whereas SNCL may present as leukemia [>25% blasts in the bone marrow] in which case it is referred to as L3 or B-ALL) and when treated inadequately relapse in the bone marrow (and CNS) occurs frequently.[73,74] Leukemia therapy was initially based on prednisone, vincristine, and antimetabolites, with maintenance therapy extending over years. In Burkitt's lymphoma, however, the demonstration of cures in the 1960s and 1970s with one or several doses of cyclophosphamide (CTX),[70–72,75–78] and the subsequent even more effective use of a small number of cycles of a combination of CTX, vincristine (VCR), and methotrexate (MTX)[77,79–81] strongly suggested that long-duration therapy would be unecessary in this disease.

In the late 1970s and early 1980s, the Children's Cancer Group (CCG) conducted a clinical trial in which two different approaches were compared.[82] Patients were randomized to receive either a four-drug regimen, COMP (CTX, VCR, MTX, and prednisone) based on regimens previously used in Burkitt's lymphoma,[77–81] but with an intermediate dose (i.d.) of MTX, following the demonstration of its activity by Djerassi and Kim,[83] or a 10-drug regimen, LSA_2L_2, based on an earlier ALL protocol used with good effect in NHLs at Memorial Sloane Kettering Cancer Institute.[84] In this comparative trial both treatment arms were made the same length—18 months—so that only the relative efficacy of the drug combinations and not the length of therapy was being addressed.

The results of this study have been particularly influential in guiding subsequent treatment approaches in the pediatric NHLs. Although both treatment arms were effective in patients with limited disease, event-free survival (EFS) was clearly superior in patients with extensive small, noncleaved cell lymphoma (SNCL) who were treated with the COMP regimen, while in patients with extensive lymphoblastic lymphoma, the LSA_2L_2 regimen gave better results.[82,85] Although these results can be applied only to the particular protocols used, many pediatric oncologists consider that LLs are better treated with an ALL type of protocol, while repeated cycles of regimens containing CTX, MTX, and VCR have provided the basis for the therapy of SNCL. Particularly important is the principle that SNCL should *not* be treated with ALL-type therapy, a conclusion that has been confirmed in other studies, including that of the BFM group who observed an EFS rate of 32% at 7 years in children with extensive B-cell lymphomas treated with the BFM's intensive regimen for ALL.[86] In the SNCL, in contrast to LL, considerable development has occurred in treatment protocols from the early three- or four-drug combinations, and therapy has been dramatically shortened—in line with the African model.

It is important to recognize that all pediatric NHLs, as is the case with ALL, require therapy directed towards the CNS to prevent CNS relapse (usually a combination of intrathecal coupled to high-dose infusions of methotrexate and Ara-C). Only patients with limited disease, particularly patients with small volume completely resected intraabdominal B-cell lymphomas and patients with early-stage lymphoblastic lymphoma not involving the head and neck, appear not to require such CNS prophylactic therapy.

B-Cell Lymphomas

Whereas some groups have chosen to treat large B-cell lymphomas differently from small, noncleaved cell lymphoma, recent European and North American studies in which these diseases are treated identically have demonstrated that the outcome is similar, that is, equally good for both histological subtypes. Thus, there seems to be no reason to continue to separate these diseases for treatment purposes, and they are dealt with together here. Children with B-cell NHL now have an

TABLE 3. Event/Progression-Free Survival Rates in Selected Studies of Patients with SNCL (St. Jude Stage)

Protocol (No. Patients[a])	Stage III	Stage IV[b]	Reference
CCG COMP (93)		50%	82, 85
CCG LSA$_2$L$_2$ (44)		29%	82, 85
BFM 86 (150)	73%	75%	97
LMB 89 (181)	87%	85%, 87%	98, 99
NCI 89-C-41	92%	81%	100
[adults/children] (38)			

[a] Children unless indicated otherwise.

[b] If only a single percentage is given, both stage IV and patients with B-ALL (if any) are included. If two percentages are given, the first is for stage IV patients, the second for B-ALL.

excellent chance, some 90% of all patients, for long-term survival, and even patients with CNS involvement have a good chance (70–80%) of cure (TABLE 3). These results are significantly better than those achieved a decade ago with regimens that included only four or five drugs, lower doses of high-dose methotrexate, and no high-dose Ara-C. It is appropriate, however, to separate the patients into risk groups, because children with the smallest tumor burdens (most have stage 1 or 2 disease) require less therapy—probably only 2 or 3 cycles—than patients with extensive disease (stage 3, i.e., extensive intraabdominal disease, or stage 4, i.e., bone marrow and/or CNS disease). The most frequently used staging system (St. Jude) is shown in TABLE 4.

Because patients with limited disease had an excellent prognosis with the earliest combination regimens—in contrast to radiation alone which, overall, gave a relapse-free survival of only approximately 20%[87–90]—attention has been focused on increasing the intensity of regimens used to treat patients with the highest tumor burdens;

TABLE 4. Staging Systems for Childhood Non-Hodgkin's Lymphoma

Staging System	Stage	Definition
St. Jude (NHL)	I	Single tumor (extranodal); Single anatomic area (nodal); Excluding mediastinum or abdomen.
	II	Single tumor (extranodal) with regional node involvement; Primary gastro-intestinal tumor with or without involvement of associated mesenteric nodes only, grossly completely resected. On same side of diaphragm: (a) Two or more nodal areas (b) Two single (extranodal) tumors with or without regional node involvement.
	III	On both sides of the diaphragm: (a) Two single tumors (extranodal) (b) Two or more nodal areas; All primary intrathoracic tumors (mediastinal, pleural, thymic); All extensive primary intraabdominal disease; All primary paraspinal or epidural tumors regardless of other sites.
	IV	Any of the above with initial CNS or bone marrow involvement (<25%).

those with St. Jude stage III (extensive abdominal tumor) or IV (bone marrow involvement <25% ± CNS involvement) disease or B-ALL. Local irradiation, which increases toxicity but is not very effective in SNCL,[92] has been eliminated, even for low-risk patients, in whom it was shown by a randomized trial to be of no benefit.[92] In children, CNS recurrence has, in the past, been a particular problem in patients with extensive disease, but the outcome of such patients has now been improved by a combination of intrathecal therapy (usually both MTX and ara-C) and high-dose systemic therapy with the same drugs. It seems likely that improved control of systemic disease, which has been accomplished by this approach, as well as by the addition of other agents such as etoposide and ifosfamide,[93–95] is a major factor in the prevention of CNS relapse. The most successful protocols for patients with extensive B-cell lymphomas include, therefore, in addition to CTX and ADR, both HDMTX (above 5 gm/m^2), and infusional or high-dose ara-C and/or VP16 or VM26.[96–100]

At the National Cancer Institute in Bethesda, Maryland, we have for many years used identical regimens for adults (18 to 60) and children (<18 years) with SNCL. In both protocols, NCI 77-04 and NCI 89-C-91, adults and children had identical EFS rates. Toxicities were similar in the two age groups. Bernstein *et al.* used a modified version of NCI 77-04 with similar results to those obtained at NCI.[102] Whereas fewer patients have been treated with protocol 89-C-41 than is the case for the highly successful BFM and SFOP protocols (in each of which, more than 400 patients have now been treated, according to reports at a recent meeting [June, 1996] in Lugano), the results appear to be similar, with approximately 90% of patients achieving prolonged disease-free survival (relapse occurs in the first year after presentation if it is going to). Whether the somewhat shorter duration

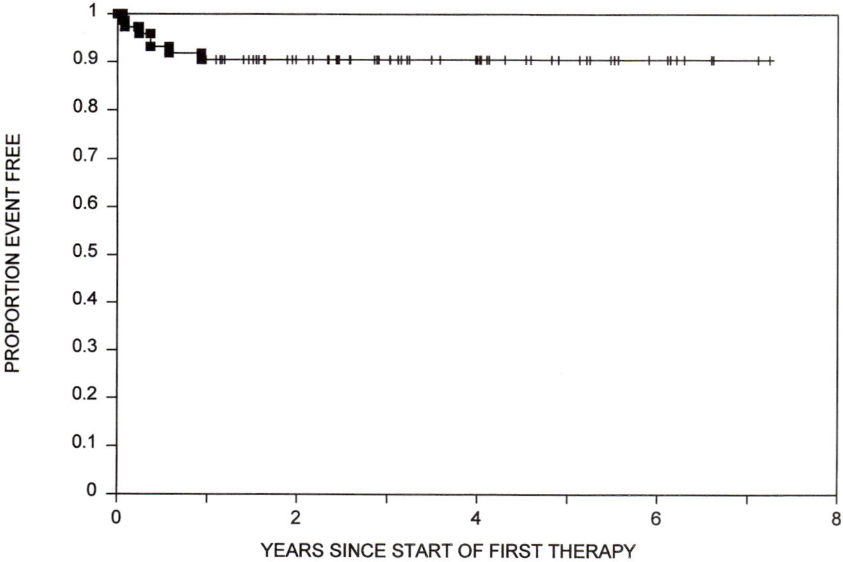

FIGURE 1. Event-free survival in 74 patients with B-cell lymphoma treated according to protocol NCI 89-C-41 and followed for at least a year (updated results).

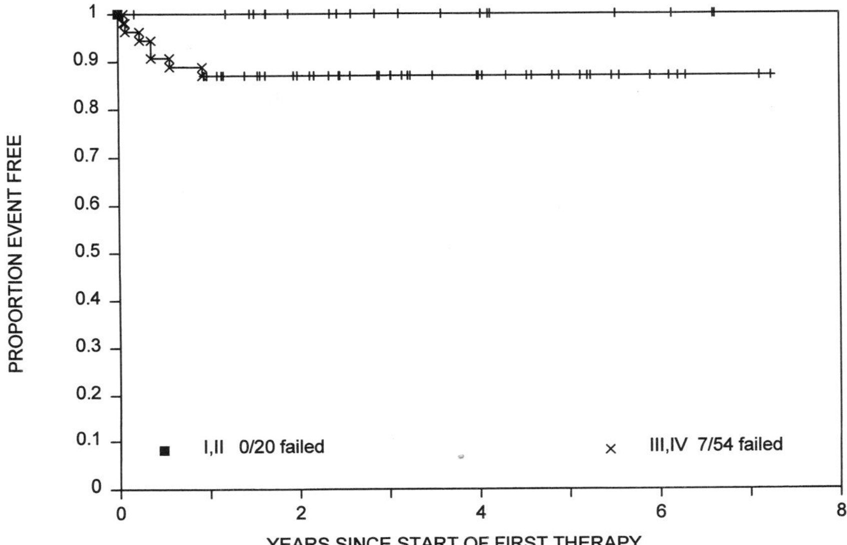

FIGURE 2. Event-free survival in the same 74 patients with B-cell lymphoma by stage. Stages I and II and III and IV are combined. When only patients with small, noncleaved lymphoma are considered, the results are essentially identical (updated results).

of 89-C-41 compared to the French and German protocols for the highest risk patients will prove to be advantageous remains to be seen. Recent results for protocol 89-C-41 are shown in FIGURES 1 and 2.

Lymphoblastic Lymphoma

In lymphoblastic lymphoma it is much less clear how patients should be divided into risk groups. As pointed out above, very few patients have localized disease, and those that do tend to have precursor B-cell disease. Although a fraction of such patients may achieve long-term survival with therapies of 9 months or less, it is not at all clear that less therapy can be safely given, and a 70% event-free survival rate in such patients is unacceptably low. Some investigators, therefore, still treat even patients with limited stage (1 or 2) lymphoblastic lymphoma with long-duration, ALL-like therapy.

Since the demonstration that LSA_2L_2 is superior to COMP, treatment for LL in children has not changed greatly. A recent update of the CCG 551 trial indicated that in the long term, EFS in patients with extensive LL treated with LSA_2L_2 was 64% at 5 years.[85] In the Pediatric Oncology Group's (POG) modified version of LSA_2L_2, EFS was approximately 55% in patients with stages III (extensive intrathoracic disease or tumor masses above and below the diaphragm) and IV (bone marrow and/or CNS) disease and mediastinal mass was a poor prognostic factor.[103] Patients with 5 to 25% lymphoblasts in the bone marrow had a similar survival to those without marrow lymphoblasts, but all eight patients with more than 25%

lymphoblasts in the marrow progressed. Apparently better results (EFS: 75% at 3 years in 84 patients) have been obtained at Institute Gustave-Roussy, where 10 additional cycles of HDMTX were incorporated into the LSA_2L_2 protocol.[104] Bone marrow disease was not a prognostic factor in this protocol (including patients with >25% marrow lymphoblasts). Some investigators have used standard ALL protocols for patients with LL and have reported EFS rates between 65% (UK Children's Cancer Study group)[105] and 82% (BFM)[97]; and although limited involvement of the bone marrow (<25%) does not appear to be a prognostic factor with such protocols, more intensive protocols may be required for patients at highest risk, and it should not be assumed that all ALL protocols are equal.

Interestingly, protocols based on a repeated cycles of a CTX-containing combination, but including ADR and/or HDMTX, have given better results than the CCG COMP regimen,[101,106,107] suggesting that ALL therapy is not necessarily essential for the successful treatment of all patients with LL. Regimens based on repeated cycles of alkylating agents, adriamycin, and vincristine, however, often give poor results in patients with bone marrow involvement and optimal therapy of these patients probably requires the addition of other agents. One important issue in the treatment of LL is the duration of treatment. Short-duration regimens have not been shown to be effective in patients with extensive disease—the shortest treatment durations of successful protocols are in excess of a year. Whether increased treatment intensity would permit shorter treatment durations is currently being tested.

REFERENCES

1. SHAD, A. & I. T. MAGRATH. 1996. Childhood non-Hodgkin's lymphomas. *In* Principles and Practice of Pediatric Oncology. Third Edition. P. A. Pizzo & D. P. Poplack, Eds. Lippincott and Raven. New York.
2. MAGRATH, I. 1997. Small noncleaved cell lymphomas. *In* The Non-Hodgkin's Lymphomas. 2nd edit. I. T. Magrath, Ed. Edward Arnold. London, U.K.
3. WRIGHT, D. H. 1985. Histogenesis of Burkitt's lymphoma: A B-cell tumour of mucosa-associated lymphoid tissue. IARC Sci. Publ. **60:** 37–45.
4. ARMITAGE, J. O., J. R. FEAGLER & D. P. SKOOG. 1977. Burkitt lymphoma during pregnancy with bilateral breast involvement. JAMA **237:** 151.
5. DURODOLA, J. I. 1976. Burkitt's lymphoma presenting during lactation. Int. J. Gynaecol. Obstet. **14:** 225–231.
6. SMITH, E. C. & B. G. T. ELMES. 1934. Malignant disease in natives of Nigeria. Ann. Trop. Med. Parasit. **28:** 461–512.
7. DAVIES, J. N. P. 1948. Reticuloendothelial tumours. East Afr. Med. J. **25:** 117.
8. EDINGTON, D. M. 1956. Malignant disease in the Gold Coast. Br. J. Cancer **10:** 41–54.
9. THIJS, A. 1957. Considérations sur les tumeurs malignes des indigénes du Congo belge et du Ruanda-Urundi. A propos de 2,536 cas. Ann. Soc. Belge Med. Trop. **37:** 483–514.
10. DE SMET, M. P. 1956. Observations cliniques de tumeurs malignes des tissus réticuloendothéliaux et des tissus hémolymphopoiétiques au Congo. Ann. Soc. Belge Med. Trop. **36:** 53–70.
11. BURKITT, D. 1958. A sarcoma involving the jaws in African children. Br. J. Surg. **46:** 218–223.
12. BURKITT, D. 1962. Determining the climatic limitations of a children's cancer common in Africa. Br. Med. J. **2:** 1019–1026.
13. BURKITT, D. 1962. A children's cancer dependent on climatic factors. Nature **194:** 232–234.
14. O'CONOR, G. T. & J. N. P. DAVIS. 1960. Malignant tumors in African children with special reference to malignant lymphomas. J. Pediatr. **56:** 526–535.
15. O'CONOR, G. 1961. Malignant lymphoma in African children. Cancer. II. A pathological entity. Cancer **14:** 270–283.

16. O'CONOR, G., H. Rappaport & E. B. Smith. Childhood lymphoma resembling Burkitt's tumor in the United States. Cancer **18:** 411–417.

17. DORFMAN, R. F. 1965. Childhood lymphosarcoma in St. Louis, Missouri, clinically and histologically resembling Burkitt's tumor. Cancer **18:** 418–430.

18. WRIGHT, D. H. 1966. Burkitt's tumor in England. A comparison with childhood lymphosarcoma. Int. J. Cancer **1:** 503–514.

19. BURKITT, D. & J. N. P. DAVIES. 1961. Lymphoma syndrome in Uganda and tropical Africa. Med. Press **245:** 367–369.

20. EPSTEIN, M. A., B. G. ACHONG & Y. M. BARR. 1964. Viral particles in cultured lymphoblasts from Burkitt's lymphoma. Lancet **i:** 702–703.

21. HENLE, G. & W. HENLE. 1966. Immunofluorescence in cells derived from Burkitt's lymphoma. J. Bacteriol. **91:** 1248–1256.

22. HENLE, W., K. HUMMELER & G. HENLE. 1966. Antibody coating and agglutination of virus particles separated from the EB3 line of Burkitt lymphoma cells. J. Bacteriol. **92:** 269–271.

23. HENLE, G., *et al.* 1969. Antibodies to EB virus in Burkitt's lymphoma and control groups. J. Natl. Cancer Inst. **43:** 1147–1157.

24. HENLE, G., *et al.* 1971. Antibodies to early Epstein-Barr virus-induced antigens in Burkitt's lymphoma. J. Natl. Cancer Inst. **46:** 861–871.

25. REEDMAN, B. M. & G. KLEIN. 1973. Cellular localizations of an Epstein-Barr virus (EBV) associated complement-fixing antigen in producer and non-producer lymphoblastoid cell lines. Int. J. Cancer **1:** 499–520.

26. DALLDORF, G., C. A. LINSELL, F. E. MARNHART & R. MARTYN. 1964. An epidemiological approach to the lymphomas of African children and Burkitt's sarcoma of the jaws. Perspect. Biol. Med. **7:** 435–449.

27. ZECH, L., *et al.* 1976. Characteristic chromosomal abnormalities in biopsies and lymphoid-cell lines from patients with Burkitt and non-Burkitt lymphomas. Int. J. Cancer **17:** 47.

28. BERNHEIM, A., R. BERGER & G. LENOIR. 1981. Cytogenetic studies on African Burkitt's lymphoma cell lines; t(8;14), t(2;8) and t(8;22) translocations. Cancer Genet. Cytogenet. **3:** 307–315.

29. TAUB, R., *et al.* 1982. Translocation of the c-*myc* gene into the immunoglobulin heavy chain locus in human Burkitt lymphoma and murine plasmacytoma cells. Proc. Natl. Acad. Sci. USA **79:** 7837.

30. DALLA-FAVERA, R., *et al.* 1982. Human c-*myc* onc gene is located on the region of chromosome 8 that is translocated in Burkitt lymphoma cells. Proc. Natl. Acad. Sci. USA **79:** 7824.

31. ADAMS, J. M., *et al.* 1982–1986. Cellular *myc* oncogene is altered by chromosome translocation to an immunoglobulin locus in murine plasmacytomas and is rearranged similarly in human Burkitt lymphomas. Proc. Natl. Acad. Sci. USA **80:** 1983.

32. MAGRATH, I. T. 1990. The pathogenesis of Burkitt's lymphoma. *In* Recent Advances in Cancer Research. G. Klein & G. Van de Woude, Eds. **55:** 133–270.

33. SANDLUND, J. T., *et al.* 1992. Theophylline induced differentiation provides direct evidence for the deregulation of c-*myc* in Burkitt's lymphoma and suggests participation of immunoglobulin enhancer sequences. Cancer Res. **53:** 127–132.

34. BHATIA, K., M. I. GUTIERREZ & I. T. MAGRATH. 1992. Burkitt's lymphoma cells frequently carry monoallelic DJ rearrangements. Curr. Top. Microbiol. Immunol. **182:** 319–324.

35. SECKER-WALKER, L., *et al.* 1987. Multiple chromosome abnormalities in a drug resistant TdT-positive B cell leukemia. Leuk. Res. **11:** 155–161.

36. DREXLER, H. G., *et al.* 1986. A case of TdT positive B-cell acute lymphoblastic leukemia. Am. J. Clin. Pathol. **85:** 735–738.

37. KANEKO, Y., *et al.* 1980. The 14q+ chromosome in pre-B-ALL. Blood **56:** 782–785.

38. GANICK, D. L. & J. L. FINLAY. 1980. Acute lymphoblastic leukemia with Burkitt cell morphology and cytoplasmic immunoglobulin. Blood **56:** 311–314.

39. OSMOND, D. G., S. PRIDDLE & S. RICO-VARGAS. 1990. Proliferation of B cell precursors

in bone marrow of pristane-conditioned and malaria-infected mice. Implications for B cell oncogenesis. Curr. Top. Microbiol. Immunol. **166:** 149–157.
40. ADAMS, J. M. & S. CORY. 1992. Oncogene co-operation in leukaemogenesis. Cancer Surv. **15:** 119–141.
41. WEDDERBURN, N. 1970. Effect of concurrent malarial infection on development of virus-induced lymphoma in Balb-c mice. Lancet **2:** 1114–1116.
42. GUNVEN, P., *et al.* 1980. Surface immunoglobulins on Burkitt's lymphoma biopsy cells from 91 patients. Int. J. Cancer **25:** 711–719.
43. GREGORY, C. D., *et al.* 1987. Identification of a subset of normal B cells with a Burkitt's lymphoma (BL)-like phenotype. J. Immunol. **139:** 313–318.
44. TAMARU, J. I., *et al.* 1995. Burkitt's lymphoma express V_H-genes with a moderate number of antigen-selected somatic mutations. Am. J. Pathol. **147:** 1398–1407.
45. BENJAMIN, D., *et al.* 1982. Immunoglobulin secretion by cell lines derived from African and American undifferentiated lymphomas of Burkitt's and non-Burkitt's type. J. Immunol. **129:** 1336–1342.
46. HENLE, G. & W. HENLE. 1970. Observations on childhood infections with the Epstein-Barr virus. J. Infect. Dis. **121:** 303–310.
47. HENLE, W. & G. HENLE. 1979. Seroepidemiology of the virus. *In* The Epstein-Barr Virus. M. A. Epstein & B. G. Achong, Eds. Springer Verlag. Berlin, Germany.
48. DIEHL, V., *et al.* 1969. Infectious mononucleosis in East Africa. East Afr. Med. J. **46:** 407–413.
49. GESER, A., *et al.* 1982. Final case reporting from the Ugandan prospective study of the relationship between EBV and Burkitt's lymphoma. Int. J. Cancer **29:** 397–400.
50. BIGGAR, R. J., *et al.* 1981. Malaria, sex and place of residence as factors in antibody response to Epstein-Barr virus in Ghana, West Africa. Lancet **2:** 115–118.
51. NIEDERMAN, J. C., *et al.* 1968. Infectious mononucleosis. J. Am. Med. Assoc. **203:** 139–143.
52. DE THÉ. G. 1977. Is Burkitt's lymphoma lymphoma related to perinatal infection by Epstein-Barr virus? Lancet **i:** 335–337.
53. WHITTLE, H. C., *et al.* 1984. T-cell control of Epstein-Barr virus infected B-cells is lost during *P. falciparum* malaria. Nature **312:** 449–450.
54. GUNAPALA, D. E., *et al.* 1990. In vitro analysis of Epstein-Barr virus: Host balance in patients with acute *Plasmodium falciparum* malaria. Parasitol. Res. **76:** 531–535.
55. LAM, K. M. C., *et al.* 1991. Circulating Epstein-Barr virus carrying B cells in acute malaria. Lancet **337:** 876–878.
56. NERI, A., *et al.* 1991. Epstein-Barr virus infection precedes clonal expansion in Burkitt's and acquired immunodeficiency syndrome-associated lymphoma. Blood **77:** 1092–1095.
57. ROWE, D. T., *et al.* 1986. Restricted expression of EBV latent genes and T-lymphocyte-detected membrane antigen in Burkitt's lymphoma cells. EMBO J. **5:** 2599–2607.
58. SCHAEFER, B. C., *et al.* 1991. Exlusive expression of Epstein-Barr virus nuclear antigen 1 in Burkitt lymphoma arises from a third promoter, distinct from the promoters used in latently infected lymphocytes. Proc. Natl. Acad. Sci. USA **88:** 6550–6554.
59. LEVITSKAYA, J., M. CORAM, V. LEVITSKY, *et al.* 1995. Inhibition of antigen processing by the internal repeat region of the Epstein-Barr virus nuclear antigen-1. Nature **375:** 685–688.
60. ROWE, M., *et al.* 1995. Restoration of endogenous antigen processing in Burkitt's lymphoma cells by Epstein Barr virus latent membrane protein-1: Coordinate upregulation of peptide transporters and HLA class I antigen expression. Eur. J. Immunol. **25:** 1374–1484.
61. TIERNEY, R. J., *et al.* 1994. Epstein-Barr latency in blood mononuclear cells: Analysis of viral gene transcription during primary infection and in the carrier state. J. Virol. **68:** 7374.
62. NIEDOBITEK, G. & L. S. YOUNG. 1994. Epstein-Barr virus persistence and virus associated tumors. Lancet **343:** 333–335.
63. BIRX, D. L., R. R. REDFIELD & G. TOSATO. 1986. Defective regulation of Epstein-

Barr virus infection in patients with acquired immunodeficiency syndrome (AIDS) or AIDS-related disorders. N. Engl. J. Med. **314:** 874–879.

64. SNUDDEN, D. K., *et al.* 1995. Alterations in the structure of the EBV nuclear antigen, EBNA-1, in epithelial cell tumors. Oncogene **8:** 1545–1562.

65. WRIGHTHAM, M. N., *et al.* 1995. Antigenic and sequence variation in the C-terminal domain of the Epstein-Barr virus nuclear antigen 1. J. Virol. **208:** 521–530.

66. BHATIA, K., *et al.* 1996. Variation in the sequence of Epstein-Barr nuclear antigen-1 in peripheral blood lymphocytes and in Burkitt's lymphoma. Oncogene **13:** 177–181.

67. WILSON, J. B., J. L. BELL & A. J. LEVINE. 1966. Expression of Epstein-Barr virus nuclear antigen-1 induces B cell neoplasia in transgenic mice. EMBO J. **15:** 3117–3126.

68. SHAD, A. 1997. Complications of therapy. *In* The Non-Hodgkin's Lymphomas. 2nd edit. I. T. Magrath, Ed.: In press. Edward Arnold. London.

69. NKRUMAH, F. K., I. V. PERKINS & R. J. BIGGAR. 1977. Combination chemotherapy in abdominal Burkitt's lymphoma. Cancer **40:** 1410–1416.

70. BURKITT, D. 1967. Long term remissions following one and two dose chemotherapy for African lymphoma. Cancer **20:** 756–759.

71. CLIFFORD, P. 1967. Long-term survival of patients with Burkitt's lymphoma; an assessment of treatment and other factors which may relate to survival. Cancer Res. **27:** 2578–2615.

72. MAGRATH, I. T. 1991. African Burkitt's lymphoma: History, biology, clinical features, and treatment. Am. J. Ped. Hematol. Oncol. **13:** 222–246.

73. MAGRATH, I. T. & J. L. ZIEGLER. 1980. Bone marrow involvement in Burkitt's lymphoma and its relationship to acute B-cell leukemia. Leuk. Res. **4:** 33–50.

74. WANATABE, A., *et al.* 1973. Undifferentiated lymphoma, non-Burkitt's type: Meningeal and bone marrow involvement in children. Am. J. Dis. Child. **125:** 57–61.

75. ZIEGLER, J., I. T. MAGRATH & C. L. M. OLWENY. 1979. Cure of Burkitt's lymphoma: 10-year follow-up of 157 Ugandan patients. Lancet **2:** 936–938.

76. ARSENEAU, J. C., *et al.* 1975. American Burkitt's lymphoma—a clinicopathological study of 30 cases. I. Clinical factors relating to long-term survival. Am. J. Med. **58:** 314–321.

77. OLWENY, C. L. M., *et al.* 1976. Treatment of Burkitt's lymphoma: Randomized clinical trials of single-agent versus combination chemotherapy. Int. J. Cancer **17:** 436–440.

78. OLWENY, C. L. M., *et al.* 1980. Long-term experience with Burkitt's lymphoma in Uganda. Int. J. Cancer **26:** 261–266.

79. ZIEGLER, J. L., *et al.* 1976. Combined modality treatment of American Burkitt's lymphoma. Cancer **38:** 2225–2231.

80. ZIEGLER, J. L. 1977. Treatment results of 54 American patients with Burkitt's lymphoma are similar to the African experience. N. Engl. J. Med. **297:** 75–80.

81. ZIEGLER, J. L., *et al.* 1978. Cancer Treat. Rep. **62:** 2031–2034.

82. ANDERSON, J. R., *et al.* 1983. Childhood non-Hodgkin's lymphoma. The results of a randomized therapeutic trial comparing a 4-drug regimen (COMP) with a 10-drug regimen (LSA$_2$L$_2$). N. Engl. J. Med. **308:** 559–565.

83. DJERASSI, I. & J. S. KIM. 1976. Methotrexate and citrovorum factor rescue in the management of childhood lymphosarcoma and reticulum cell sarcoma (non-Hodgkin's lymphomas). Cancer **38:** 1043–1051.

84. WOLLNER, N., *et al.* 1975. Non-Hodgkin's lymphoma in children. Med. Ped. Oncol. **1:** 235–263.

85. ANDERSON, J. R., *et al.* 1993. Long-term follow-up of patients treated with COMP or LSA$_2$L$_2$ therapy for childhood non-Hodgkin's lymphoma: A report of CCG-551 from the Children's Cancer Group. J. Clin. Oncol. **11:** 1024–1032.

86. MÜLLER-WEIHRICH, S., *et al.* 1982. BFM-Studie 1975–81 zur Behandlung der Non-Hodgkin-Lymphoma hoher Malignität bei Kinder und Jugendlichen. Klin. Pädiatr. **194:** 219–225.

87. JENKIN, R. D., *et al.* 1984. The treatment of localized non-Hodgkin's lymphoma in children: A report from the Children's Cancer Study Group. J. Clin. Oncol. **2:** 88–97.

88. MURPHY, S. B., *et al.* 1983. End results of treating children with localized non-Hodgkin's lymphoma with a combined modality approach of lessened intensity. J. Clin. Oncol. **1:** 326–330.

89. MURPHY, S. B. & H. O. HUSTU. 1980. A randomized trial of combined modality therapy of childhood non-Hodgkin's lymphoma. Cancer **45:** 630–637.
90. NELSON, D. F., *et al.* 1977. The role of radiation therapy in localized, resectable intestinal non-Hodgkin's lymphoma in children. Cancer **39:** 89–97.
91. NORIN, T., *et al.* 1971. Conventional and superfractionated radiation therapy in Burkitt's lymphoma. Acta Radiol. **10:** 545–557.
92. LINK, M. P., *et al.* 1990. Results of treatment of childhood localized non-Hodgkin's lymphoma with combination chemotherapy with or without radiotherapy. N. Engl. J. Med. **322:** 1169–1174.
93. JONES, G. R. & L. J. ETTINGER. 1985. Continuous infusion of high-dose cytosine arabinoside for treatment of childhood acute leukemia and non-Hodgkin's lymphoma in relapse. Semin. Oncol. **12:** 150–154.
94. GENTET, J. C., *et al.* 1990. Phase II study of cytarabine and etoposide in children with refractory or relapsed non-Hodgkin's lymphoma: A study of the French Society of Pediatric Oncology. J. Clin. Oncol. **8:** 661–665.
95. MAGRATH, I. T., *et al.* 1991. Ifosfamide in the treatment of high-grade recurrent non-Hodgkin's lymphomas. Hematol. Oncol. **9:** 267–274.
96. REITER, A., *et al.* 1992. Favorable outcome of B-cell acute lymphoblastic leukemia in childhood: A report of three consecutive studies of the BFM group. Blood **80:** 2471–2478.
97. REITER, A., *et al.* 1995. Non-Hodgkin's lymphomas of childhood and adolescence: Results of a treatment stratified for biological subtypes and stage. A report of the BFM group. J. Clin. Oncol. **7:** 186–193.
98. PATTE, C., *et al.* 1991. High survival rate in advanced-stage B-cell lymphomas and leukemias without CNS involvement with a short intensive polychemotherapy: Results from the French Pediatric Oncology Society of a randomized trial of 216 children. J. Clin. Oncol. **9:** 123–132.
99. PATTE, C., *et al.* 1993. High cure rate in B-cell (Burkitt's) leukemia in the LMB 89 protocol of the SFOP (French Pediatric Oncology Society) (Abstr.). Proc. Am. Soc. Clin. Oncol. **12:** 1050.
100. MAGRATH, I. T., *et al.* 1996. Adults and children with small noncleaved cell lymphoma have a similar excellent outcome when treated with the same chemotherapy regimen. J. Clin. Oncol. **14:** 925–934.
101. MAGRATH, I. T., *et al.* 1984. An effective therapy for both undifferentiated (including Burkitt's) lymphomas and lymphoblastic lymphomas in children and young adults. Blood **63:** 1102–1111.
102. BERNSTEIN, J. I., *et al.* 1988. Combined modality therapy for adults with small noncleaved cell lymphoma (Burkitt's and non-Burkitt's types). J. Clin. Oncol. **4:** 847–858.
103. SULLIVAN, M. P., *et al.* 1985. Pediatric oncology group experience with modified LSA$_2$L$_2$ therapy in 107 children with non-Hodgkin's lymphoma (Burkitt's lymphoma exluded). Cancer **55:** 323–336.
104. PATTE, C., *et al.* 1992. Results of the LMT81, a modified LSA2L2 protocol with high-dose methotrexate, on 84 children with non-B cell (lymphoblastic) lymphoma. Med. Ped. Oncol. **20:** 105–113.
105. EDEN, O. B., *et al.* 1992. Treatment of advanced stage T cell lymphoblastic lymphoma: Results of the United Kingdom Children's Cancer Study Group (UKCCSG) protocol 8503. Br. J. Haematol. **82:** 310–316.
106. HVIZDALA, E. V., *et al.* 1988. Lymphoblastic lymphoma in children—a randomized trial comparing LSA$_2$L$_2$ with the A-COP+ therapeutic regimen: A Pediatric Oncology Group study. J. Clin. Oncol. **6:** 26–33.
107. TUBERGEN, D. G., *et al.* 1995. Comparison of treatment regimens for pediatric lymphoblastic non-Hodgkin's lymphoma: a Children's Cancer Group Study. J. Clin. Oncol. **13:** 1368–1376.

Biologic Prognostic Factors in Acute Lymphoblastic Leukemia

J. OTTEN, J. VAN DER WERFF TEN BOSCH,
AND N. BALDUCK

Department of Pediatrics, Pediatric Oncology and Hematology
Academisch Kinderziekenhuis
Vrije Universiteit Brussel (VUB)
Laarbeeklaan 101
B 1090 Brussels, Belgium

For the last 20 years, the prognosis of acute lymphoblastic leukemia (ALL) in children has been markedly improved, mainly as a result of the intensification of chemotherapy. On the other hand, ALL is now considered to be a heterogeneous group of discrete entities characterised by clinical, immunophenotypic, and genetic correlations as well as by their prognosis. Considering the treatment intensity of most current protocols, evaluation of prognosis has become important in order not to overtreat patients with very good prognosis and to identify those patients whose very small chances of cure with conventional treatment regimens justify experimental therapeutic approaches.

Nowadays, most treatment protocols stratify ALL patients in two or more different risk groups taking known clinical and hematologic prognostic features into account such as peripheral leukocyte or blast count at diagnosis, age, gender, and evaluation of liver and spleen enlargement. Recently, early response to treatment has also been found to be very predictive of outcome.[1-3] In order to further refine the evaluation of prognosis, many authors have analyzed the predictive value of biologic characteristics such as immunophenotype and cytogenetic or molecular genetic markers of the leukemic clone. The practical utility of such prognostic factors, however, if used as a patient's stratification tool in large co-operative studies, depends on the following prerequisites: easy and early identification, reliable evaluability, and independence with regard to other known prognostic factors.

Immunophenotyping of leukemic cells is easy, fast, and generally quantitatively well defined. It has helped in identifying particular subsets of ALL such as FAB L3 or Burkitt ALL for which specific and successful therapies have been designed.[4,5] Precursor T-cell ALL has long been recognized as an immunophenotypic variant with a very bad prognosis. However, it is very often associated with other unfavorable features, such as high leukocyte counts, and multivariate analyses have shown that T-cell lineage per se was not the predominant determinant of a worse prognosis.[6] This is illustrated in FIGURE 1, which shows the distribution of leukocyte counts in T- and B-cell lineage ALL, respectively, as observed in the EORTC 58881 study.

Due to the predominance of high leukocyte counts in precursor T-cell ALL, a comparison of event-free survival (EFS) curves indicates a poorer outcome for the overall population of patients with T-cell ALL as compared to precursor B-cell ALL (FIG. 2). However, when patients with similar leukocyte counts are compared, the impact of the immunophenotype on EFS completely disappears. This observation does not imply that the optimal treatment of T- versus B-cell lineage ALL should necessarily be identical. On the basis of a historical comparison between

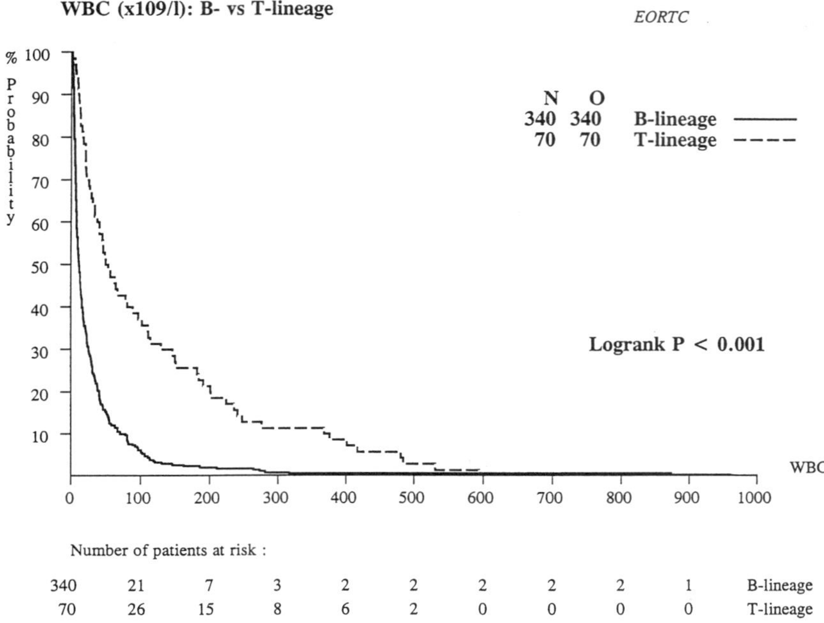

FIGURE 1. Distribution of leukocyte counts at diagnosis in B-cell lineage (—) and T-cell lineage (----) ALL, respectively.

two successive trials (BFM 83 and BFM 96), the German group concluded that the introduction of high doses of methotrexate (5 g instead of 0.5 g) during interim therapy has improved the EFS of T-lineage ALL patients with good response to the initial prednisone + i.t. methotrexate prephase from 62% to 84% at 5 years from diagnosis.[7] In the EORTC 58881 trial, ALL patients were initially treated with a prephase consisting of prednisolone for one week and one intrathecal injection of methotrexate. During the first year of the study, this intrathecal injection was given on day 8. Evaluation of the response on day 8 showed a poor response in 22% of the patients, that is, the persistence of >1,000 blast cells/mm³ in the peripheral blood. This percentage of poor responders was twice has high as that reported by the German BFM group who used to administer the first intrathecal methotrexate on day 1 of the prephase.[1] Therefore, in the subsequent part of the EORTC 58831 study, intrathecal methotrexate was given on day 1 as well, or as soon as possible during the prephase depending on feasibility. The percentage of poor response fell to 11%, in keeping with the BFM data. This serendipitous observation revealed the importance of the systemic effect of one low dose of intrathecal methotrexate. However, this effect was restricted to precursor B-cell ALL patients only. In T-cell-lineage ALL, no significant difference in response rate was observed whether the intrathecal injection was given on day 1, on day 8, or in between (TABLE 1). The observations are in keeping with the results of Barredo *et al.*,[8] who reported very low methotrexate–polyglutamate accumulation in T-lineage leukemic blasts after exposure to usual, low doses of the drug. These authors also showed that high

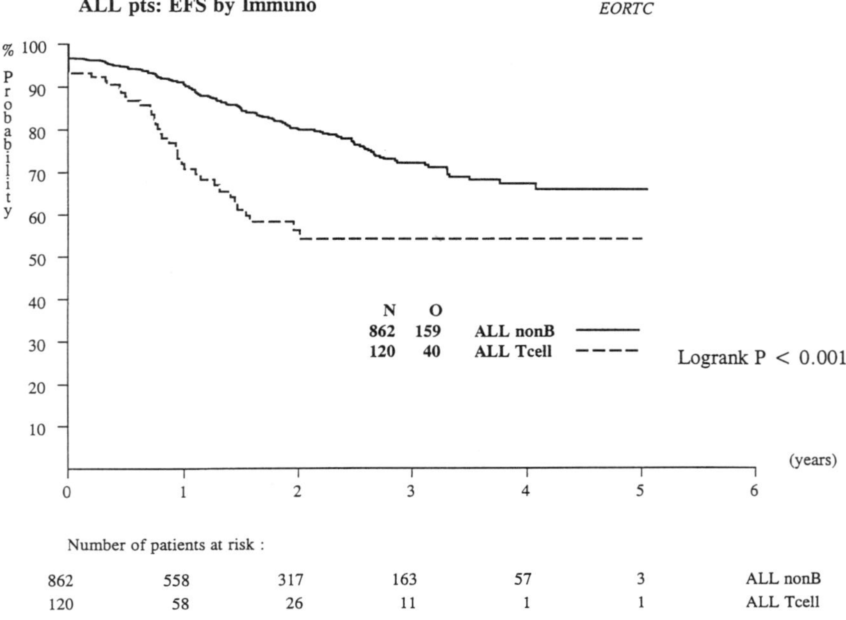

FIGURE 2. Event-free survival of B-cell lineage (—) and T-cell lineage (----) ALL patients.

doses of methotrexate were required for intracellular methotrexate polyglutamate concentrations to reach those achieved in B-lineage leukemic blasts exposed to low doses of methotrexate.[8]

On the basis of these clinical and laboratory studies, one might question the efficiency, if any, of usual ($\pm$20 mg/m^2) methotrexate dosing as commonly used in continuation therapy of precursor T-cell ALL. The example of T-lineage ALL shows that immunophenotype may not be the best guide for deciding about intensity of treatments, although its specific pharmacodynamics may provide hints with regard to qualitative adjustment of the chemotherapy regimen.

The immunophenotype may nevertheless be the easiest criterion to identify

TABLE 1. Effect of Timing of First Intrathecal Methotrexate Injection on Response Rate to the Prephase in B-Cell Lineage and T-Cell Lineage ALL[a]

		Gp1 (%)	Gp2 (%)	Gp3 (%)	*p* Value
Non T ALL	R	434 (96)	217 (90)	112 (79)	<0.001
	N	18 (4)	24 (10)	27 (19)	
T ALL	R	40 (75.5)	23 (53.5)	23 (60.5)	0.10
	N	13 (24.5)	20 (46.5)	15 (39.5)	

[a] R, responder to the prephase; N, nonresponder to the prephase; Gp1, patients who received the intrathecal injection on day 1 of the prephase; Gp3, received injection on day 8; Gp2, received injection between day 1 and day 8.

discrete subsets of ALL requiring specific adjustments of therapy. ALL in infants is considered to have a very bad prognosis. In EORTC studies 58831 and 58881, the EFS of 68 infants treated over the last 10 years is 42%, compared to 68% for the overall population of children with ALL. This poor outcome is at least partially related to the frequent occurrence of rearrangements of the MLL gene at the 11Q23 locus in the leukemic cell clone in infant ALL.[12]

The great majority of these cases lack CD10 (CALLA) expression at the surface of the blasts. Accordingly CD10 negativity may be used as a marker, albeit indirect and imperfect, of the particular ALL subtype. In EORTC studies, the 6-year survival of infants with CD10 + ALL has been 81%, compared to 38% for CD10-negative ALL ($p = 0.006$). In this particular age group, this one particular marker, identifiable at diagnosis, discriminates between two populations of patients at very different risks of treatment failure. Use of this marker should permit an adjustment of the therapy from the very first day of the treatment.

CONCLUSION

Immunophenotypic markers, identifying lineage and maturation stage of ALL blasts, do not usually add much discriminating power to stratification systems based on a combination of clinical and hematological factors and on early response to therapy. However, they may help identify ALL variants characterized by specific pharmacodynamic properties at the leukemic cell level which could be exploited for optimization of therapy. In some subgroups of patients, some immunophenotypic markers offer an easy tool to identify subsets of patients at very different risks of relapse.

REFERENCES

1. RIEHM, H., *et al.* 1987. Die Corticosteroid abhängige Dezimierung der Leukämiezellzahl im Blut als Prognosefactor bei akuten lymphoblastischen Leukämie im Kindesalter. Klin. Pädiat. **199:** S151–160.
2. ARICO, M., *et al.* 1995. Good steroid response in vivo predicts a favorble outcome in children with T-cell acute lymphoblastic leukemia. Cancer **75:** S1684–1693.
3. STEINHERZ, P. G., *et al.* 1996. Cytoreduction and prognosis in acute lymphoblastic leukemia. The importance of early marrow response. J. Clin. Oncol. **14:** 385–398.
4. PATTE, C., *et al.* 1991. High survival rate in advanced stage B cell lymphomas and leukemias without CNS involvement with a short intensive polychemotherapy. J. Clin. Oncol. **9:** 123–132.
5. REITER, A., *et al.* 1992. Favourable outcome of B cell ALL in childhood, a report of three consecutive studies of the BFM group. Blood **80:** 2471–2478.
6. SHUSTER, J. J., *et al.* 1990. Prognostic factors in childhood T-cell acute lymphoblastic leukemia: A pediatric oncology study group. Blood **75:** 166–171.
7. FIECKERT, H. J., *et al.* 1993. Event-free survival of children with T-cell acute lymphoblastic leukemia after introduction of high dose methotrexate in multicenter trial ALL-BFM 86. Am. Soc. Clin. Oncol. Abstr. **12:** 317.
8. BARREDO, J. C., *et al.* 1994. Difference in constitutive and post-methotrexate folylpolyglutamate synthetase activity in B-lineage and T-lineage leukemia. Blood **84:** 464–569.
9. CHESSELLS, J. M., *et al.* 1992. ALL in infancy: Experience in UK national trials. Med. Pediat. Oncol. **20:** 372.
10. CRIST, W., *et al.* 1986. Clinical and biologic features predict a poor prognosis in acute lymphoid leukemias in infants: A Pediatric Oncology Group study. Blood **167:** 135–140.

11. HILDEN, J. M., *et al.* 1995. Molecular analysis of infant acute lymphoblastic leukemia: MLL gene rearrangement and reverse transcriptase–polymerase chain reaction for t(4;11)(q21;q23). Blood **86:** 3876–3882.
12. CHIEN-SHING, CHEN, *et al.* 1993. Molecular rearrangements on chromosome 11Q23 predominate in infant acute lymphoblastic leukemia and are associated with specific biologic variables and poor outcome. Blood **81:** 2386–2393.

Arvid Lindau's Cerebellar Hemangioblastoma 70 Years Later

Some Pediatric Aspects[a]

JOHN J. KEPES[b] AND FELICIA SLOWIK[c]

[b]Department of Pathology and Laboratory Medicine and
Division of Neurosurgery
University of Kansas Medical Center
3901 Rainbow Boulevard
Kansas City, Kansas 66160-7410

[c]National Institute of Neurosurgery
Budapest, Hungary

It is with a combination of pride and humility that I am undertaking the task of reviewing the present state of understanding of that most interesting, and in some respects downright mysterious, neoplasm, the cerebellar capillary hemangioblastoma that Dr. Arvid Lindau described 70 years ago, in 1926. The present session is part of the meeting graciously sponsored by the New York Academy of Sciences, a rousing finale to five years of fruitful cooperation under the aegis of the U.S. State Department's Agency for International Development (AID) program, Pediatric Oncology Outreach to Hungary (P.O.O.H.), between the University of Kansas Medical Center and two Hungarian Institutions with pediatric patients: the Second Department of Pediatrics of the Semmelweis University Medical School of Budapest and the Section of Pediatric Neurosurgery of the National Institute of Neurosurgery, also of Budapest. I have had a long-standing interest and fascination with everything and anything connected to hemangioblastomas, but I certainly owe this audience some explanation why I have selected for my subject Lindau's cerebellar hemangioblastomas, which as every textbook will readily tell you, manifests itself most commonly in adulthood. Nevertheless, this tumor and related manifestations of the von Hippel–Lindau (VHL) syndrome may present clinically at a very young age. Cerebellar hemangioblastoma has been found even in a newborn,[1] and there are several reports of VHL or even sporadic cerebellar hemangioblastomas encountered at an early age.[2] Cases have been reported by Pasztor *et al.* as well as Julow *et al.* from the National Institute of Neurosurgery, one of the active participants in the P.O.O.H. program.[3,4] Not only cerebellar hemangioblastomas, but also other manifestations of VHL, may present early: A case of a sellar paraganglioma forming part of VHL was recently reported to have occurred in a 14-year-old boy.[5]

The main connection to pediatrics, however, stems from the fact that members of families with VHL have a life-long *potential* to develop one or more of the manifestations of this disease complex, and early screening will help to diagnose and treat these at a phase when, hopefully, they are still curable and need less radical intervention. For example, a 1994 International Symposium on Von Hippel–Lindau

[a] As presented by Dr. Kepes at the conference Challenges and Opportunities in Pediatric Oncology in Budapest, Hungary.

disease[6] recommended that clinical screening of individuals at risk should start at age six with ophthalmoscopy, evaluation of neurological status, abdominal ultrasound examination, and a 24-hour catecholamines study, with MRI study of the central nervous system recommended to begin at age 10. It follows then, that VHL definitely should be regarded as at least a potential pediatric problem in both the therapeutic and the preventive sense. This paper is not about the genetics of VHL: suffice it to state that dramatic advances have been made in molecular genetic analysis through which it has been established that the VHL gene is localized on the short arm of chromosome 3. To the uninitiated, this may sound like this gene in the 3p25-26 location is the villain responsible for this malady, but in fact in its normal state this gene's job is to *protect* one against the ravages of VHL, and trouble develops when this gene becomes deleted or undergoes one or several mutations. In other words it appears that the VHL gene is a tumor suppressor gene.[7]

Regarding what constitutes a case or a patient said to have the von Hippel–Lindau syndrome, two classes of individuals qualify: (1) A person who is known to harbor a central nervous system hemangioblastoma *plus* one or several other manifestations of VHL, be it retinal angiomatosis, pancreatic cyst, or any of the other components of this entity. It is not necessary for that person to have a known relative with VHL syndrome. (2) A person with hemangioblastoma (even without other manifestations) who has at least one relative with VHL syndrome. In this respect, some analogy exists with another syndrome inherited as an autosomal dominant: neurofibromatosis. A person with a large number of neurofibromas of the skin and nerve trunks and multiple café-au-lait spots is considered to be suffering from neurofibromatosis even without a known relative with the same disease. Conversely, a patient with a single, positively identified neurofibroma who has relatives with neurofibromatosis would also be considered to be one suffering from that disease complex.

Turning now to historic aspects, we will focus on Dr. Arvid Lindau, his times, and his contributions. Arvid Lindau was born on July 23, 1892 in the city of Malmö, Sweden, son of a regiment physician, Dr. Anders Lindau.[8] Dr. Arvid Lindau received his medical education at the University of Lund and graduated in 1914. He trained in pathology and bacteriology at both the Karolinska Institut in Stockholm and at the University of Lund. In 1925 he undertook an extensive journey through Europe. This was at least partly motivated by his great interest in certain cerebellar cystic lesions, in particular one often cystic vascular tumor that he felt bore some similarities to the lesion known as "angiomatosis retinae" described by von Hippel of Göttingen in 1904.[9]

So, in addition to many other European centers of pathology, Lindau visited Dr. von Hippel, and the conclusion was reached that angiomatosis retinae has very similar histological features to cerebellar hemangioblastomas. Thus, the von Hippel–Lindau syndrome was born, the hyphen between the two names indicating the 22 years that elapsed before the connection was made. Upon his return to Sweden, Dr. Lindau completed his historical opus in the form of a dissertation, sufficiently long and detailed to deserve a special issue, a supplement to the *Acta Pathologica and Microbiologica Scandinavica*: 126 pages, plus an added 29 pages with 71 illustrations (gross and microscopic photographs of excellent quality).[10] Although the title referred to the composition of cerebellar cysts, their pathogenesis and relationship to angiomatosis retinae, the paper also discussed other types of cerebellar cysts, for example, parasitic cysts, cysts related to gliomas, epidermoid cysts, and even pseudocysts that were consequences of cerebellar infarction or hemorrhage. The emphasis, however, was on cystic hemangioblastomas. This dissertation made Dr. Lindau famous in a very short time. He was able to spend three months in Germany with Professor Aschoff, and in 1931 received a Rockefeller fellowship to Harvard University, where he came into close contact with Harvey Cushing and Percival Bailey, who supported their young colleague by providing world-wide recognition for him.

To turn now to the title of this presentation, what important developments can one point to in our understanding of this unusual tumor, cerebellar hemangioblastoma, and the von Hippel–Lindau complex? Topographically, hemangioblastomas themselves have been discovered in locations other than the cerebellum: in the brain stem, spinal cord, cerebral and spinal meninges, but, importantly, only in tissues that actually form part of or are directly derived from nervous tissue. Spinal nerve roots have been found to be the site of hemangioblastomas; the location farthest from the spinal cord came to be the radial nerve between the biceps and triceps muscles as reported by Brodkey et al.[11]

Other important manifestations of VHL include angiomatosis of the retina (in reality retinal hemangioblastomas themselves), pancreatic cysts and tumors, renal cysts and renal cell carcinomas, neoplasms of the epididymis, and pheochromocytomas (and as mentioned earlier, extraadrenal paragangliomas as well). It is certainly possible for any of those lesions to become clinically manifest before any CNS hemangioblastoma presented itself. In one of our patients, a 42-year-old woman not known to suffer from VHL, a nephrectomy was done for renal cell carcinoma; and when 1½ years later a symptomatic round mass appeared in the cerebellum, it was naturally considered to represent a metastasis, except that upon surgical removal it turned out to be a classical hemangioblastoma, thereby placing the patient into the VHL category.[12]

With regard to the hemangioblastomas themselves, as we all know, they have a vascular and a stromal cell component leading some to consider this tumor a "hamartomatous" growth.[13] As far as the vascular component is concerned, the term commonly used is "capillary" hemangioblastoma, although capillaries are certainly not the only type or caliber of vessels present in these lesions. Venules and, often, also very large and wide venous channels are often seen in this tumor. In our classification of "ordinary" hemangiomas, we would likely call such lesions "capillary and venous hemangiomas," but in the case of hemangioblastomas somehow the newly formed venous channels have not received much attention, only the capillaries. Regarding stromal cells, I shall endeavor to address three problems that keep challenging all of us who hold a steady fascination for this unique tumor: (1) Are these cells simply providing a "stroma" for a neoplasm composed primarily of blood vessels, or do they form an integral or perhaps even principal contingent of this neoplasm? (2) What is the identity and histogenesis of stromal cells? and (3) Is there a functional relationship between stromal cells and the proliferating capillaries?

Regarding the first question, it is our observation that stromal cells in hemangioblastomas have characteristics worthy of neoplastic cells in their own right. They are not relegated to "accompany" capillaries or larger vessels, but are often seen to form poorly vascularized, or even entirely avascular cell clusters, sometimes of respectable size. Whereas the heavily vascularized areas of hemangioblastomas are very rich in reticulin fibers (thanks to the basement membranes of the capillaries), stromal cells forming avascular clusters are mostly free of reticulin and on low-power view on reticulin stain come quite close to imitating a nodular area of a glioma. As to nuclear characteristics, they very often contain bizarre, hyperchromatic nuclei (Fig. 1), such as one encounters in neoplasms with an active growth potential, but admittedly also in so-called "ancient" schwannomas and in angiomatous meningiomas, in those latter instances without signifying malignant potential. Perhaps more

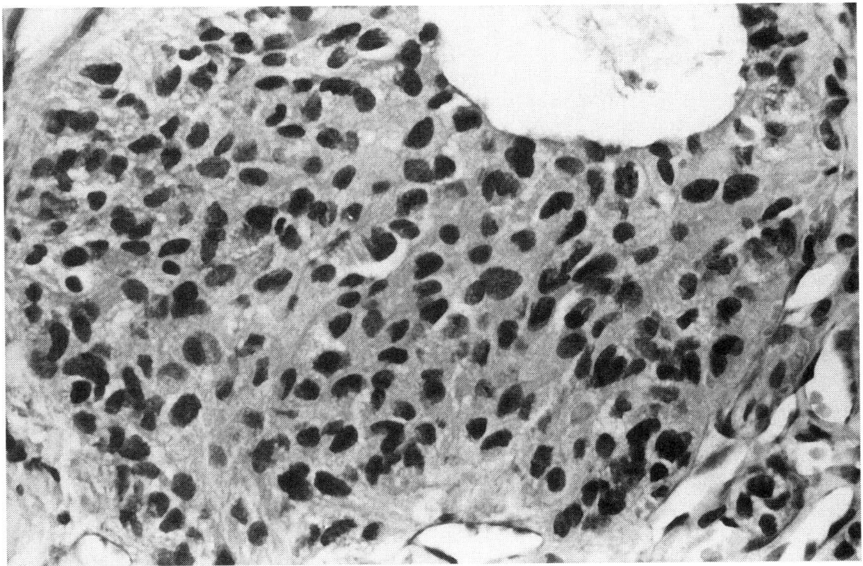

FIGURE 1. Hemangioblastoma of cerebellum. Cluster of stromal cells is seen with only a few capillaries at the periphery of the cluster. Stromal cell nuclei show marked pleomorphism and hyperchromasia. H&E × 160.

important in this respect is the often-encountered growth pattern of stromal cell clusters with relation to neighboring capillaries. Here, one frequently observes clusters of stromal cells compressing capillary walls, protruding/invaginating into their lumina by forming cell-rich "peninsulas" (FIG. 2). The compressed vessels often passively assume a "staghorn" pattern, closely reminiscent of the vascular alterations seen in hemangiopericytomas, where of course it is universally accepted that the actively proliferating neoplastic pericytes are responsible for this pattern by "pushing" from the outside into the capillary lumina. Examining proliferative tendencies with the MIB-1 immunostain (paraffin section equivalent of the Ki-67 stain), it rapidly becomes obvious that the stromal cells have many more positively staining nuclei than the vascular endothelial cells (FIG. 3), although the latter obviously also proliferate (but perhaps not in a neoplastic manner). Drs. Diane Persons and Susan Venuti, in our Department of Pathology at Kansas University Medical Center, analyzed the ploidy in hemangioblastomas and found that practically all endothelial cells were diploic, whereas many stromal cells showed aneuploidy, mostly tetraploidy (FIG. 4). The above-listed criteria of stromal cells: foci of proliferation independent of the vascular component; the presence of bizarre, hyperchromatic, atypical nuclei; a MIB-1 proliferative nucleus index much higher than is seen in vascular endothelium; frequent polypoidy; and, in terms of histologically visible proliferative activity, the commonly seen indentation of capillary walls from the outside creating a "staghorn" pattern similar to that seen in hemangiopericytomas all suggest that the active neoplastic component of hemangioblastomas is likely the stromal cell population.

This recognition gradually found expression in the attempts to find a proper "niche" for hemangioblastomas in classification schemes of CNS neoplasms. In the first edition of the WHO histopathological classification of tumors of the central nervous system (1972), hemangioblastomas were classified as tumors of the vascular system.[14] In the fifth (last) edition of Russell and Rubinstein's *Pathology of Tumours of the Nervous System* (1989), we still find "capillary" hemangioblastomas listed as the leading example of tumors of vascular origin.[15] In striking contrast, both the second edition of the WHO brain tumor classification[16] and Burger and Scheithauer's *AFIP Atlas of Tumor Pathology*[17] list hemangioblastomas as "tumors of uncertain origin," with the aforementioned latest WHO classification actually referring to the stromal cells as "the principal cells." It is of historic interest that Castaigne *et al.* used this term for stromal cells as early as 1968(!), almost 30 years ago, in their study of ultrastructural features of hemangioblastomas.[18]

The next question we all would like to have an answer for is *who* are these "stromal cells" whom we increasingly consider to be the "principal cells" of hemangioblastomas? Unfortunately, we don't know the answer to this seemingly simple question!

In 1930, Lindau himself considered them to be detached endothelial cells that frequently undergo lipidization.[19] The latter part of his observation is certainly true, and since the biochemical studies of Jakobiec *et al.* on cases of retinal hemangioblastomas[20] we also learned that the lipid contained within stromal cells has the same chemical composition as plasma lipids do, making it very likely that thin-walled blood vessels in the tumor allow plasma components (lipids as well as some proteins) to diffuse into the interstitium and gain ingress into stromal cells. As for these cells representing endothelial cells "away" from blood vessels, this assumption seems

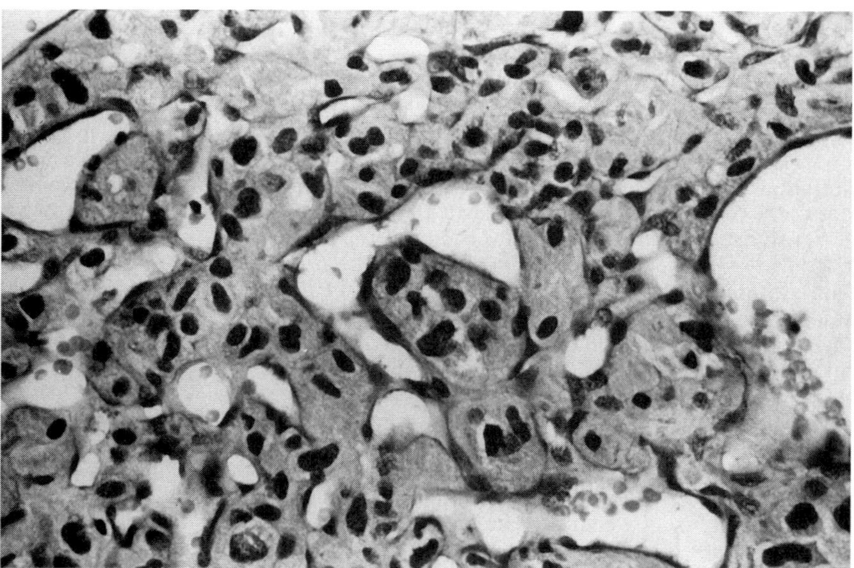

FIGURE 2. A more vascularized portion of the tumor shown in FIGURE 1. Proliferation of stromal cells causes indentation deformities of capillaries by external pressure. H&E × 160.

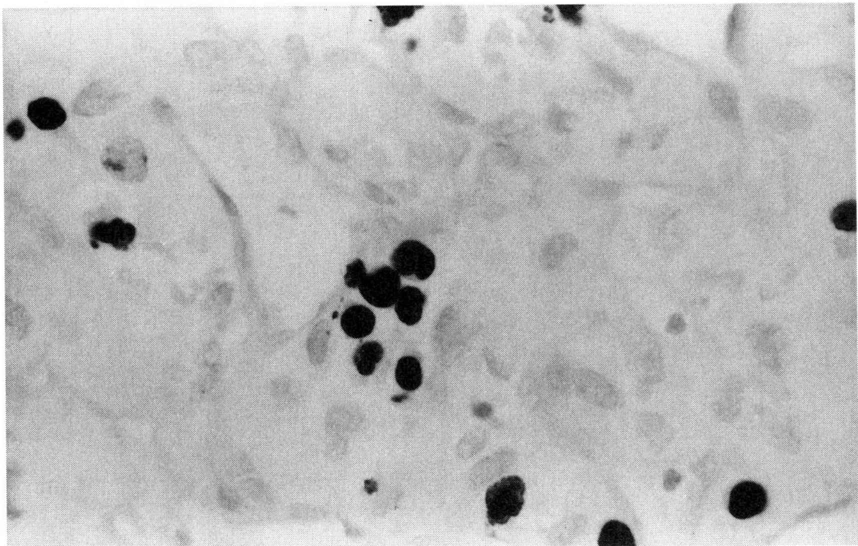

FIGURE 3. Same tumor as in FIGURES 1 and 2. MIB-1 stain (paraffin section application for Ki-67 stain for proliferative phase nuclei) shows positive staining in many stromal cells, but at least in this area not in nuclei of capillary endothelium. MIB-1 immunostain ×160.

highly unlikely: Stromal cells have no counterpart in vascular tumors of any kind outside the central nervous system, its covering, and derivates. By immunohistology, they are negative for factor VIII, CD-34, and other endothelial markers; and Ho *et al.*'s 1984 observations[21] regarding the presence of Weibel-Palade bodies (important ultrastructural markers of endothelial cells) in stromal cells of hemangioblastomas could not be confirmed by later studies.[22] The suggestion to regard stromal cells as modified pericytes also runs counter to the fact that, whereas extraneural capillary and other hemangiomas certainly possess pericytes, they do not contain stromal cells. Our immunohistological studies of hemangioblastomas for smooth muscle actin (HHF-35) showed strong presence of that substance in both endothelial cells and the pericytes surrounding them in hemangioblastomas, but were utterly negative in the stromal cells (FIG. 5). Conversely, stromal cells typically are quite positive for γ-enolase (neuron-specific enolase).[23] They also often express S-100 protein, and although this is not restricted to neuroectodermal cells (it is, among other cells, typically expressed by Langerhans cells of the epidermis and in Langerhans cell histiocytosis), it is regularly present in astrocytes, Schwann cells, and melanocytes, all of neuroectodermal descent. In addition, Becker, Paulus, and Roggendorf found synaptophysin and neuropeptide Y in stromal cells,[24] which is also in line with their having some relationship to neural or neuroendocrine functions. In the latter sense (also regarded by some as "paracrine" functions), one should also refer to the known production of erythropoietin by stromal cells and the presence of vascular endothelial growth factor (VEGF) in these cells.[25,26] The finding of neurosecretory granules in stromal cells by electron microscopy[27,28] may well represent the ultrastructural basis for such activities. Unfortunately, to this day, no *normal* component of

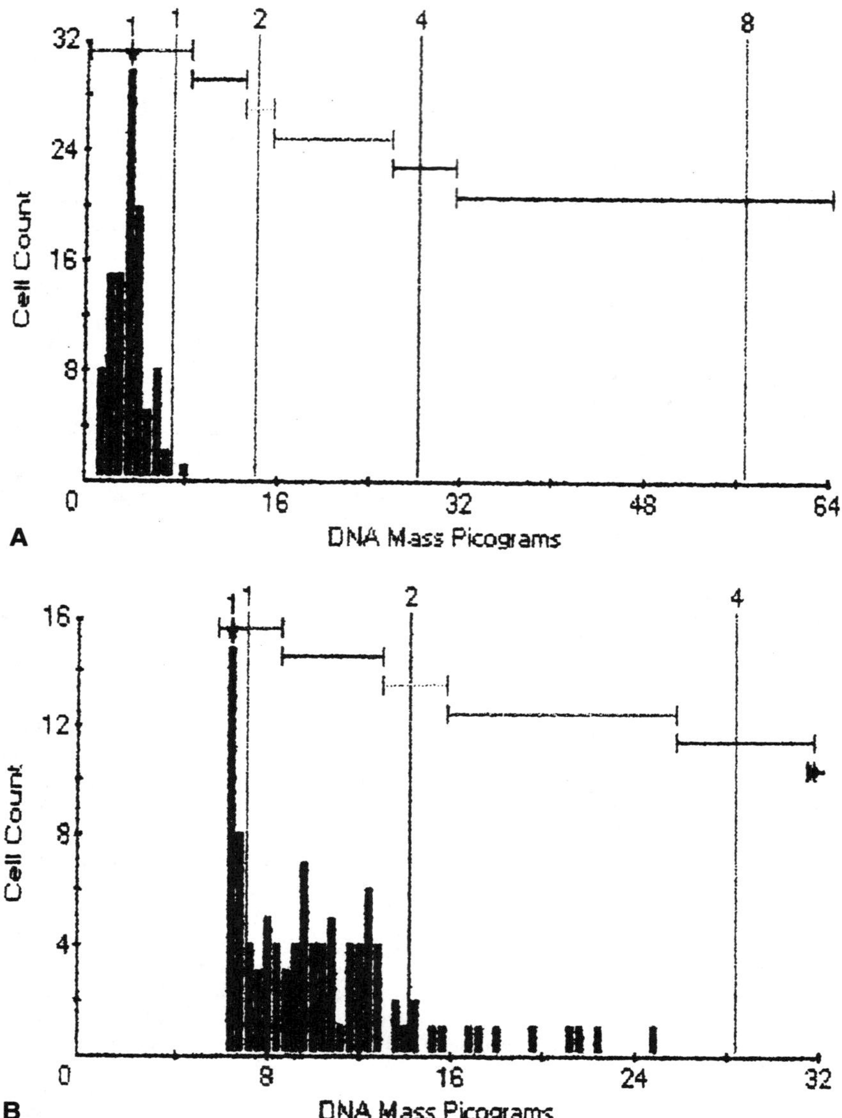

FIGURE 4. DNA mass picograms showing distribution of DNA mass in various cell nuclei in cerebellar hemangioblastoma case. (A) In endothelial cells nearly all nuclei have a DNA mass of 8 picograms or less, consistent with diploic nuclei. **(B)** In stromal cells of the same tumor, DNA mass in nuclei shows much greater variation, extending from 7 to 25 picograms per nucleus, indicating aneuploidy.

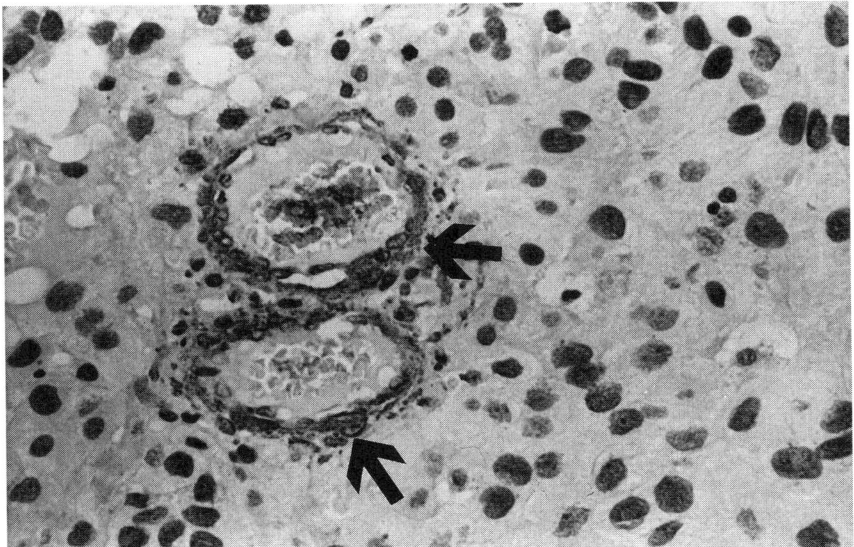

FIGURE 5. Immunostain for smooth muscle actin (SMA) is positive for cytoplasm of endothelial cells and pericytes of venules in tumor (*arrows*), but entirely negative for stromal cells, making pericytic derivation of stromal cells unlikely. Immunoperoxidase stain for SMA, ×180.

the central nervous system or its coverings (the meninges) has been identified as being the precursor of this cell; in other words we are discussing a cell that so far has been recognized and identified only in its neoplastic form, certainly a unique situation in the entire history of pathology in general and the study of neoplasms in particular.

In addition to vascular channels and stromal cells (as well as some scattered mast cells), hemangioblastomas may also contain islands of hematopoietic cells of the erythropoietic series.[27] These are supposed to represent the "local" or intratumoral result of the production of erythropoietin by stromal cells, a substance that, of course, is also produced by neoplastic epithelial cells of renal cell carcinomas. But the difference is that erythropoietin production is a well-known function of the normal kidney as well, but is not recognized as an activity of any of the known normal elements of the central nervous system.

One should also mention that, based on ultrastructural examination of actual hemangioblastomas, astrocytes have been suspected to be closely related to stromal cells;[20] and, indeed, lipidized cells within hemangioblastomas may have positive immunostaining for glial fibrillary acidic protein. This is due to the fact that hemangioblastomas of the nervous parenchyma (cerebellum or other sites), while rather well circumscribed, are not truly encapsulated; and portions of neighboring brain tissue with reactive astrocytic gliosis quite often become, at least superficially, incorporated in the tumor mass. Such cells are naturally GFAP positive, but being exposed to the same leaked-out plasma elements as the real tumor cells, often become lipidized, and while some still maintain their well-recognizable contours of trapped astrocytes, others, secondary to high lipid content, have more rounded outlines and come to resemble stromal cells with a similarly high degree of lipidization. Such cells nevertheless cannot be considered to be actual stromal cells. A vivid proof of this is the fact that in the pure meningeal examples of hemangioblastomas many lipidized stromal cells may be encountered, but, with no trapped astrocytes from neighboring nervous parenchyma, none of the stromal cells in the meningeal examples ever express GFAP.[29] Deck and Rubinstein also strongly felt that the presence of GFAP in cells of hemangioblastomas does not indicate astrocytic nature of stromal cells, but they preferred to regard GFAP-positive lipidized cells in these tumors as "real" stromal cells that happened to ingest or incorporate glial fibrillary acidic protein from gliotic tissue around the tumor.[30]

Finally, the question of a possible causal role of stromal cells in bringing about the proliferation of vascular channels has recently been the focus of great interest. In our and other authors' material, stromal cells of hemangioblastomas were found to contain significant amounts of VEGF (FIG. 6) and other authors found corresponding VEGF receptors in vascular endothelial cells of these tumors.[25,26] This would appear to be a process similar to that seen in often highly vascularized gliomas, particularly glioblastomas, where the malignant neoplastic glial cells were found to contain demonstrable VEGF.[31] Because in malignant gliomas the endothelial proliferation is often so pronounced that "glomeruloid" formations with endothelial cells piling up within lumina are regularly observed features, this type of secondary vascular proliferation is regarded by some as a histological and to some extent diagnostic hallmark of neoplasms of neural origin.[32] In our own cases of CNS hemangioblastomas, we were able to find similar excessive proliferation of

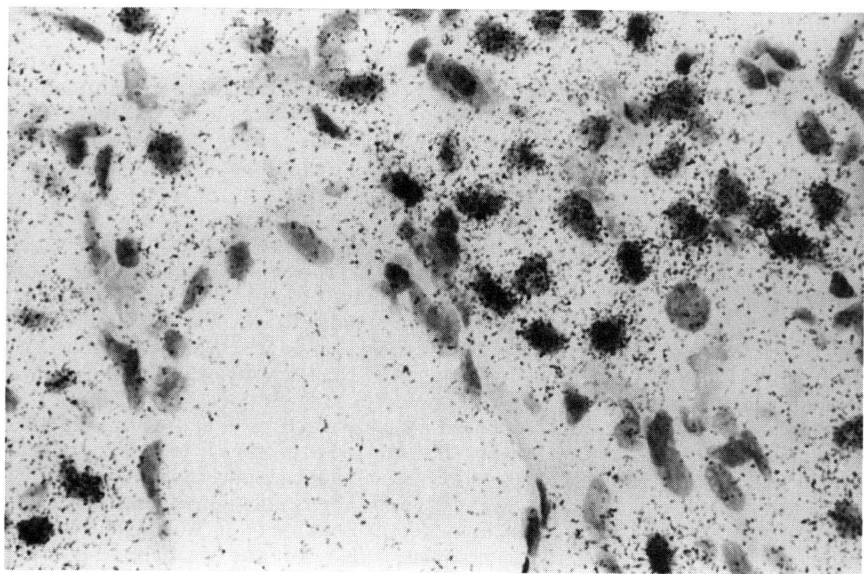

FIGURE 6. Cerebellar hemangioblastoma with *in situ* hybridization using ^{35}S-labeled antisense riboprobes for vascular endothelial growth factor (VEGF). Many positive (*black*) dots mark the cytoplasm of stromal cells indicating the highest concentration of VEGF within the cell population of this tumor. ×180. (Courtesy of Dr. S. Y. Leung, Hong Kong.)

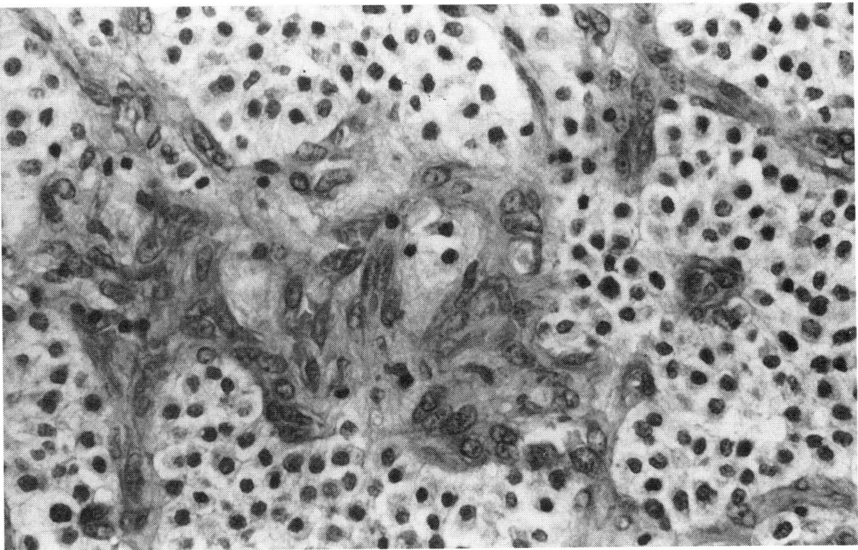

FIGURE 7. Another case of cerebellar hemangioblastoma. Stromal cells form compact clusters; capillary proliferation is very active with endothelial cells piling up, creating "glomeruloid" formations similar to those seen in malignant gliomas. H&E × 150.

endothelial lining cells resulting in glomeruloid formations, very similar to those found in gliomas (FIG. 7).

In summary, it now appears that capillary hemangioblastomas (which as stated above are not restricted to capillaries in their vascular composition) may actually represent tumors of a heretofore unidentified cell of central nervous and/or meningeal tissue, which in its neoplastic form is recognized as the so-called stromal cell. This cell is capable of independent proliferation (as in the vessel-free portions of hemangioblastomas) but is also able to stimulate in a paracrine fashion vascular proliferation, similar to, but in excess of what may be observed in gliomas. It is therefore possible that, lacking a better term, hemangioblastomas in the future may come to be called "highly vascularized stromomas" of the central nervous system and its coverings.

ACKNOWLEDGMENTS

The authors wish to thank Drs. Diane Persons and Susan Venuti for providing DNA analysis of the tumors examined, Dr. Suet Yi Leung for performing antisense riboprobe studies for vascular endothelial growth factor, Mr. Dennis Friesen for photographic help, and Mrs. Edna Esfandiary for the careful typing of the manuscript.

REFERENCES

1. BERGSTRAND, H., H. OLIVECRONA & H. TÖNNIS. 1936. Gefässmissbildungen und Gefässgeschwülste des Gehirns. G. Thieme Leipzig. cit. *In* Pathology of Tumours of the Nervous System. 5th edit. D. S. Russell & L. J. Rubinstein, Eds.: 640. 1989. Williams & Wilkins. Baltimore, MD.

2. BOURDILLON, P. J. & R. C. HICKMAN. 1967. Von Hippel–Lindau's disease presenting at an early age. J. Neurol. Neurosurg. Psychiat. **30:** 559–562.

3. PÁSZTOR, A., L. SZTENÁK, I. DOBRONYI, J. SZÉNÁSY & G. HARMAT. 1985. Angioblastoma of the posterior fossa in children. J. Pediatr. Neurosci. **1:** 4.

4. JULOW, J., K. BÁLINT, P. GORTVAI & E. PÁSZTOR. 1994. Posterior fossa haemangioblastomas. Acta Neurochir. (Wien) **128:** 109–114.

5. SCHEITHAUER, B. W., A. PARAMESWARAN & B. BURDICK. 1996. Intrasellar paraganglioma: Report of a case in a sibship of von Hippel–Lindau disease. Neurosurgery **38:** 395–399.

6. NEUMANN, H. P. H. & O. D. WIESTLER. 1994. Report of the International Symposium von Hippel–Lindau Syndrome: Molecular Basis and Clinical Management. Freiburg, May 27–28, 1994. Personal Communication.

7. LATIF, F., et al. 1993. Identification of the von Hippel–Lindau disease tumor suppressor gene. Science **260:** 1317–1320.

8. NEUMANN, H. P. H. 1993. Arvid Lindau zum 100. Geburtstag. Pathol. **14:** 178–180.

9. HIPPEL, E. VON. 1904. Über eine sehr seltene Erkrankung der Netzhaut. A. von Graefe's Arch. Ophthalmol. **59:** 83–86.

10. LINDAU, A. 1926. Studien über Kleinhirncysten. Bau, Pathogenese und Beziehungen zur Angiomatosis Retinae. Acta Pathol. Microbiol. Scand. Suppl. **I:** 1–128.

11. BRODKEY, J. A., J. A. BUCHINGANI & T. F. O'BRIEN. 1995. Hemangioblastoma of the radial nerve: Case report. Neurosurgery **36:** 198–201.

12. KEPES, J. J. & W. L. YARDE. 1994. Renal cell carcinoma followed by a cerebellar mass. Kansas Med. **95:** 15–17.

13. SEYAMA, S., M. OHTA, S. NISHIO, et al. 1982. Cells constituting cerebellar hemangioblastomas. Ultrastructural study. Acta Pathol. Jpn. **32:** 399–413.

14. ZÜLCH, K. J. 1979. Histological Typing of Tumours of the Central Nervous System. 2nd edit. World Health Organization. Geneva. p. 62.

15. RUSSELL, D. S. & L. J. RUBINSTEIN. 1989. Pathology of Tumours of the Nervous System. 5th edit. Williams and Wilkins. Baltimore, MD. p. 639.

16. KLEIHUES, P., P. C. BURGER & B. W. SCHEITHAUER. 1994. *In* World Health Organization. International Histological Classification of Tumours of the Central Nervous System. Springer Verlag. Geneva. 41–42.

17. BURGER, P. E. & B. W. SCHEITHAUER. 1994. Tumors of the central nervous system. *In* AFIP Atlas of Tumor Pathology. 3rd edit. Series No. 10. Armed Forces Institute of Pathology. Washington DC. p. 239.

18. CASTAIGNE, P., M. DAVID, B. PERTUISET, R. ESCOUROLLE & J. POIRIER. 1968. L'ultrastructure des hemangioblastomes du system nerveux central. Rev. Neurol. (Paris) **118:** 5–26.

19. LINDAU, A. 1930. *In* Discussion on vascular tumors of the brain and spinal cord. Proc. R. Soc. Med. **24:** 363–370.

20. JAKOBIEC, F. A., R. L. FONT & F. B. JOHNSON. 1976. Angiomatosis retinae. An ultrastructural study and lipid analysis. Cancer **38:** 2042–2056.

21. HO, K-L. 1984. Ultrastructure of cerebellar capillary hemangioblastoma. I. Weibel-Palade bodies and stromal cell histogenesis. J. Neuropathol. Exp. Neurol. **43:** 592–608.

22. SHIMURA, T., A. HIRANO & J. F. LLENA. 1985. Ultrastructure of cerebellar hemangioblastoma. Some new observations on the stromal cells. Acta Neuropathol. (Berlin) **67:** 6–12.

23. FELDENZER, J. A. & P. E. MCKEEVER. 1987. Selective localization of gamma enolase in stromal cells of cerebellar hemangioblastomas. Acta Neuropathol. **72:** 281–285.

24. BECKER, I., W. PAULUS & W. ROGGENDORF. 1989. Histogenesis of stromal cells in cerebellar hemangioblastomas. An immunohistochemical study. Am. J. Pathol. **134:** 271–275.

25. BÖHLING, T., E. HATVA, M. KUJALA, L. CLAESSONWELSH, K. ALITALO & K. HALTIA. 1996. Expression of growth factors and growth receptors in capillary hemangioblastoma. J. Neuropathol. Exp. Neurol. **55:** 522–527.

26. WIZIGMANN-VOOS, S., G. BREIER, W. RISAU & K. H. PLATE. 1995. Up-regulation of vascular endothelial growth factor and its receptors in von Hippel–Lindau disease-associated and sporadic hemangioblastomas. Cancer Res. **55:** 1358–1364.

27. ISHWAR, S., R. M. TANIGUCHI & F. S. VOGEL. 1971. Multiple supratentorial hemangioblastomas. Case study and ultrastructural characteristics. J. Neurosurg. **35:** 396–404.

28. LINARES, J., V. DIAZ-FLORES, V. ARJONA, D. AGUILAR & A. R. LUCAS. 1978. Hemangioblastoma cerebeloso. Bases ultraestructurales de secrecion endocrina. Morfol. Norm. Patol. B. **2:** 777–799.

29. KEPES, J. J., S. S. RENGACHARY & S. H. LEE. 1979. Astrocytes in hemangioblastomas of the central nervous system and their relationship to stromal cells. Acta Neuropathol. **47:** 99–104.

30. DECK, J. H. N. & L. J. RUBINSTEIN. 1981. Glial fibrillary acidic protein in stromal cells of some capillary hemangioblastomas: Significance and possible implications of an immunoperoxidase study. Acta Neuropathol. **54:** 173–181.

31. LEUNG, S. Y., A. S. Y. CHAN, M. P. WONG, S. T. YUEN, N. CHEUNG & L. P. CHUNG. 1996. Vascular endothelial growth factor expression in gliomas. Abstract 140. AANP 72nd Annual Meeting. J. Neuropathol. Exp. Neurol. **55:** 640.

32. GAUDIN, P. B. & J. ROSAI. 1995. Florid vascular proliferation associated with neural and neuroendocrine neoplasms. A diagnostic clue and potential pitfall. Am. J. Surg. Pathol. **19:** 642–652.

Immunobiology of Gliomas:
New Perspectives for Therapy[a]

PIERRE-YVES DIETRICH,[b] PAUL R. WALKER,[b]
PHILIPPE SAAS,[b] AND NICOLAS DE TRIBOLET[c]

[b]Division of Oncology
Laboratory of Tumor Immunology
[c]Division of Neurosurgery
Hôpital Universitaire de Genève
Rue Micheli-du-Crest 24
CH-1211 Genève 14, Switzerland

INTRODUCTION

Brain tumors constitute the second leading cause of cancer death in children younger than 15 years of age. The commonest tumors derive from glial precursors (astrocytes, oligodendrocytes, ependymocytes) and, among them, malignant gliomas (anaplastic astrocytoma and glioblastoma) are generally lethal soon after diagnosis. In addition, benign gliomas often develop in sites with critical neurological function, limiting the opportunity for complete surgical removal or high-dose irradiation. Furthermore, the rare survivors are at high risk for immediate and long-term toxicities induced by conventional therapies. Innovative ways to treat gliomas are therefore being intensively explored, including new drugs, radiosurgery, and various approaches based on gene therapy technology (suicide gene, anti-angiogenesis, insertion of wild-type p53, etc.).

A major impediment to the success of therapy is the propensity of glioma to microscopically infiltrate normal structures at an early stage of their development. This renders the specific targeting of all tumor cells and the sparing of normal cells particularly difficult. Therefore, eliciting an efficient immune response against glioma appears to be an attractive alternative treatment strategy, taking advantage of the natural circulatory properties of immune cells and their anti-tumor activities. Until the 1990s, this had been attempted mainly by nonspecific stimulation using interleukin-2 (IL-2) alone or with lymphokine-activated killer (LAK) cells. The therapeutic benefit for glioma patients is very limited, and treatment-related toxicities (e.g., capillary leak syndrome) are particularly severe in the CNS. Recent advances in basic and tumor immunology suggest that at least in some circumstances, glioma cells may interact with immune cells, despite the location of glioma in the CNS, a long-believed immune-privileged site. Therefore, it may be possible to selectively stimulate immune cells with anti-tumor function, avoiding the drawbacks of global immune activation. This approach should lead to a better therapeutic window, with an increase in efficacy and a decrease in toxicity.

In this review, we will first summarize the current understanding of the possible dialogue between glioma cells and immune cells with particular emphasis on T cells,

[a] This work was supported by the Swiss National Foundation (No. 31-40704.94), la Ligue Genevoise contre le Cancer (P.R.W.), la Fondation de France (P.S.), and les Fondations Terwindt and Spinola.

before discussing in a second part the most promising treatment strategies designed to increase the immune response against glioma.

HOW AN IMMUNE RESPONSE AGAINST GLIOMA MAY OCCUR IN THE CNS

There is now cumulative evidence that T lymphocytes play a critical role in the anti-tumor response. The initiation or priming of a specific T-cell response requires an antigen-specific interaction between tumor cells and T lymphocytes, probably with the help of professional antigen-presenting cells (APCs). In the case of glioma, the physical isolation provided by the blood–brain barrier and the limited lymphatic drainage of the CNS[1] suggest that the immune cells involved in these interactions may differ from those in other sites. Thus, major efforts are currently made to elucidate the mechanisms of antigen presentation in the CNS, to characterize T cells involved in tumor recognition, and to identify glioma antigens. In addition to this antigen-specific interaction, other factors can influence the outcome of the response. The importance of co-stimulatory signals and mechanisms contributing to immunosuppression in the tumor microenvironment will be discussed.

Antigen Presentation

Self and foreign antigenic proteins are continuously presented to both CD8[+] and CD4[+] T cells as peptides associated with class I and II MHC molecules, respectively (FIG. 1). Which cells in the CNS have the ability to present antigen? The classical APCs present in peripheral lymphoid tissue are generally absent in the CNS, a factor that may contribute to low immunoreactivity in such an "immunologically privileged site."[1]

Candidate APCs for the brain must express appropriate cell surface MHC molecules as a prerequisite for antigen presentation to T cells. Microglial cells are particularly attractive candidates that fulfill this basic requirement for such a function. They correspond to 5–15% of the total cellular composition of brain tissue and are distributed throughout the CNS.[2] Microglial cells possess many characteristics of dendritic cells, such as dendritic morphology, intracytoplasmic Birbeck granules, expression of CD1a, MHC class I and class II molecules, and expression of accessory molecules such as B7 (CD80), LFA3 (CD58), and ICAM-1 (CD54). Their hematopoietic origin was recently directly confirmed in transplantation studies in animals, in which the grafting of bone marrow induced a new microglia population of donor origin in the host.[3]

Other APC candidates in the brain are endothelial cells and capillary pericytes, on which class II molecule and adhesion molecule expression is inducible by IFN-γ, TNF-α, and IL-1.[1] The perivascular location of pericytes seems to be particularly appropriate for APC function, being accessible to resident CNS cells and blood-derived immune cells. Indeed, to elicit an immune response, they could migrate to lymphoid organs through the recently demonstrated lymphatic vessels connecting the brain to draining cervical lymph nodes.[4] This would be comparable to the well-documented migration of Langerhans cells or dendritic cells to regional lymph nodes and spleen from skin, cardiac, or renal transplants.[5]

The role of both normal and tumoral astrocytes in antigen presentation is not clear. However, a positive feature when considering the immune control of glioma

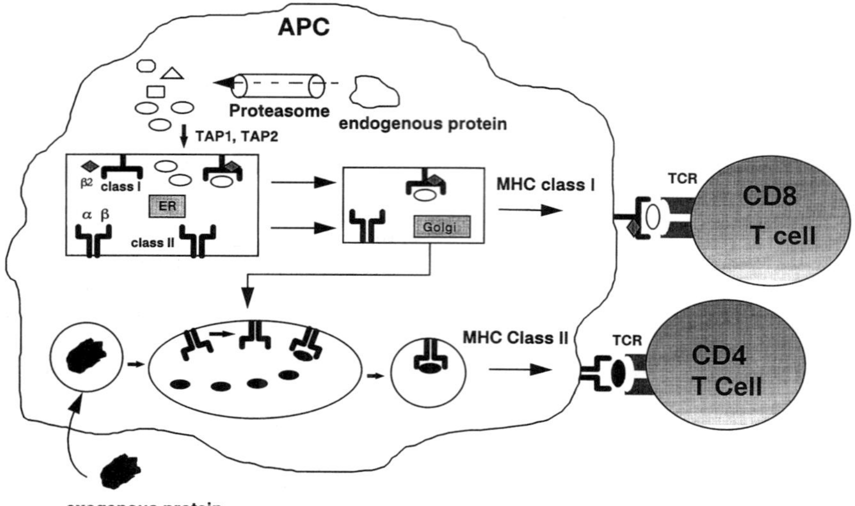

FIGURE 1. Simplified view of processing and presentation of antigens to T cells. Concerning antigen presentation by class I molecules, endogenous proteins are first degraded into small peptides by the proteasomes that are able to unfold the proteins and engulf them into a degradative tunnel. Whereas the majority of peptides degraded in proteasomes are rapidly hydrolyzed to amino acids by cytosolic exopeptidases, a minority of them, having an appropriate size, are transported into the endoplasmic reticulum (ER) by peptide transporter (TAP molecules) associated with antigen processing.[73] In the ER, peptides bind to and stabilize MHC class I molecules. The MHC–peptide complexes are then transported through the Golgi apparatus to the cell surface, allowing a specific recognition by the T-cell receptor (TCR) of CD8[+] T cells. The processing of proteins presented by class II molecules follows a different and complex pathway. Briefly, antigen is taken up by endocytosis and then proteolyzed in successive intracellular compartments (endosomes, lysosomes). At some stage, the peptides interact with MHC class II molecules (previously assembled in the ER). They are then transported as MHC–peptide complexes to the cell surface, where recognition by the TCR of CD4[+] T cells can occur.

It is clear from experimental data that some endogenous and exogenous antigens do not always follow pathways described above. Indeed, endogenous antigens may also be presented by class II molecules and exogenous antigens by class I molecules. This suggests that both tumor cells and other APC can be implicated in tumor antigen presentation and the induction of an anti-tumor response.

is that astrocytes appear to have the potential to present their antigens to T cells, since MHC molecules are expressed under some circumstances. Considering normal astrocytes in the first instance, divergent data emerge from *in vivo* and *in vitro* studies. MHC class I and II molecules are either undetectable or weakly expressed by normal astrocytes *in vivo*,[6] whereas cultured astrocytes can express a high level of MHC class I and II molecules, particularly after incubation with IFN-γ.[7,8] Primary cultures of astrocytes can present foreign antigens to class I and class II restricted CTL,[7–9] but may be unable to trigger a complete T-cell activation program.[8]

Similar results were reported for tumor cells derived from astrocytic lineage, with generally low levels of MHC expression by malignant glioma *in vivo*,[6] but *in vitro* MHC molecule expression that was usually constitutive for MHC class I and inducible for class II.[7-9] Furthermore, antigen presentation of both endogenous antigens and native exogenous proteins to both class I and class II restricted T cells was shown for some glioma cell lines, suggesting that antigen was processed as well as presented to T cells.[7,9,10] These apparently conflicting *in vivo* and *in vitro* data may be reconciled if one considers that the anti-tumor response *in vivo* is a dynamic process in which there may be a transient expression of MHC molecules induced by local cytokine secretion.

Do Glioma Antigens Exist?

The existence of antigens at the surface of tumor cells that can be recognized by immune cells is the *sine qua non* condition for the generation of a specific immune response against a tumor. For many years, it was thought that tumor antigens demonstrated in experimental tumors were due to the artificial induction of such tumors (by oncogenic viruses, ultraviolet irradiation, or chemical carcinogens) and were unlikely to be found in human cancer. However, the identification of MAGE-1 (melanoma-antigen) by T. Boon and coworkers in 1991[11] has forced a reassessment of this assumption. MAGE-1 is the first identified gene encoding a human *tumor-specific antigen* recognized by T cells (FIG. 2). It belongs to a gene family of several members that are silent in normal cells (with the exception of testis), but expressed as a nonameric peptide in a significant proportion of melanoma and other neoplasms (e.g., breast tumors, non-small-cell lung tumors, head and neck carcinomas). This pioneering work gave a new impetus to tumor immunology research, leading to the characterization of different types of tumor antigens (FIG. 2).

To date, no tumor antigen able to elicit an immune response has been characterized in glioma *in vivo*. MAGE family members are expressed by some glioblastoma cell lines,[10] but not in uncultured tumors.[12] This discrepancy could be due to a different level of DNA methylation induced by culture, since this regulates MAGE expression.[13] Perhaps the most attractive tumor antigen candidates in the case of glioma are the proteins structurally altered during malignant transformation of glial cells (or contributing to this). For example, the frequent p53 alterations observed early in the carcinogenesis of glioma could provide new antigenic peptides that may trigger an immune response. Indeed, specific cytotoxic T-cell clones have been generated *in vitro* against mutated p53 protein,[14] and *in vivo* immunization with a mutated p53 peptide induced specific CTL clones that lysed MHC-matched tumor cells expressing the mutated p53 gene.[15] Another frequent molecular event observed in glioma is the amplification and mutation of the epidermal growth factor receptor (EGF-R) gene,[16] creating new epitopes identifiable by monoclonal antibodies (for diagnostic purposes). Whether these altered proteins can induce an immune response or serve as immune targets when expressed on glioma cells is still unknown.

Encouraging results emerged from a recent study suggesting that an antibody response against glioma had been elicited *in vivo*. The screening of one astrocytoma cDNA expression library by autologous serum identified specific IgG reactivity with the expression product of TEGT, a gene that is developmentally regulated in the testis. Antibodies to this protein were detected in 2 out of 13 glioma patient sera, but were never found in healthy controls or patients with other cancers.[17] This antibody response also suggests that there may be CD4[+] T helper cell activity in the anti-glioma immune response. In the future, the identification of glioma

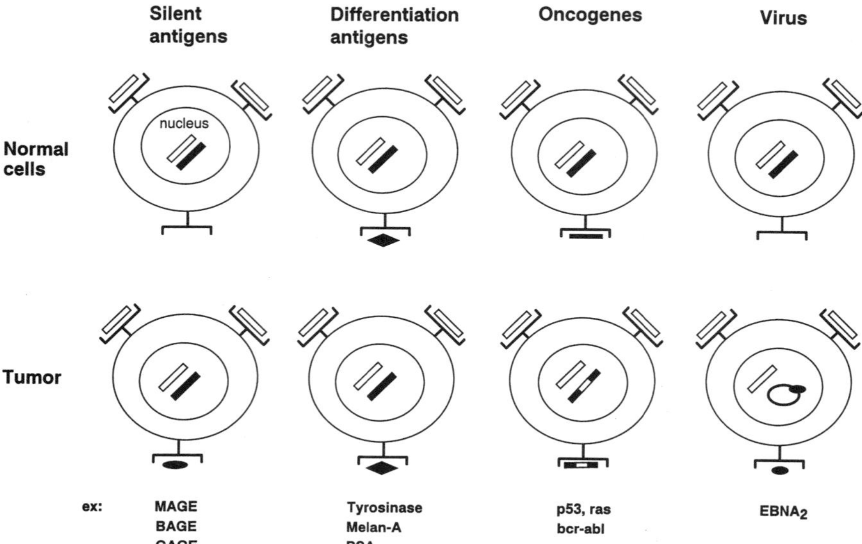

FIGURE 2. Schematic representation of tumor antigens able to elicit a T-cell response. *Tumor-specific antigens*: The prototype is MAGE-1, with expression restricted to tumor cells (see text). *Differentiation antigens*: mainly described for melanoma, these antigens are cell lineage specific. They are expressed by tumor cells as well as by the normal cell counterpart (i.e., the melanocyte). *Oncogene and tumor suppressor gene products, fusion proteins*: Peptides derived from mutated proto-oncogene or tumor suppressor gene, or encoded by a fusion gene resulting from a chromosomal translocation can be recognized by T cells and may be immunogenic. *Viral oncoproteins*: the immunogenic properties of E6 and E7 viral peptides of human papilloma virus-16 justify the current development of immunization strategies for patients with cervical cancer. The open rectangles in the nucleus and bound to MHC molecules in all cells represent normal self genes and peptides, respectively.

antigens would aid our understanding of the interaction between glioma cells and the immune system, and would facilitate the design of new treatment strategies for glioma.

Immunosuppression in the Microenvironment of Gliomas and Function of T Cells

A high proportion of gliomas are infiltrated by lymphocytes; these are mostly T cells, whereas B cells and NK cells are more scattered. The level of T lymphocyte infiltration is apparently not related to a good prognosis.[18] This apparent paradox implies that tumor-infiltrating lymphocytes (TIL) are functionally compromised, and, indeed, a range of immunological defects have been identified in glioma patients. These include abnormal delayed hypersensitivity responses, low numbers of circulating T cells, depressed mitogen responsiveness, decreased antibody responses (probably due to defective CD4$^+$ T helper cell activity), and impaired T-cell cytotoxicity. An underlying cause may be that T cells (particularly TIL) of glioma patients express a defective high-affinity IL-2 receptor (IL-2R). This may explain, at least partly, the poor proliferative abilities of glioma TIL in culture, despite the addition of recombinant IL-2; their low IL-2 production after mitogen stimulation; and the difficulty in generating T-cell clones with MHC-restricted cytotoxicity against tumor cells.[19]

During the last few years, it has become evident that soluble factors produced by glioma cells may hinder an adequate immune response. Indeed, addition of culture supernatants obtained from glioblastoma cell lines or freshly dissociated gliomas inhibits several lymphocyte functions (see above). Moreover, T lymphocytes from normal individuals exhibit similar immunologic abnormalities when cultured in the presence of glioma supernatant.[19] The most extensively studied soluble suppressor factor is TGF-β2, originally called glioblastoma cell–derived T-cell suppressor factor (G-Tsf), that was first identified in the supernatant of a human glioblastoma cell line that suppressed T-cell growth.[20] TGF-β2 exerts multiple and complex immunosuppressive effects. However, as with the other isoforms, TGF-β2 is secreted as a large and inactive precursor acquiring biological properties only after cleavage by proteases such as plasmin or cathepsin. Interestingly, there are differences in TGF secretion between normal and tumoral astrocytes. Normal astrocytes in culture express three TGF-β isoform mRNAs (for TGF-β1, 2, and 3) but secrete TGF-β2 in its inactive *latent* form, whereas glioblastoma cell lines secrete TGF-β2 isoform mainly in its *active cleaved* form. This was elegantly confirmed in studies showing that the T-cell suppression mediated by TGF-β2 was inhibited when glioma-derived proteolytic enzymes were blocked by protease inhibitors.[21] The critical role of TGF-β2 in inducing immunosuppression was further underlined in experiments in which anti-sense TGF-β2 phosphorothioate oligonucleotides inhibited TGF-β2 secretion by glioblastoma cell lines, restoring the proliferative and cytotoxic functions of autologous glioma lymphocytes.[22]

Other soluble factors produced by malignant glioma could contribute to the immunosuppression associated with this tumor. They include prostaglandin E_2 (PGE_2), IL-1-receptor (IL-1R) antagonist, and interleukin-10 (IL-10),[23-25] but the *in vivo* biological relevance of these factors was not fully investigated in these studies. Considering IL-10, while this cytokine inhibits certain aspects of the cellular immune response, its potential role in glioma is not clear because its activities are not solely immunosuppressive.[26] Indeed, recent studies demonstrated that murine and human IL-10 can stimulate the acquisition of an effective and specific anti-tumor immune response in different animal models.[27] Furthermore, although IL-10 release in or by glioma has been suggested, this was only characterized at the mRNA level using RT-PCR methods.[28,29] Normal brain tissue also yields positive signals when tested under such conditions, and thus the biological significance of these data is questionable. More recently, an IL-10 activity in glioma cell line supernatant was suggested to abrogate the release of IFN-γ by lymphocytes and to be partly responsible for inhibition of MHC class II expression by monocytes.[25] However, the contribution of IL-10 was only indirectly studied, using an antibody to block glioma supernatant activity. Further studies are thus clearly warranted to better define the precise role of IL-10 in the immune response against glioma.

T-Cell Receptor of Lymphocytes Infiltrating Glioma

Most T lymphocytes recognize antigenic peptides presented by MHC molecules using a heterodimeric α/β TCR composed of an α and a β chain (FIG. 3). The

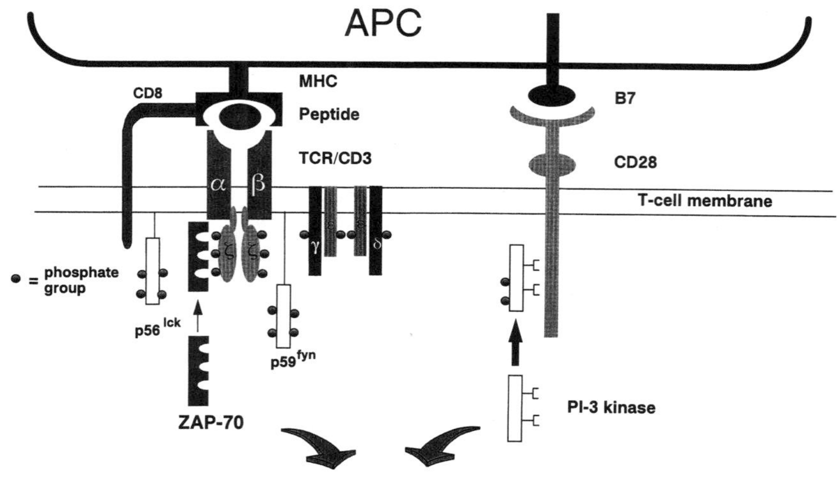

FIGURE 3. Antigen recognition and activation of T cells. Schematically, the activation of T cells requires two signals. The first signal is provided by an antigen-specific interaction between MHC–peptide complex and TCR (see text). The TCR of most mature lymphocytes is a heterodimeric structure (the association of one α and one β chain), which is noncovalently linked to the CD3 molecular complex involved in intracellular signal transduction. The second signal can be provided by costimulatory molecules, such as B7 (see text). This allows the generation of an effective response consisting of cytokine production, cellular proliferation, and effector function.

molecular analysis of TCR can provide useful information concerning the specific T-cell response that may take place *in vivo* against tumors. Indeed, the expression of unique rearranged TCR gene products determines the specificity of a given T cell, and the identification of recurrent TCR transcripts (same rearrangement and same junctional region) in large T-cell populations generally indicates antigen-driven expansion of the corresponding T-cell clones. PCR-based methods have been developed in recent years,[30] allowing the detection of T-cell clonal expansions having occurred *in vivo* in various tumor types. Such a molecular analysis is not informative per se concerning the significance of the detected clonal expansions, but can be a powerful tool when coupled with subsequent analyses of function and specificity. Indeed, the recent demonstration (in a case of melanoma and a case of RCC) that T-cell clonal expansions detected by TCR molecular analysis were actually cytolytic against autologous tumor cells in a MHC-restricted fashion highlights the interest of this approach.[31,32]

The TCR of TIL-infiltrating glioma have not been extensively studied to date. A limited usage of TCR Vα and Vβ gene segments was reported in a series of Japanese glioma patients.[33] In a series of Caucasian patients, we also found an elevated expression of certain Vβ transcripts with conserved structural features of the junctional region (the most critical region for peptide recognition) (manuscript submitted). These data offer compelling evidence for T-cell involvement in glioma, possibly reflecting a response against still-putative glioma antigens. However, it is also apparent that such a TCR molecular analysis must be combined with studies to define the activation state and the biological significance of the *in vivo* expanded T-cell populations. This may then offer the possibility of characterizing the glioma antigens recognized by such cells.

Costimulatory Molecules

In order to elicit complete T-cell effector function, antigen presentation must also deliver activation signals to the T cell, particularly through a subgroup of surface molecules termed costimulatory molecules. Schematically, antigen-specific T-cell clonal activation requires two signals (FIG. 2). The absence of a second signal results in the specific unresponsiveness of T cells, or anergy. Cumulative data suggests that B7–CD28 interaction plays a pivotal role in determining immune reactivity versus anergy.[34] To date, two members of the B7 family have been cloned (B7.1 and B7.2), and there are two counter-receptors on T cells, namely CD28 and CTLA-4. Recent results suggest that costimulation pathways are more complex than initially thought. It appears that CD28 mediates stimulatory effects, whereas CTLA-4 may be a negative regulator of T-cell responses. In accordance with this assumption, impressive anti-tumor responses have been recently reported in animal models in which CTLA-4 function was blocked by antibodies.[35]

Glioma cells do not constitutively express B7.1 and B7.2 molecules, as demonstrated by Northern blot analysis and immunocytochemistry.[36] Moreover, B7 expression is not inducible by IFN-γ. *In vivo*, glioblastoma cells and monocytes infiltrating the tumor are also B7 negative, whereas monocytes at the tumor boundary remain positive.[36] This suggests that a local phenomenon may take place in glioma to downregulate the B7 expression, impeding efficient T-cell priming, or even favoring T-cell anergy. Furthermore, the B7–CD28 interactions in the CNS were recently shown to be important in generating an efficient CTL response to viral antigens.[37] Overall, these data encourage future examination as to whether B7 gene transfer can restore or elicit an appropriate immune response against glioma.

NEW STRATEGIES FOR IMMUNOTHERAPY

Vaccination Strategy

There is nowadays an increasing body of evidence that tumor cells can communicate with immune cells, with or without the help of professional APC. Such interactions are mediated not only by receptor–ligand pairing during cell–cell contact but also a plethora of cytokines. The growth of aggressive tumors such as malignant glioma suggests that there is an imbalance in the tumor–host relationship that could be the result of deficiencies in many possible components in the response (TABLE 1). It has been proposed that immunizations using either tumor antigens themselves or genetically modified tumor cells as immunogens could tilt the balance in favor of an efficient immune response. It has now been shown in *in vivo* models that such immunogens could induce not only a local immune response at the site of injection, but also systemic effects.[38,39] In the particular case of glioma, these vaccination approaches must take into account the accessibility of effector cells to their target in the CNS. It is now clear that there is significant trafficking of activated T cells through the CNS[40,41] and that T cells primed by glioma cells in the periphery can recirculate and reach the brain to mediate their anti-tumor effects.[42–45]

TABLE 1. Possible Mechanisms of Immune-Escape by Glioma

Tumor antigens are not presented to immune cells
Lack of tumor antigen
Lack of MHC class I or II molecules
Lack of TAP molecules
Lack of professional APC in the local environment of glioma
Tumor antigens are presented, but do not elicit an immune response
Tumor peptide density below the threshold level required for T-cell activation
Onco-fetal antigen for which immune tolerance has been established
Costimulatory signal absent or with inhibitory effect
Appearance of new epitopes that are tolerogenic rather than immunogenic
Inefficacy of immune response
Limited accessibility of immune cells to the CNS
Abnormalities in intracellular signal transduction of T cells
Inappropriate immune stimulation (e.g., solely T helper and antibody response)
Immunosuppressive microenvironment
Soluble factors (TGF-β2, IL-10, PGE2, IL-1R antagonist)
Cell–cell contact (FasL)

Vaccines Using Tumor Antigens

The prototype for this approach is melanoma, where the identification of a series of melanoma-associated antigens allowed the investigation of the immunogenic properties of these antigens. Several types of vaccines are currently under evaluation: (i) peptides alone or in conjunction with immune-boosting adjuvants; (ii) recombinant proteins; (iii) dendritic cells pulsed with tumor peptides or injected with tumor lysates; (iv) naked DNAs encoding tumor antigens. For glioma, the absence of well-characterized tumor antigen prevents most immunizations of this kind. The identification of glioma antigens is thus an important objective, not only for direct therapeutic purposes, but also to aid our understanding of the dialogue between glioma cells and immune cells.

Vaccines Using Genetically Modified Tumor Cells

Recent experimental data demonstrate that the immunogenicity of malignancies can be enhanced by transfecting tumor cells with diverse cytokine genes (e.g., IL-2, IL-4, IL-7, IL-12, GM-CSF, IFN-γ) or with genes encoding surface molecules (e.g., B7) acting as costimulatory molecules for T-cell activation.[38] This type of immunization is an attractive strategy to treat glioma, since it does not depend on the identification of tumor antigens. Specific examples are discussed according to the presumed area of activity for each proposed therapy.

Inhibition of Local Immunosuppression (TGF-β2 Antisense)

Recent *in vivo* experiments confirm the critical immunosuppressive effect of TGF-β2 (see above) and suggest that its inhibition may be a useful objective in glioma treatment.[46] Using TGF-β2 antisense to inhibit the TGF-β2 expression of rat 9L gliosarcoma, it was shown that transfected cells, when inoculated subcutaneously, were highly immunogenic and induced the eradication of an established wild-type (wt) tumor. Unfortunately, as in most studies using fast-growing experimentally induced tumors, the wt tumor was established only a short time (5 days) before immunization with transduced tumor cells. However, the data did convincingly demonstrate that a memory response giving partial protection (tested at 12 weeks) had been induced. Enhanced cytotoxic effector function was observed in these animals, although its specificity or *in vivo* role was not thoroughly investigated.

Enhancement of Lymphocyte Responsiveness (Interleukin-2, Interleukin-7, Interleukin-12)

The immunogenicity of IL-2-transfected tumor cells has been extensively examined (e.g., Refs. 47 and 48). In most *in vivo* assays performed with neoplasms of different origin, IL-2-producing tumor cells induced a T-cell-mediated, specific immune response leading to protection against wt tumor and/or its regression. Based on these experimental results, numerous human gene therapy programs using IL-2-transduced tumor cells are under way. In glioma, this approach with tumor cells genetically modified to produce IL-2 has yielded mixed results with an immune response generally insufficient to mediate tumor rejection or to prevent tumor development,[49] except for a slight anti-tumor effect obtained in an allogeneic model.[50] IL-2 transfection was even found to be deleterious when combined with TGF-β2 antisense.[46] Such results could be due to the decreased expression of high-affinity IL-2R on glioma lymphocytes.[19] An alternative explanation could be the level of IL-2 production by the transduced clones. Indeed, it was reported in a model of melanoma that the most successful immunizations were achieved with vaccines producing intermediate IL-2 levels, whereas vaccines producing low or high levels of IL-2 were ineffective in protecting against parental tumor challenge.[51]

Interleukin-7 (IL-7) is a 152 amino acid glycoprotein that exhibits several immunostimulatory properties including the induction of activated T-cell proliferation, enhancement of CTL- and LAK-cell-mediated cytotoxicity, and increase of perforin mRNA expression in the CD8$^+$ subset. The anti-tumor effects of cells genetically modified to produce IL-7 have been investigated in a murine glioma (ependymoblastoma 203) model,[42] where tumor cells were transfected with an expression vector containing murine IL-7 cDNA. IL-7-transfected glioma cells were vigorously rejected by a CD8$^+$ T-cell-mediated immune response that was proportional to the level of IL-7 production. In addition, the anti-tumor effect observed was tumor-specific, since there was no protection against other syngeneic tumors (melanoma B-16 and fibrosarcoma YM-12). These *in vivo* data suggest that IL-7 should be further investigated. In addition, some immunologic effects mediated by IL-7 render this cytokine particularly interesting in the setting of malignant glioma. Indeed, IL-7 increases IL-2R α chain expression in CD4$^+$ cells[52] and inhibits TGF-β mRNA expression and its production by murine macrophages.[53]

IL-12 is a heterodimeric protein (35-kDa and 40-kDa subunits) with pleiotropic properties. These include the promotion of both NK and cytotoxic T lymphocyte activity and growth, as well as the induction of a Th1 cytokine secretion phenotype. A promising enhancement of anti-tumor responses has been observed in various cancer models,[54] and in addition IL-12 has nonimmune properties such as anti-angiogenic effects.[55] The activity of IL-12 in glioma has yet to be reported, but it is clearly an additional attractive cytokine candidate to be tested in future studies.

Enhancement of Antigen Presentation and T-Cell Costimulation

As has been discussed, the fundamental requirement for antigen presentation to T cells is the expression of MHC molecules. IFN-γ and TNF-α are cytokines that are particularly active in augmenting both class I and class II MHC molecule expression.[1,56] Indeed, a variety of tumor cells transduced with genes encoding these factors exhibited reduced tumorigenicity *in vivo* (e.g., Refs. 57 and 58). However, given the pleiotropic nature of these cytokines, it is not clear whether the upregulation of MHC molecules is the critical mechanism leading to the anti-tumor effect. A more direct way of manipulating only the MHC expression of a tumor cell would be with the gene transfer of the MHC class II transactivator CIITA, which plays a critical role for the regulation of MHC class II genes.[59] To date, this has been only attempted *in vitro*, and its application as a single therapy may be limited by the fact that its effects are too restricted: CIITA gene transfer induced MHC class II expression on tumor cells, but not antigen-processing function.[60] Nevertheless, in combination with other genes, this may be a useful way of manipulating the immune response in a manner that can be subtly controlled and "fine-tuned." MHC Class II expression and the subsequent CD4$^+$ helper T-cell response that it may augment could be useful in the induction and expansion phases of an immune response. However, MHC class I restricted CD8$^+$ cells are generally considered to be the most potent immune effectors. Strategies to specifically enhance MHC class I expression of tumor cells have not been examined for effects on antigen presentation. However, transient transfections of mouse astrocytes with constructs containing an MHC class I promoter sequence showed enhanced surface expression of the H-2L^d molecule.[61] If the same effects are found in astrocytoma cells, this could offer a way of independently modulating MHC class I expression by glioma cells and to determine if its expression level is a limiting factor in an anti-tumor immune response.

The interest in examining whether B7 gene transfer can restore or elicit an appropriate immune response against glioma has been discussed in Section I, but to date, no data are available.

Miscellaneous Gene Transfers Exhibiting Immunological "Bystander" Effects

IGF-1 antisense: Insulin-like growth factor-1 (IGF-1) is a polypeptide crucially involved in normal growth and development. It acts as a growth factor for a variety of cells including fibroblasts, epithelial cells, smooth muscle cells, osteoclasts, and bone marrow stem cells. Its vital role in development has been highlighted by the observation that IGF-1 or IGF-1 receptor (IGF-1R) knock-out mice had profound fetal growth retardation.[62] IGF-1 can also function as a growth factor in a variety of tumors.[63] For example, neuroblastomas produce IGF-1, which promotes autocrine growth.

In glioma, the role of IGF-1 has been investigated using the rat C6 model. C6 cells express both IGF-1 and IGF-1R and are therefore an appropriate model to examine whether the inhibition of IGF-1 expression can influence tumorigenicity. Trojan *et al.* observed that C6 glioma cells transfected with IGF-1 antisense continued growing *in vitro* but had lost their tumorigenicity, apparently because of an immune response.[43] Indeed, an intense CD8$^+$ T-cell infiltrate was observed at the site of injection of transfected glioma cells. Furthermore, subcutaneous injection of IGF-1 antisense-transfected C6 cells prevented the development of tumors induced by wt C6 cells and induced the regression of pre-established wt gliomas. The anti-tumor immune response was shown to be specific for C6 cells, since co-injected neuroblastoma cells were not affected. Overall, these data suggested that the inhibition of IGF-1 expression restores or induces an adequate and specific immune response. However, several important immunological parameters, such as whether a memory response had been elicited and the absolute requirement for CD8[+] cells, have yet to be investigated.

The role of IGF-1 in rat C6 glioma growth was independently examined using an antisense RNA to IGF-1R.[44] Results similar to those reported by Trojan were obtained, suggesting that IGF-1 could act as an immunosuppressive factor. However, the activities of IGF-1 are complex, since immunostimulatory properties have also been demonstrated, such as stimulating the repopulation of the atrophied thymus in diabetic rats[64] and the stimulation of lymphopoiesis.[65] Whatever the mechanisms involved, this approach appears to be an exciting new therapeutic strategy. If antisense treatment specifically inhibits the production of IGF-1, the only obligatory condition for clinical application would be that human glioma produce IGF-1 *in vivo*, but this has yet to be fully assessed. Alternatively, it cannot be excluded that the impressive results observed in this model were due to nonspecific and unexpected effects of the antisense procedure, such as the modulation of surface molecules involved in T-cell activation. This has to be defined before applying this strategy in human therapy.

S-myc: The *S-myc* gene belongs to the *myc* family, encoding proteins that have important roles in cellular proliferation, transformation, and apoptosis. The *s-myc* protein was shown to inhibit the progression of glioma cell cycle towards the S phase and to induce apoptosis of tumor cells. In *in vivo* models (rat 9L and C6 gliomas), the co-administration of *s-myc* gene (linked to a viral promoter) with the glioma cell lines prevented the formation of tumors when injected either subcutaneously or intracerebrally. Moreover, the inhibitory effect was clearly restricted to the glioma tested, further supporting the view that a specific immune response was taking place.[45] Here also, the immune mechanisms responsible for this anti-tumor effect have to be fully characterized in order to refine and optimize any strategy based on these observations.

Tumor suppressor genes: Proteins derived from oncogenes and tumor suppressor genes participate in the regulation of cytokine expression. The transfection of wt tumor suppressor genes into glioma may therefore be accompanied by significant modification of tumor immunogenicity. For example, the wt p53 represses the expression of genes having a TATA-box in their promoter region, such as TGF-$\beta 2$.[66,67] In contrast, the mutant p53 protein upregulates the secretion of TGF-$\beta 2$, and also that of vascular endothelial growth factor, an important angiogenic factor in glioma. It is likely that p53 and other genes involved in glioma pathogenesis exert numerous other effects on cytokine or surface molecule expression. Because p53 alteration is an early event in glioma pathogenesis, gene therapy of glioma using wt p53 transfection is currently being investigated. This single gene modification might induce a series of positive events (immunoenhancement and anti-angiogenesis) that may synergize to produce a therapeutic effect.

INTRATUMORAL DELIVERY OF CYTOKINES USING GENE THERAPY

The strict localization of glioma in the CNS, with only exceptional metastases, may be particularly appropriate for approaches employing direct intratumoral delivery. Genetically modified cells can be implanted directly in the tumor (or the tumor bed after surgical removal) in order to sustain a local secretion of cytokines in the tumor microenvironment. This has the advantage of more closely mimicking the natural biology of cytokine action (paracrine effect) than does the intravenous administration of recombinant cytokines. Cell lines producing a viral vector containing the gene of interest can also be placed intratumorally, leading to the *in vivo* infection of tumor cells.[68] However, the locoregional gene therapy for glioma has major limitations. First, genetic material is injected in a site where glioma-associated immunosuppression is probably the strongest; and second, only a small percentage of cells are actually transfected *in vivo* with current gene delivery techniques. This latter problem is a major impediment for a tumor that has infiltrated normal brain structures at an early stage of development.

In spite of these theoretical disadvantages, these approaches are being experimentally investigated in glioma, particularly with interleukin-4 (IL-4). This cytokine, mainly derived from CD4[+] T cells, exhibits pleiotropic functions, including the increase of T-cell proliferation and cytotoxicity, and enhancement of eosinophil and B-cell proliferation and differentiation. IL-4 also exerts direct antiproliferative effects *in vitro* against diverse tumor cell lines originating from lung cancer, gastric cancer, renal cell carcinoma, or multiple myeloma.[69] *In vivo*, the anti-tumor effects induced by IL-4-transfected cells were also observed in nude mice, suggesting a T-cell-independent mechanism.[69] In glioma, an anti-tumor response was also reported, using a nude mice model where a human glioma cell line was co-injected (subcutaneously or intracerebrally) with a transfected IL-4-producing plasmocytoma cell line.[70] This induced a strong recruitment of eosinophils with subsequent growth inhibition of the tumor. However, a direct antiproliferative effect mediated by IL-4 cannot be totally excluded, since no eosinophil depletion was performed. The non-T-cell-dependent nature of the anti-tumor effect induced by IL-4 could be an advantage in glioma patients showing several T-cell abnormalities, although whether this benefit will be counteracted by poor memory T-cell induction remains to be clarified and cannot be assessed until studies are performed in immunocompetent animals.

Killing Glioma Cells by Induction of Fas-Mediated Apoptosis

Recently, a new, exciting way to treat malignant glioma emerged from the discovery that these tumors often express Fas (CD95), whereas normal cells in CNS do not.[71] Fas is a transmembrane glycoprotein belonging to the nerve growth factor/ TNF receptor superfamily: Like TNF-R1 (TNF-receptor 1), it can transduce an apopoptic signal through its cytoplasmic domain. Apoptosis is triggered by the binding of Fas to its natural ligand (FasL) or by cross-linking with anti-Fas antibodies. A high proportion of human glioma cell lines are sensitive to apoptosis mediated by anti-Fas antibody *in vitro*. Others are resistant, but may be rendered sensitive after stable transfection of a human Fas cDNA.[72] These results offer new possibilities to treat glioma with anti-Fas antibodies or soluble FasL. One possible drawback of such an approach is that other Fas-positive cells may be affected. Thus, infiltrating leukocyte activity may be reduced, restricting strategies relying on simultaneous immunoenhancement.

CONCLUSION

Cumulative data suggest that a limited immune response against malignant glioma has taken place *in vivo*, although this is clearly insufficient in the case of tumor growth. The thorough characterization of this spontaneous response occurring in the CNS will assist the subsequent design of rational glioma therapies. The pre-existing elements of the response can be targeted for reinforcement, while complementary components that may be lacking can be induced.

Finally, it must be emphasized that the growth of a tumor results from numerous factors, such as multiple genetic events, neoangiogenesis, and escape from immune control. The current experience in cancer treatment shows that several components in this complex process should be targeted to provide maximal chances of cure. Therefore, therapies based on immunoenhancement should be combined with other approaches such as those designed to inhibit neoangiogenesis, to restore the control of cell cycle or to induce apoptosis. Such an integrated treatment strategy combining nonantagonistic, low-toxicity procedures that attack different components of tumor growth present an opportunity to improve the disastrous prognosis of patients bearing malignant glioma.

ACKNOWLEDGMENT

We would like to thank Melissa Morawitz for her help in preparing the manuscript and the figures.

[**Note added in proof:** A new mechanism of immunosuppression has recently been proposed to play a role in tumor immune escape: tumor cell expression of Fas ligand (FasL) (discussed in Ref. 74). FasL is a membrane protein of the TNF family that exerts biological function by cross-linking its receptor Fas, a member of the tumor necrosis factor receptor (TNFR)/nerve growth factor receptor (NGFR) family. The role of FasL was initially elucidated in the immune system, where it is used as a cytotoxic effector molecule by T cells to induce apoptosis in Fas-expressing targets and as a means of limiting the pool size of activated T cells after antigen elimination.[75] More recently, FasL was also shown to be expressed by non-lymphoid tissues such as the eye and testis, where the immune-privileged status of these organs may be maintained by the elimination of infiltrating Fas[+] leukocytes.[76,77]

Concerning glioma, we have now demonstrated the *in vivo* expression of Fas ligand (FasL) by human glioma, and the efficient killing of Fas-bearing cells by glioma lines (human and rodent) *in vitro* and by human tumor cells *ex vivo*.[78] In addition, CD4[+] and CD8[+] T-cell lines derived from tumor-infiltrating lymphocytes were killed *in vitro* by autologous glioma cells.[74] FasL expression is not restricted to glioma and was also reported in large granular lymphocytic leukemia of T or NK origin,[79] colon carcinoma,[80] melanoma,[81] and hepatocellular carcinoma.[82] Further experiments to resolve the *in vivo* significance of these findings are necessary, since FasL expression may act synergistically with other immunosuppressive factors to limit the potential success of immunoenhancing therapies.

In addition, Fas/FasL interactions may occur between glioma cells in the absence of immunocytes, since there is the potential for both Fas and FasL expression in the same tumor. This raises the possibility of autocrine suicide or "fratricide," a factor to be considered in therapeutic strategies designed to exploit the Fas death pathway to directly reduce tumor growth. To optimize glioma therapies (and indeed those destined for many other cancers), it will thus be important to understand how the components of the Fas pathway are regulated not only in normal immune homeostasis, but also in pathological situations.]

REFERENCES

1. FABRY, Z., C. S. RAINE & M. N. HART. 1994. Immunol. Today **15:** 218–224.
2. DAVIS, E. J., T. D. FOSTER & W. E. THOMAS. 1994. Brain Res. Bull. **34:** 73–78.
3. KRIVIT, W., J. H. SUNG, E. G. SHAPIRO & L. LOCKMAN. 1995. Cell Transplant. **4:** 385–392.
4. CSERR, H. F. & P. M. KNOPF. 1992. Immunol. Today **13:** 507–512.
5. AUSTYN, J. M. & C. LARSEN. 1990. Transplantation **49:** 1–7.
6. LAMPSON, L. A. & W. F. HICKEY. 1986. J. Immunol. **136:** 4054–4062.
7. DHIB-JALBUT, S., C. V. KUFTA, M. FLERLAGE, N. SHIMOJO & H. F. MCFARLAND. 1990. J. Neuroimmunol. **29:** 203–211.
8. WEBER, F., E. MEINL, F. ALOISI, C. NEVINNY-STICKEL, E. ALBERT, H. WEKERLE & R. HOHLFELD. 1994. Brain **117:** 59–69.
9. DAUBENER, W., S. S. ZENNATI, P. WERNET, T. BILZER, H. G. FISCHER & U. HADDING. 1992. J. Neuroimmunol. **41:** 21–28.
10. RIMOLDI, D., P. ROMERO & S. CARREL. 1993. Int. J. Cancer **54:** 527–528.
11. VAN DER BRUGGEN, P., C. TRAVERSARI, P. CHOMEZ, C. LURQUIN, E. DE PLAEN, B. VAN DEN EYNDE, A. KNUTH & T. BOON. 1991. Science **254:** 1643–1647.
12. DE SMET, C., C. LURQUIN, P. VAN DER BRUGGEN, E. DE PLAEN, F. BRASSEUR & T. BOON. 1994. Immunogenetics **39:** 121–129.
13. DE SMET, C., S. J. COURTOIS, I. FARAONI, C. LURQUIN, J. P. SZIKORA, O. DE BACKER & T. BOON. 1995. Immunogenetics **42:** 282–290.
14. HOUBIERS, J. G., H. W. NIJMAN, S. H. VAN DER BURG, J. W. DRIJFHOUT, P. KENEMANS, C. J. VAN DE VELDE, A. BRAND, F. MOMBURG, W. M. KAST & C. J. MELIEF. 1993. Eur. J. Immunol. **23:** 2072–2077.
15. NOGUCHI, Y., Y. T. CHEN & L. J. OLD. 1994. Proc. Natl. Acad. Sci. USA **91:** 3171–3175.
16. LIBERMANN, T. A., H. R. NUSBAUM, N. RAZON, R. KRIS, I. LAX, H. SOREQ, N. WHITTLE, M. D. WATERFIELD, A. ULLRICH & J. SCHLESSINGER. 1985. Nature **313:** 144–147.
17. SAHIN, U., O. TURECI, H. SCHMITT, B. COCHLOVIUS, T. JOHANNES, R. SCHMITS, F. STENNER, G. LUO, I. SCHOBERT & M. PFREUNDSCHUH. 1995. Proc. Natl. Acad. Sci. USA **92:** 11810–11813.
18. ROSSI, M. L., J. T. HUGHES, M. M. ESIRI, H. B. COAKHAM & D. BROWNELL. 1987. Acta Neuropathol. **74:** 269–277.
19. ROSZMAN, T., L. ELLIOTT & W. BROOKS. 1991. Immunol. Today **12:** 370–374.
20. FONTANA, A., H. HENGARTNER, N. DE TRIBOLET & E. WEBER. 1984. J. Immunol. **132:** 1837–1844.
21. HUBER, D., J. PHILIPP & A. FONTANA. 1992. J. Immunol. **148:** 277–284.
22. JACHIMCZAK, P., U. BOGDAHN, J. SCHNEIDER, C. BEHL, J. MEIXENSBERGER, R. APFEL, R. DORRIES, K. H. SCHLINGENSIEPEN & W. BRYSCH. 1993. J. Neurosurg. **78:** 944–951.
23. SAWAMURA, Y., A. C. DISERENS & N. DE TRIBOLET. 1990. J. Neurooncol. **9:** 125–130.
24. TADA, M., A. C. DISERENS, I. DESBAILLETS, R. JAUFEERALLY, M. F. HAMOU & N. DE TRIBOLET. 1994. J. Neuroimmunol. **50:** 187–194.
25. HISHII, M., T. NITTA, H. ISHIDA, M. EBATO, A. KUROSU, H. YAGITA, K. SATO & K. OKUMURA. 1995. Neurosurgery **37:** 1160–1167.
26. MOSMANN, T. R. 1994. Adv. Immunol. **56:** 1–26.
27. BERMAN, R. M., T. SUZUKI, H. TAHARA, P. D. ROBBINS, S. K. NARULA & M. T. LOTZE. 1996. J. Immunol. **157:** 231–238.
28. NITTA, T., M. HISHII, K. SATO & K. OKUMURA. 1994. Brain Res. **649:** 122–128.
29. MERLO, A., A. JURETIC, M. ZUBER, L. FILGUEIRA, U. LUSCHER, V. CAETANO, J. ULRICH, O. GRATZL, M. HEBERER & G. SPAGNOLI. 1993. Eur. J. Cancer **29A:** 2118–2125.
30. PANNETIER, C., J. EVEN & P. KOURILSKY. 1995. Immunol. Today **16:** 176–181.
31. MACKENSEN, A., G. CARCELAIN, S. VIEL, M. C. RAYNAL, H. MICHALAKI, F. TRIEBEL, J. BOSQ & T. HERCEND. 1994. J. Clin. Invest. **93:** 1397–1402.
32. CAIGNARD, A., M. GUILLARD, C. GAUDIN, B. ESCUDIER, F. TRIEBEL & P. Y. DIETRICH. 1996. Int. J. Cancer **66:** 564–570.
33. EBATO, M., T. NITTA, H. YAGITA, K. SATO & K. OKUMURA. 1993. Immunol. Lett. **39:** 53–64.

34. JUNE, C. H., J. A. BLUESTONE, L. M. NADLER & C. B. THOMPSON. 1994. Immunol. Today **15:** 321–331.
35. LEACH, D. R., M. F. KRUMMEL & J. P. ALLISON. 1996. Science **271:** 1734–1736.
36. TADA, M., A. C. DISERENS, M. F. HAMOU, R. JAUFEERALLY, E. VAN MEIR & N. DE TRIBOLET. 1996. Brain Tumor Res. Ther. 327–337.
37. KUENDIG, T. M., A. SHAHINIAN, K. KAWAI, H. W. MITTRUECKER, E. SEBZDA, M. F. BACHMANN, T. W. MAK & P. S. OHASHI. 1996. Immunity **5:** 41–52.
38. PARDOLL, D. M. 1995. Annu. Rev. Immunol. **13:** 399–415.
39. ZITVOGEL, L., J. I. MAYORDOMO, T. TJANDRAWAN, A. B. DE LEO, M. R. CLARKE, M. T. LOTZE & W. J. STORKUS. 1996. J. Exp. Med. **183:** 87–97.
40. OWENS, T., T. RENNO, V. TAUPIN & M. KRAKOWSKI. 1994. Immunol. Today **15:** 566–571.
41. HAFLER, D. A. & H. L. WEINER. 1987. Ann. Neurol. **22:** 89–93.
42. AOKI, T., K. TASHIRO, S. MIYATAKE, T. KINASHI, T. NAKANO, Y. ODA, H. KIKUCHI & T. HONJO. 1992. Proc. Natl. Acad. Sci. USA **89:** 3850–3854.
43. TROJAN, J., T. R. JOHNSON, S. D. RUDIN, J. ILAN, M. L. TYKOCINSKI & ILAN. 1993. Science **259:** 94–97.
44. RESNICOFF, M., C. SELL, M. RUBINI, D. COPPOLA, D. AMBROSE, R. BASERGA & R. RUBIN. 1994. Cancer Res. **54:** 2218–2222.
45. ASAI, A., Y. MIYAGI, H. HASHIMOTO, S. H. LEE, K. MISHIMA, A. SUGIYAMA, H. TANAKA, T. MOCHIZUKI, T. YASUDA & Y. KUCHINO. 1994. Cell Growth Diff. **5:** 1153–1158.
46. FAKHRAI, H., O. DORIGO, D. L. SHAWLER, H. LIN, D. MERCOLA, K. L. BLACK, I. ROYSTON & R. E. SOBOL. 1996. Proc. Natl. Acad. Sci. USA **93:** 2909–2914.
47. FEARON, E. R., D. M. PARDOLL, T. ITAYA, P. GOLUMBEK, H. I. LEVITSKY, J. W. SIMONS, H. KARASUYAMA, B. VOGELSTEIN & P. FROST. 1990. Cell **60:** 397–403.
48. GANSBACHER, B., K. ZIER, B. DANIELS, K. CRONIN, R. BANNERJI & E. GILBOA. 1990. J. Exp. Med. **172:** 1217–1224.
49. RAM, Z., S. WALBRIDGE, J. D. HEISS, K. W. CULVER, R. M. BLAESE & E. H. OLDFIELD. 1994. J. Neurosurg. **80:** 535–540.
50. LICHTOR, T., R. P. GLICK, T. S. KIM, R. HAND & E. P. COHEN. 1995. J. Neurosurg. **83:** 1038–1044.
51. SCHMIDT, W., T. SCHWEIGHOFFER, E. HERBST, G. MAASS, M. BERGER, F. SCHILCHER, G. SCHAFFNER & M. L. BIRNSTIEL. 1995. Proc. Natl. Acad. Sci. USA **92:** 4711–4714.
52. ARMITAGE, R. J., A. E. NAMEN, H. M. SASSENFELD & K. H. GRABSTEIN. 1990. J. Immunol. **144:** 938–941.
53. DUBINETT, S. M., M. HUANG, S. DHANANI, J. WANG & T. BEROIZA. 1993. J. Immunol. **151:** 6670–6680.
54. TAHARA, H. & M. T. LOTZE. 1995. Gene Ther. **2:** 96–106.
55. VOEST, E. E., B. M. KENYON, M. S. O'REILLY, G. TRUITT, R. J. D'AMATO & J. FOLKMAN. 1995. J. Natl. Cancer Inst. **87:** 581–586.
56. FABRY, Z., M. M. WALDSCHMIDT, D. HENDRICKSON, J. KEINER, L. LOVE-HOMAN, F. TAKEI & M. N. HART. 1992. J. Neuroimmunol. **36:** 1–11.
57. GANSBACHER, B., R. BANNERJI, B. DANIELS, K. ZIER, K. CRONIN & E. GILBOA. 1990. Cancer Res. **50:** 7820–7825.
58. TJUVAJEV, J., B. GANSBACHER, R. DESAI, B. BEATTIE, M. KAPLITT, C. MATEI, J. KOUTCHER, E. GILBOA & R. BLASBERG. 1995. Cancer Res. **55:** 1902–1910.
59. MACH, B. 1995. N. Engl. J. Med. **332:** 120–122.
60. SIEGRIST, C. A., E. MARTINEZ-SORIA, I. KERN & B. MACH. 1995. J. Exp. Med. **182:** 1793–1799.
61. MASSA, P. T., K. OZATO & D. E. MCFARLIN. 1993. Glia **8:** 201–207.
62. LIU, J. P., J. BAKER, A. A. PERKINS, E. J. ROBERTSON & A. EFSTRATIADIS. 1993. Cell **75:** 59–72.
63. LE ROITH, D., R. BASERGA, L. HELMAN & C. T. ROBERTS, JR. 1995. Ann. Intern. Med. **122:** 54–59.
64. BINZ, K., P. JOLLER, P. FROESCH, H. BINZ, J. ZAPF & E. R. FROESCH. 1990. Proc. Natl. Acad. Sci. USA **87:** 3690–3694.
65. CLARK, R., J. STRASSER, S. MCCABE, K. ROBBINS & P. JARDIEU. 1993. J. Clin. Invest. **92:** 540–548.

66. MACK, D. H., J. VARTIKAR, J. M. PIPAS & L. A. LAIMINS. 1993. Nature **363:** 281–283.
67. MCCARTNEY-FRANCIS, N. L. & S. M. WAHL. 1994. J. Leuk. Biol. **55:** 401–409.
68. WEI, M. X., T. TAMIYA, R. K. HURFORD, JR., E. J. BOVIATSIS, R. I. TEPPER & E. A. CHIOCCA. 1995. Hum. Gene Ther. **6:** 437–443.
69. TEPPER, R. I. 1993. Res. Immunol. **144:** 633–637.
70. YU, J. S., M. X. WEI, E. A. CHIOCCA, R. L. MARTUZA & R. I. TEPPER. 1993. Cancer Res. **53:** 3125–3128.
71. WELLER, M., K. FREI, P. GROSCURTH, P. H. KRAMMER, Y. YONEKAWA & A. FONTANA. 1994. J. Clin. Invest. **94:** 954–964.
72. WELLER, M., U. MALIPIERO, A. RENSING-EHL, P. J. BARR & A. FONTANA. 1995. Cancer Res. **55:** 2936–2944.
73. SUH, W. K., M. F. COHEN-DOYLE, K. FRUH, K. WANG, P. A. PETERSON & D. B. WILLIAMS. 1994. Science **264:** 1322–1326.
74. WALKER, P. R., P. SAAS & P.-Y. DIETRICH. 1997. J. Immunol. In press.
75. LYNCH, D. H., F. RAMSDELL & M. R. ALDERSON. 1995. Immunol. Today **16:** 569–574.
76. GRIFFITH, T. S., T. BRUNNER, S. M. FLETCHER, D. R. GREEN & T. A. FERGUSON. 1995. Science **270:** 1189–1192.
77. BELLGRAU, D., D. GOLD, H. SELAWRY, J. MOORE, A. FRANZUSOFF & R. C. DUKE. 1995. Nature **377:** 630–632.
78. SAAS, P., P. R. WALKER, M. HAHNE, A.-L. QUIQUEREZ, V. SCHNURIGER, G. PERRIN, L. FRENCH, E. G. VAN MEIR, N. DE TRIBOLET, J. TSCHOPP & P. Y. DIETRICH. 1997. J. Clin. Invest. **99:** 1173–1178.
79. TANAKA, M., T. SUDA, K. HAZE, N. NAKAMURA, K. SATO, F. KIMURA, K. MOTOYOSHI, M. MIZUKI, S. TAGAWA, S. OHGA, K. HATAKE, A. H. DRUMMOND & S. NAGATA. 1996. Nat. Med. **2:** 317–322.
80. O'CONNELL, J., G. C. O'SULLIVAN, J. K. COLLINS & F. SHANAHAN. 1996. J. Exp. Med. **184:** 1075–1082.
81. HAHNE, M., D. RIMOLDI, M. SCHRÖTER, P. ROMERO, M. SCHREIER, L. FRENCH, P. SCHNEIDER, T. BORNAND, A. FONTANA, D. LIENARD, J.-C. CEROTTINI & J. TSCHOPP. 1996. Science **274:** 1363–1366.
82. STRAND, S., W. J. HOFMANN, H. HUG, M. MÜLLER, G. OTTO, D. STRAND, S. M. MARIANI, W. STREMMEL, P. H. KRAMMER & P. R. GALLE. 1996. Nat. Med. **2:** 1361–1366.

Supra and Infra Tentorial Gliomas in Children

D. A. BRUCE

Department of Neurosurgery
University of Texas Southwestern Health Sciences Center
Columbia Children's Hospital at
Medical City Dallas
Children's Medical Center of Dallas
1935 Motor Street
Dallax, Texas 75235-7794

Gliomas are the most common pediatric brain tumor, with frequency varying between 30 and 50% of tumors: 1.7/100,000 out of a total of 3.8/100,000 every year (33%). From 1974 to 1985 the frequency of astrocytoma at the Children's Hospital of Philadelphia was 49% of all brain tumors: 66% were benign, 14% malignant, and 21% brain stem. Combining the brain stem with the malignant lesions gives a 35% incidence of malignantly behaving tumors. Mixed gliomas, oligodendrogliomas, gangliogliomas, and xanthoastrocytomas account for about 10% of brain tumors.[1] This article will focus on astrocytomas of the brain.

Pathological classification has veered away from the Kernohan classification into four grades, and now tumors are low grade or malignant. The former includes grades I and II, and the latter, grades III and IV.[2] Pilocytic, fibrillary, mixed gliomas, and gangliogliomas account for the low-grade tumors, and anaplastic and glioblastoma account for the malignant grades. Excluding brain stem tumors, malignant lesions account for 14 to 20% of gliomas in children. With the advent of MRI scanning, there is now a tendency to group astrocytomas as diffuse and focal, with the former being more likely to be fibrillary and the latter pilocytic. This is probably truer in the brain stem than elsewhere.[3] For the malignant tumors, there is a definite correlation with length of survival and pathology, with the anaplastic lesions doing much better than the glioblastoma multiforme (GBM). (See TABLE 1.)

In addition to pathologic diagnosis, the location of the tumor is an important factor for therapy and outcome. Cerebellar astrocytomas account for one-third of posterior fossa tumors. These may be cystic or solid and are located in the vermis or the cerebellar hemispheres. With current imaging and surgical techniques, both of these varieties of tumor seem to do equally well. The aim of surgery is total resection. If the tumor is cystic and the cyst wall enhances, then resection of the cyst wall must be performed, whereas if there is no wall enhancement only the nodule must be removed. If there is significant, >1.5 cm, tumor left on the postoperative scan, a second operation is performed for complete resection if that portion of tumor was left accidentally. If the tumor was left because of its location, as a surgical decision, then repeat surgery makes no sense. Complete resection or minimal residual tumor requires no additional therapy and can be followed for evidence of regrowth. Radiation therapy, especially in children under 5 years is not usually recommended, although long-term survival in benign astrocytomas following radiation therapy occurs.[4,5] Significant residual tumor is treated either by observation and re-operation if the tumor grows; with chemotherapy, usually carboplatinum, vincristine, and VP 16; or in the older children with radiation. Usually, we follow

TABLE 1. Incidence of Pediatric Astrocytomas

	Ages (years)			
	0–15	0–4	5–9	10–14
Astrocytoma	2.1	0.7	0.6	0.8
Pilocytic	1.5	0.5	0.6	0.4
Anaplastic	0.4	0.1	0.1	0.2
Mixed and others	1.0	0.4	0.3	0.3
Total	5.0	1.7	1.6	1.7
Incidence/100,000	1.7	0.6	0.5	0.6
Total all tumors	11.5	4.1	3.7	3.7
Incidence	3.8	1.4	1.2	1.2
Relative percent astrocytoma	33%	34%	32%	32%

the children with scans and no adjuvant therapy until growth occurs. The use of focused beam radiation is being studied, but significant results are not yet available. The decision about additional therapy should be an individual one, based on the age of the child and size, pathology, and location of residual tumor. The rate of long-term tumor-free survival with total resection is 90 to 95%, and this is also true for minimal residual disease. Astrocytomas that occur in the superior vermis or in the hemisphere with extension into the subarachnoid space and around the vessels are most often fibrillary tumors and have a higher incidence of recurrence, often requiring radiation therapy, chemotherapy, or both despite benign histology.

Brain stem gliomas must be subdivided both by location and by imaging. Those diffuse tumors of the pons, while often having benign histology on biopsy, behave like malignant tumors, and mean survival is around 10 months with all modalities of therapy. Biopsy is rarely recommended unless there is a therapeutic option. Biopsy is difficult with a high incidence of no definitive pathology. Most benign biopsy tumors still behave in a malignant fashion with early death.[6,7] Most of the diffuse pontine tumors are fibrillary astrocytomas that may or may not be malignant.[8,9] Therapy, including escalating doses of hyperfractionated radiation, although showing excellent early responses on MRI, has not resulted in prolonged survival, and 80% of children die within 2 years. The addition of chemotherapy, cisplatinum, taxol, and bone marrow transplantation have all failed to demonstrate significant benefit.[10–13] New strategies are required for this tumor.

Focal tumors occur in the medulla, tectum, and occasionally in the pons. The later are usually exophytic into the fourth ventricle or cystic in the pons. Recent studies have suggested that local, cystic, enhancing lesions on MRI are likely to be pilocytic astrocytomas and lend themselves to surgical resection, whereas the fibrillary astrocytomas do not enhance unless they are malignant. Although this may generally be true, I have seen small cystic enhancing lesions at the obex that were GBMs. Surgery and attempted resection of medullary and exophytic pontine tumors are the best current therapy, since often no further adjunctive therapy is required. Great care is required to avoid producing new and significant neurological deficits, and the use of intraoperative ultrasound, the cavitron, operating microscope, and experience are required to get good results. Tectal gliomas are often very indolent and can often be observed as long as the only problem is hydrocephalus. The treatment of the hydrocephalus is either by shunt or by third ventriculostomy. These tumors may take many years before growing and requiring therapy. Surgical resection may be adequate, but if tumor regrowth occurs radiation therapy is the next line of treatment. All of these pilocytic tumors can be successfully treated and a cure can be expected in most cases.

OPTIC PATHWAY GLIOMAS

Optic Nerve Lesions

Although reports in the past have suggested that pure optic nerve gliomas are hamartomatous and rarely grow, current experience demonstrates a different picture, and progressive proptosis and visual loss do occur with these lesions. If vision is better than 20/200 and proptosis is not so bad that corneal ulceration is occurring, local radiation therapy is the best therapy. This will often shrink the tumor and preserve vision. Chemotherapy with carboplatinum, vincristine, and/or VP16 has shown some response in the purely intraorbital tumors and can be tried in the younger children. When vision is less than 20/200, the best treatment is surgical resection of the tumor, taking a portion of the intracranial nerve, if the MRI scan shows no chiasmatic involvement.

Optic Chiasm and Hypothalamus

The pathology of these lesions is usually pilocytic astrocytoma, but the more posterior lesions in the hypothalamus may be malignant, and spongioblastoma polare can also be found. If the lesion involves the nerves and chiasm with preserved vision, biopsy is usually performed to be certain of the diagnosis; however, if there is no safe area from which to take a biopsy, treatment can be begun without biopsy. Radical surgery is rarely appropriate in this area unless there is a clear exophytic portion or vision is lost. In the younger children the first line of adjuvant therapy is chemotherapy. The drugs currently being used are carboplatinum, vincristine, and VP16. This tumor and those in the hypothalamus have responded well to chemotherapy, with a better than 50% partial response rate and 5-year relapse-free survival of 80%.[17,18] Radiation therapy, although resulting in 50% 10-year survival, is associated with many complications: cognitive dysfunction, pituitary insufficiency, Moya-Moya disease, and second tumors. Thus, this is reserved for older patients, 9 years and up, who have failed surgery and chemotherapy.

Tumors that are exophytic in the third ventricle are often amenable to extensive surgical resection. The dangers are damage to the septal area and or fornices with memory loss, supra optic injury with diabetes insipidus, damage to the middle cerebral or thalamic perforating vessels with resultant stroke, and injury to the hypothalamus with loss of temperature control and either coma or prolonged lethargy. Malignant tumors require chemotherapy and/or radiation dependent on age.

HEMISPHERIC GLIOMAS

The majority of hemispheric gliomas in children are benign, and the ideal therapy is complete surgical excision with no adjutant therapy. If there are small pieces of residual tumor, careful follow-up with repeat surgery if growth occurs is the best therapy. If there is significant residual tumor, repeat surgery to complete the resection should be done unless the tumor was left because of surgical difficulties and further resection is believed to be impossible without producing unacceptable neurological damage. Depending on the age of the child, observation or chemotherapy should be used.[19-21] Radiation therapy can be of benefit in this setting and is still of value in the older children.[4,22] Deep-seated tumors in the thalamus are usually just biopsied and radiated. Radiation has a definite role in prolonging survival in these deeper lesions.[4] It is now possible to do quite extensive resections of these deep tumors with minimal morbidity, and residual tumor is treated as it would be for the more superficial lesions.

Anaplastic Glioma and Glioblastoma

In children older than 3 years, the degree of surgical resection has been shown to correlate with improved outcome especially in the anaplastic lesions. In younger children no such correlation has been shown.[23-25] Chemotherapy is of proven value in the treatment of malignant gliomas, and when combined with radiation therapy the five-year event-free survival is 46% versus 18% for those treated with radiation therapy alone. In children under 2 years with 8 in 1 therapy, progression-free survival (PFS) at three years is 36% and survival 51%, not dissimilar to the results in the older children when radiation therapy was also used. As in the older children, those with anaplastic astrocytomas did considerably better than those with glioblastoma—PFS of 44% vs. 0%. The older children with glioblastoma who received radiation therapy had a 16% long-term survival rate, obviously better than the infants, and this could be the result of the radiation therapy.

Recurrence in most of these lesions is local, and attempts are under way to prevent local recurrence. Chemotherapy delivered by slow-release wafers has achieved some early success, as has the use of Hypercin and fluorescence.[26]

Biological response modifiers have not yet found a place in the treatment of gliomas. There is no information on the use of active immunotherapy. Adaptive immunotherapy has been tried, with some success reported with the use of IL-2.[27] Passive immunotherapy with monoclonal antibodies has had its best success via intrathecal injection for meningeal tumor.[28] Restorative or nonspecific immunotherapy has been shown to have beneficial effects on tumor shrinkage (β-interferon,[29] γ-interferon,[30] and Imuvert[31]). Combinations of these therapies with radiation and chemotherapy and with each other are being studied and are likely to be routes for the future.

GENE THERAPY

New studies are under way to find the effectivenesss of the thymidine kinase blockade by ganciclovir via a retrovirus vector. Laboratory studies *in vitro* and *in vivo* have shown that it is possible using a cell cycle–blocking gene strategy, MAD-MAX complex P21 and P27, to inhibit glioma growth. These genes are delivered using an adenovirus vector.[32,33] In the near future there will be other trials of gene therapy in brain tumors. Gamma knife and stereotaxic radiation therapy are likely to decrease the radiation toxicity. Strategies to block tumor resistance to drugs and to examine the value of such drugs as toremifene, an angiogenesis factor blocker, in combination with radiation sensitizers, are being explored. At the same time local delivery systems for drugs or virus vector into the tumors are being examined.

CONCLUSIONS

The current best therapy for low-grade gliomas in the brain is surgical resection wherever possible. Residual tumor may be followed by MRI, repeat surgery, or treated by chemotherapy. Radiation is usually saved for recurrent or deeply seated tumors but is effective in prolonging survival. Even deep thalamic and hypothalamic tumors can often have extensive resection with minimal new neurological problems. Implanted radiation has some role in treating recurrent disease.

Malignant gliomas also should be as completely resected as possible. Chemotherapy improves survival over radiation alone, and it is unclear how much additional benefit is obtained from the radiation therapy. In children under two, radiation is rarely used. Children with anaplastic gliomas have a close to 50% 3- to 5-year survival rate, whereas infants with glioblastomas have essentially 0% long-term survival. In older children who are treated with surgery, radiation therapy, and chemotherapy, there is a 16% long-term survival with glioblastoma.

Newer therapies with monoclonal antibodies, immune modulation, and gene therapy via virus vectors; newer drugs; and more focused radiation are all still in their infancy but show promise as potential adjuvant agents.

REFERENCES

1. POLLOCK, I. F. 1994. Brain tumors in children. N. Engl. J. Med. **331:** 1500–1507.
2. RORKE, L. B., F. M. GILLES, R. L. DAVIS & L. E. BECKER. 1985. Revision of the World Health Organization classification of brain tumors for childhood brain tumors. Cancer **56:** 1869–1886.
3. FISCHBEIN, N. J., M. D. PRADOS, W. WARA, et al. 1996. Radiological classification of brain stem tumors: Correlation of magnetic resonance imaging appearance with clinical outcome. Pediatr. Neurosurg. **24:** 24–29.
4. WEST, C. G. H., R. GATTAMANENI & V. BLAIR. 1995. Radiotherapy in the treatment of low grade astrocytomas I. A survival analysis. Child. Nerv. Syst. **11:** 438–442.
5. ALVORD, E. C., JR. & S. LOFTON. 1988. Gliomas of the optic nerve and chiasm: Outcome by patient's age, tumor site and treatment. J. Neurosurg. **68:** 85–98.
6. ALBRIGHT, A. L., A. N. GUTHKELCH, R. J. PACKER, et al. 1986. Prognostic factors in brain-stem gliomas. J. Neurosurg. **65:** 751–755.
7. GRANEL, M. T. J., G. BENZ-BOHM, R. SCHRODER, et al. 1995. Prognostic factors in brain stem gliomas: A retrospective analysis in 79 children. Med. Pediatr. Oncol. **24:** 234–243.
8. MANTRAVADI, R. V. P., R. PHATAK, S. BELLUR, et al. 1982. Brain stem gliomas: An autopsy study of 25 cases. Cancer **49:** 1294–1296.
9. BURGER, P. C. 1996. Pathology of brain stem astrocytomas. Pediatr. Neurosurg. **24:** 35–40.
10. KAPLAN, A. M., A. L. ALBRIGHT, R. A. ZIMMERMAN, et al. 1996. Brainstem gliomas in children. Pediatr. Neurosurg. **24:** 185–192.
11. FREEMAN, C. R. 1996. Hyperfractionated radiotherapy for diffuse intrinsic brain tumors in children. Pediatr. Neurosurg. **24:** 103–110.
12. FREEMAN, C. R., J. P. KRISCHNER, R. A. SANFORD, et al. 1993. Final results of a study of escalating doses of hyperfractionated radiotherapy in brain stem tumors in children: A Pediatric Oncology Group Study. Int. J. Radiat. Oncol. Biol. Phys. **27:** 197–206.
13. PACKER, R. J., J. M. BOYETT, R. A. ZIMMERMAN, et al. 1994. Outcome of children with brain stem gliomas after treatment with 7,800 cGy of hyperfractionated radiotherapy. A Children's Cancer group Phase I/II Trial. Cancer **74:** 1827–1834.
14. FULTON, D. S., V. A. LEVIN, W. M. WARA, et al. 1981. Chemotherapy of pediatric brain-stem tumors. J. Neurosurg. **54:** 721–725.
15. PACKER, R. J., H. S. NICHOLSON, L. G. VEZINA, et al. 1992. Brain-stem gliomas. Pediatr. Neurooncol. **4:** 863–879.
16. DUNKEL, I., J. GARVIN, S. GOLDMAN, et al. 1994. High dose chemotherapy with autologous

      bone marrow rescue does not cure children with brain-stem tumors. Pediatr. Neurosurg. **21:** 219–225.

17. PACKER, R. J., L. N. SUTTON, L. T. BILANIUK, *et al.* 1988. Treatment of chiasmatic/ hypothalamic gliomas of childhood with chemotherapy: an updated report. Ann. Neurol. **23:** 79–85.

18. RODRIGUEZ, L. A., M. S. B. EDWARDS & V. A. LEVIN. 1990. Management of hypothalamic gliomas in children: An analysis of 33 cases. Neurosurgery **26:** 242–247.

19. NORTH, C. A., R. B. NORTH, J. A. EPSTEIN, *et al.* 1990. Low-grade astrocytomas: Survival and quality of life after radiation therapy. Cancer **66:** 6–14.

20. PACKER, R. J., B. LANGE & J. ATER. 1993. Carboplatin and vincristine for recurrent and newly diagnosed low-grade gliomas of childhood. J. Clin. Oncol. **11:** 850–856.

21. ALLEN, J. & J. SIFFERT. 1996. Contemporary chemotherapy issues for children with brainstem gliomas. Pediatr. Neurosurg. **24:** 98–102.

22. LAWS, E. R., JR., W. F. TAYLOR & M. B. CLIFTON. 1984. Neurosurgical management of low grade astrocytomas of the cerebral hemispheres. J. Neurosurg. **61:** 665–673.

23. SPOSTO, R., I. J. ERTEL, R. D. JENKIN, *et al.* 1989. The effectiveness of chemotherapy for treatment of high grade astrocytomas in children: Results of a randomized trial; a report from the Children's Cancer Study Group. J. Neuro. Oncol. **7:** 165–177.

24. GEYER, J. R., J. L. FINLAY, J. M. BOYETT, *et al.* 1995. Survival of infants with malignant astrocytomas. Cancer **75:** 1045–1050.

25. DUFFNER, P. K., M. E. HOROWITZ, J. P. KRISCHER, *et al.* 1993. Post-operative chemotherapy and delayed radiation in children less than 3 years of age with malignant brain tumors. N. Engl. J. Med. **328:** 1725–1731.

26. AGOSTINIS, P., A. VANDENBOGAERDE, A. DONELLA-DEANA, *et al.* 1995. Photosensitized inhibition of growth factor regulated protein kinases by Hypercin. Biochem. Pharmacol. **49:** 1615–1622.

27. PACKER, R. J., E. D. KRAMMER & J. A. RYAN. 1991. Biological and immune modulating agents in the treatment of childhood brain tumors. Neurol. Clin. **9:** 405–422.

28. LASHFORD, L., R. P. MOSLEY, J. C. BENJAMIN, *et al.* 1989. Antibody targeted irradiation for leptomeningeal neoplasm: Current status. J. Neuro. Oncol. **7:** 517–519.

29. PACKER, R. J., J. C. ALLEN, W. A. BLEYER, *et al.* 1989. Efficacy and toxicity of recombinant beta-interferon (IFN-Bser) in childhood brain tumors. Neurology **39:** 264–268.

30. MAHALEY, M. S., M. B. URSO, F. WHALEY, *et al.* 1985. Immunobiology of primary intracranial tumors Part 10: Therapeutic efficacy of interferon in recurrent gliomas. J. Neurosurg. **63:** 719–725.

31. JAECKLE, K. A., A. MITTLEMAN & F. HILL. 1990. Phase II trial of imuvert in recurrent malignant astrocytomas. J. Clin. Oncol. **8:** 1408–1418.

32. CHEN, J., T. WILLINGHAM, L. R. MARGRAF, *et al.* 1995. Effects of the Myc oncogene antagonist, MAD, on proliferation, cell cycling and the malignant phenotype of human brain tumors. Nature Med. **1:** 638–643.

33. CHEN, J., T. WILLINGHAM, M. SHUFORD, *et al.* 1996. Effects of ectopic overexpression of p21$^{WAF1/CIP1}$ on aneuploidy and the malignant phenotype of human brain tumor cells. Oncogene **13:** 1395–1403.

ADDITIONAL READING

PACKER, R. J., E. D. KRAMER & J. A. RYAN. 1991. Biological and immune modulating agents in the treatment of childhood brain tumors. Neurol. Clin. **9:** 405–422.

HANAUSKE, A. R. 1993. New approaches to chemotherapy of brain tumors with and without radiation. Crit. Rev. Neurosurg. **3:** 45–49.

ALLEN, J., R. PACKER, A. BLEYER, *et al.* 1991. Recombinant interferon beta: A phase I–II trial in children with recurrent brain tumors. J. Clin. Oncol. **9:** 783–788.

FOLKMAN, J. & D. E. INGBER. 1992. Inhibition of angiogenesis. Sem. Oncol. Biol. **3:** 89–96.

REDEKOP, G. J. & C. C. G. NAUS. 1995. Transfection with bFGF sense and antisense cDNA resulting in modification of malignant glioma growth. Neurosurgery **82:** 83–90.

TISHLER, R. B., C. R. P. B. GEARD, E. J. HALL & P. B. SCHIFF. 1992. Taxol sensitizes human astrocytoma cells to radiation. Cancer Res. **52:** 3495–3497.

ALLEN, J. C. & J. SIFFERT. 1996. Contemporary chemotherapy issues for children with brainstem gliomas. Pediatr. Neurosurg. **24:** 98–102.

LEFKOWITZ, I. B., R. J. PACKER, L. N. SUTTON, *et al.* 1988. Results of treatment of children with recurrent gliomas with lomustine and vincristine. Cancer **61:** 896–902.

PACKER, R. J., P. J. SAVINO, L. T. BILANIUK, *et al.* 1983. Chiasmatic gliomas of childhood. A reappraisal of natural history and effectiveness of cranial radiation. Child's Brain **10:** 393–403.

WISOFF, J. H. 1992. Management of optic pathway tumors of childhood. Pediatr. Neurooncol. Neurosurg. Clin. N. Am. **3:** 791–802.

Brain Tumors during the First Year of Life

LÁSZLÓ BOGNÁR

Pediatric Department
Hungarian National Institute of Neurosurgery
1145, Amerikai út 57
Budapest, Hungary

INTRODUCTION

Brain tumors are among the most common early childhood neoplasms, second only to neuroblastomas according to Farwell *et al.*[1] The current literature holds little information concerning brain tumors in newborns and infants. It has been reported that brain tumors in neonates and infants occur more frequently in the supratentorial compartment than in the infratentorial compartment.[2–4] Others observed that this supratentorial–infratentorial ratio was reversed during the first year of life.[3,5,6] Childhood tumors tend to be located along the neural axis according to Koos and Miller[3] and Matson.[7] Jooma *et al.*[2] found 69% of the lesions along the midline. Brain tumors during the first year of life are clinically silent because of the lack of symptoms and verbal complaints. The elasticity of the infant's skull and the brain's capacity for functional adaptation can delay the manifestation of clinical signs. Furthermore, the vulnerability of the immature brain to raised intracranial pressure can result in severe retardation even after successful surgical treatment.

With the advent of computed tomography (CT), magnetic resonance, and ultrasonographic (US) imaging, the early diagnosis of intracranial lesions is more easily accomplished, thus allowing earlier surgical intervention. The purpose of this study is to compare the brain tumors of neonates, infants, and children and to discuss the preoperative evaluation and current therapy.

MATERIALS AND METHOD

During the 40-year period from 1954 to 1995, 1728 infants and children with intracranial tumors were treated in the Pediatric Department of the Hungarian National Institute of Neurosurgery in Budapest. Since the introduction of CT, MRI, and US diagnostic imaging during the period from 1984 to 1995, the total number of children treated for brain tumors was 723, out of which 51 were diagnosed under one year of age, compared with only 32 patients diagnosed under one year of age from 1954 to 1984. We consider the pool of 51 patients as the statistical basis for our presentation, and all the percentages are derived from them.

In this study we define neonates to be under 2 months of age, infants to be 2 to 12 months of age, and children to be 1 to 16 years of age. The records of all the pediatric patients with a clinical or histological diagnosis of brain tumor were reviewed. In our analysis the average percentage of tumors discovered for neonates was approximately 6%, which is comparable to the rate found for infants. The low percentage observed in the adolescent age group is explained by referral differences,

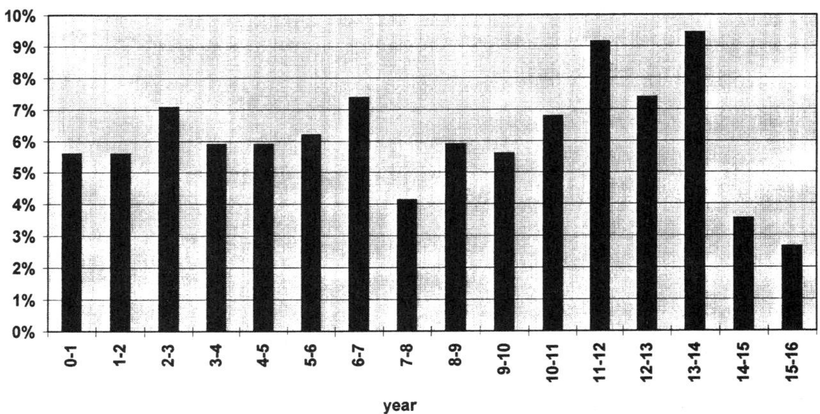

FIGURE 1. Age distribution of children (1984–1995; patient number 723).

because in the past we admitted children into our department up to 14 years of age (FIG. 1).

The sex distribution was nearly equal with a slight male predominance of 58%. This was more marked in the infant group, in which 63% were males. The little group of five neonates is statistically insignificant.

TUMOR LOCATION AND CLINICAL SIGNS

The proportion of infratentorial and supratentorial tumors was the same for both infants and children. The tumor location, axial or hemispheric, was also equally divided; the infants had slightly more hemispheric lesions than axial. In children the most frequently seen clinical signs were vomiting, 23%; headache, 24%; gait disturbances, 18%; vision problems, 18%; and vertigo, 12%. In infants the clinical signs were vomiting, 23%; macrocephaly, 18%; disordered eye movements, 16%; and delayed development, 10%. In all five neonates we found the same signs: vomiting, head tilting, seizures, and developmental arrests.

HISTOLOGY

The histological distribution of malignant to benign tumors is higher in neonates (100%) and infants (53%) than in children (43%). FIGURE 2 shows the most common histologic tumor types in order of frequency. For children, benign gliomas, malignant gliomas, and medulloblastomas were the most frequent histological diagnoses; whereas in infants malignant gliomas, benign gliomas, and medulloblastomas were the most frequent.

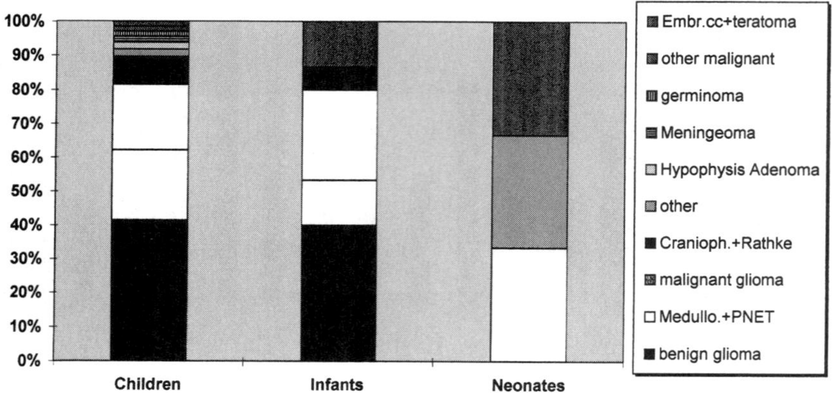

FIGURE 2. Distribution of histological result.

TREATMENT AND RESULTS

Children in the series received various modes of treatments, as depicted in FIGURE 3. Craniotomies were performed in 85% of the children, whereas the remaining 15% of the cases were considered inoperable (FIG. 4). Shunt implantation as a palliative surgical solution was implemented in 21% of the inoperable group and 12% of the directly operated group in children. Twenty percent of the infants needed shunt implantation, and 33% of the neonates required shunts (FIG. 5). In the craniotomy group, the surgeon indicated that the tumor was radically excised in 70%, subtotally excised in 22%, and partially excised in 5% of the cases; whereas a biopsy only had been performed in 3% of the children. In the infant group, a

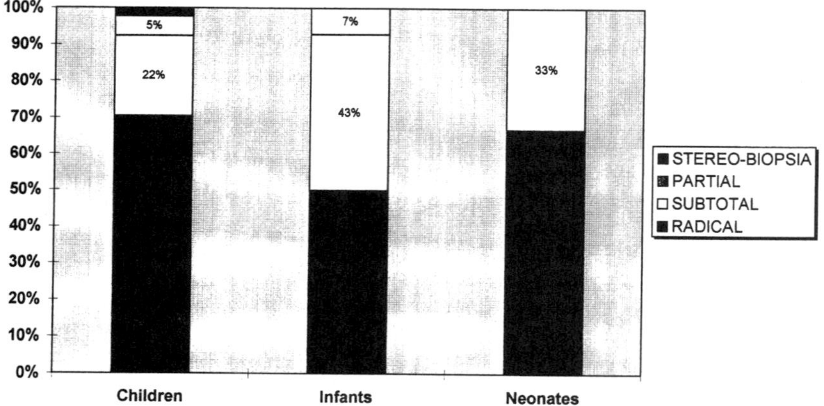

FIGURE 3. Surgical therapy.

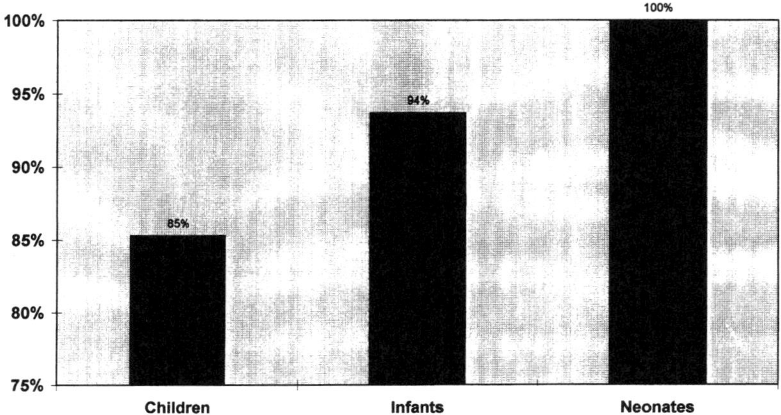

FIGURE 4. Proportion of patients with tumor surgery.

direct surgical procedure was performed in 94% of the cases with radical resection of the tumor in 50%, subtotal resection in 43%, and partial resection in 7% of the patients. All five neonates had undergone craniotomy, with radical surgery in three and subtotal in two cases. Among the directly operated patients, shunt implantation was needed for 12% of the children, 20% of the infants, and 33% of the neonates.

The surgical mortality rate, defined as occurring within the first two months post-operatively, was 5% for the children and 13% for the infant group. All five neonates survived the surgical procedure (FIG. 6). The surgical mortality rate varied depending on tumor location. In children it was 10% for axial and 6% for hemispherical tumors, whereas in infants it was 14% and 22%, respectively (FIG. 7).

Radiation therapy was given in 29% of the children and in 7% of the infants. The tumor dose of radiation therapy varied from 3000 to 5400 rad while spinal and

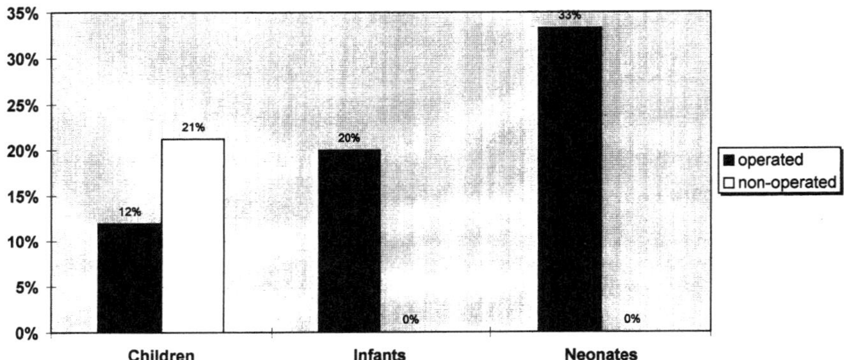

FIGURE 5. Proportion of shunted patients in operated and nonoperated groups.

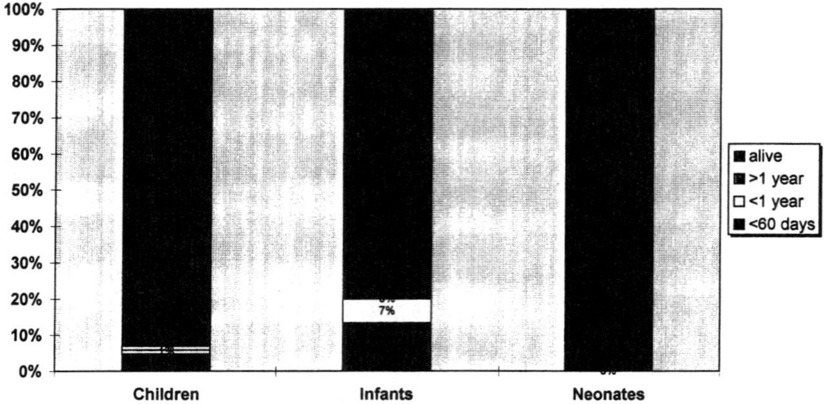

FIGURE 6. Surgical outcome.

whole-brain dosage ranged from 2500 to 3200 rad. Radio- and chemotherapy was administered to 11% of the children. Chemotherapy was used as the single post-operative treatment in 4% of the children, 20% of the infants, and 33% of the neonates (FIG. 8).

DISCUSSION

We found the incidence of pediatric brain tumors during the first year of life to be approximately 6%, which agrees with the 7–11% average Gjerris[8] reported in 1976. In the last decade the average number of admitted children with brain

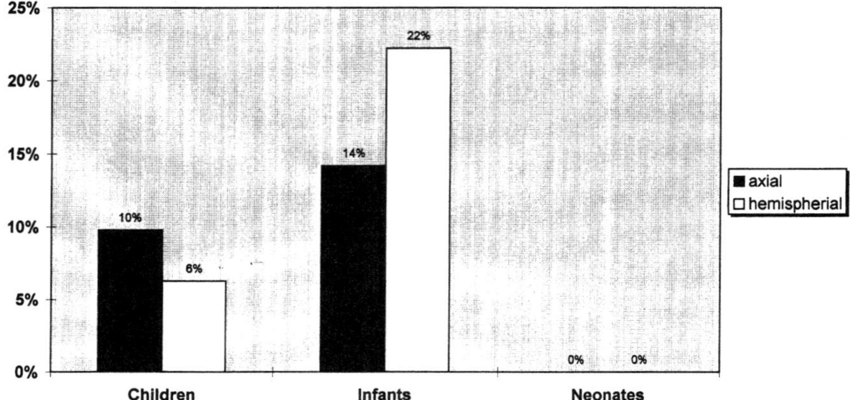

FIGURE 7. Mortality rate depending on tumor location.

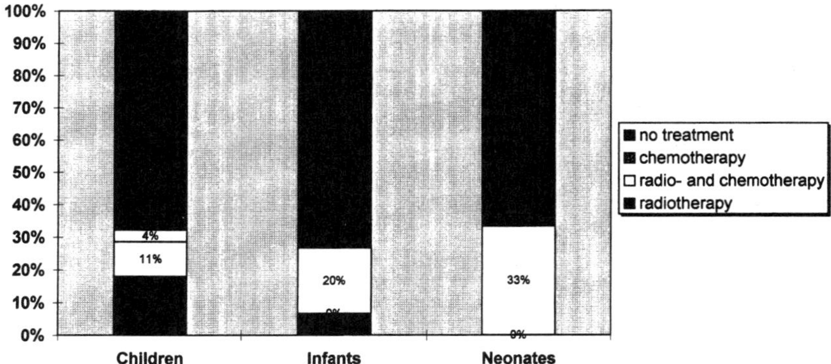

FIGURE 8. Proportion of patients with postoperative treatment.

tumors in our department has almost doubled to 66 patients per year compared to 34 patients per year in the pre-imaging period of 1954–1983, clearly demonstrating the importance of CT and MRI imaging for early diagnosis. The average number of tumors discovered under one year of age is four times greater now than during the pre-imaging period. Currently, we know that brain tumors are not uncommon in infants.

We found an equal distribution between axial-hemispheric and also between supratentorial–infratentorial tumor localization. Thirty years ago, Matson[7] reported that 40% of the infants with brain tumors were suitable candidates for operation. The percentage of operable infants has increased since that time. In 1984 Jooma et al.[2] published his series in which 68% of the infants underwent surgery. In our center 85% of the children and 94% of the infants had undergone surgical intervention, which is consistent with the common trend of increasing operability as reported by Lapras et al.[9] who stated that the best treatment of infantile brain tumor is radical surgery.

Patients with malignant or unresectable tumors require radiotherapy for tumor control, but one should be aware of the serious adverse effect. Clinical observation and experimental data suggest that the immature brain is much more susceptible to irradiation. Intellectual retardation, delayed hypopituitarism, growth retardation, and occlusive neurovascular complications are reported.[2,11–19]

In our study radiotherapy was used for infants with malignant tumors until 1990. We consider 3–5 years as the lower limit for radiotherapy, even though Radcliffe et al.[10] documented a mean IQ loss of 27 points in children younger than 7 years after whole-brain RT. For the younger children, chemotherapy is the only adjuvant post-operative treatment, but its efficacy is still controversial. The long-term effects are still under investigation.

SUMMARY

Of 1728 childhood brain tumors treated at the National Institute of Neurosurgery, Budapest during the years from 1954 until 1995, 83 of the affected children were younger than one year of age. Because of the advent of the CT and MRI scans in the last 11 years, 51 out of the 83 are presented, these being patients treated since these technological advances have been available. There was a male predominance, with 30 boys and 21 girls. Five of the 51 infants were diagnosed before two months of age. The ratio of supratentorial to infratentorial tumors was almost 1:1. Vomiting, alteration of psychomotor development, and macrocrania were the most common presenting features. Craniotomy and tumor debulking was performed in 85% of the children and 94% of the infants. The most frequent histological diagnosis was benign glioma, PNET, malignant glioma, and craniopharyngioma. The surgical mortality rate was 5% for the children and 13% for the infant group. All five neonates survived the surgical procedure. Radiation therapy was given in 29% of the children and in 7% of infants.

REFERENCES

1. FARWELL, J. R., G. J. DOHRMANN, FLANNERY. 1977. Central nervous systems tumors in children. Cancer **40:** 3123.
2. JOOMA, R., R. D. HYWARD & D. N. GRANT. 1984. Intracranial neoplasms during the first year of life: Analysis of one hundred consecutive cases. Neurosurgery **14:** 31–41.
3. KOOS, W. T. M. & H. MILLER. 1971. Intracranial Tumors of Infants and Children. Georg Thieme Verlag. Stuttgart.
4. SUNDER-PLASSMAN, M. & K. JELLINGER. 1971. Neuroektodermale Hirngeschwulste im ersten lebensjahr. Acta Neurochir. (Wien) **24:** 107–120.
5. FESSARD, C. 1968. Cerebral tumors in infancy: 66 clinicoanatomical case studies. Am. J. Dis. Child. **115:** 302.
6. HARWOOD-NASH, D. C. & C. R. FITZ. 1976. Neuroradiology in Infants and Children. Mosby. St. Louis. pp. 668–788.
7. MATSON, D. D. 1969. Neurosurgery of Infancy and Childhood. 2nd edit. Charles C Thomas. Springfield, Illinois.
8. GJERRIS, F. 1976. Clinical aspects and long-term prognosis of intracranial tumours in infancy and childhood. Dev. Med. Child. Neurol. **18:** 145–159.
9. LAPRAS, C., J. N. GUILBURD, J. GUYOTAT & J. D. PATET. 1988. Brain tumors in infants: A study of 76 patients operated upon. Child. Nerv. Syst. **4:** 100–103.
10. RADCLIFFE, J., R. J. PACKER, T. E. ATKINS, G. R. BUNIN, L. SCHUT, J. W. GOLDWEIN & L. N. SUTTON. 1992. Three- and four-year cognitive outcome in children with noncortical brain tumors treated with whole-brain radiotherapy. Ann. Neurol. **32:** 551–554.
11. CLEMENTE, C. D., J. N. YAMAZAKI, L. R. BENNETT & R. A. McFALL. 1960. Brain radiation in newborn rats and differential effects of increased age: II. Microscopic observations. Neurology (NY) **10:** 669–675.
12. DEUTSCH, M. 1982. Radiotherapy for primary brain tumors in very young children. Cancer **50:** 2785–2789.
13. HIRSCH, J. F., D. RENIER, P. CZERNICHOW, L. BENVENISTE & A. PIERRE-KAHN. 1979. Medulloblastoma in childhood: Survival and functional results. Acta Neurochir. (Wien) **48:** 1–14.
14. SHALET, S. M. 1982. Growth and hormonal status of children treated for brain tumors. Child. Brain **9:** 284–293.
15. RICHARDS, G. E., W. M. WARA, M. M. GRUMBACH, S. L. KAPLAN, G. E. SHELINE & F. A. CONTE. 1976. Delayed onset of hypopituitarism: Sequelae of therapeutic irradiation of central nervous system, exe, and middle ear tumors. J. Pediatr. **89:** 553–559.
16. PARK, T. S., H. J. HOFFMAN, E. G. HENDRICK, R. P. HUMPHREYS & L. E. BECKER. 1983. Medulloblastoma: Clinical presentation and management. Experience at the Hospital for Sick Children, Toronto, 1950–1980. N. Neurosurg. **58:** 543–552.
17. RAIMONDI, A. J. & T. TOMIAT. 1979. The disadvantages of prophylactic whole CNS postoperative radiation therapy for medulloblastoma. *In* Multidisciplinary Aspects of Brain Tumor Therapy. P. Paoletti, W. B. Butti & R. Knerich, Eds.: 209–218. Elsevier/North Holland Biomedical Press. Amsterdam.

18. NEUHAUSER, E. B. D., M. H. WITTENBORG, C. Z. NERMAN & J. COHEN. 1939. Irradiation effects of roentgen therapy on the growing spine. Radiology **59:** 637–650.
19. GREEN, O. C. 1983. Endocrinology: Hormonal involvement in children with brain tumors. *In* Brain Tumors in the Young. L. Amador, Ed.: 785–821. Charles C Thomas. Springfield, IL.

Adjuvant Chemotherapy of Pediatric Brain Tumors

T. S. VATS

Texas Tech University Health Sciences Center
1400 Wallace Boulevard
Amarillo, Texas 79106

INTRODUCTION

Brain tumors account for 20% of all pediatric malignancies and are the most common solid tumors seen in children.[1,2] Nearly two-thirds of brain tumors are infratentorial in children older than one year of age. In infants and children less than one year of age, the supratentorial tumors are as common as infratentorial ones.[2] There are various histological subtypes, and their relative frequencies are listed in TABLE 1.

The major treatments for brain tumors are surgery, radiation therapy (RT), and recently chemotherapy for certain tumors. Surgery remains the front-line treatment for low-grade astrocytomas of the cerebellum and cerebral hemispheres. Total resection of these tumors is associated with a high rate of cure. Total resection in some of the other brain tumors, such as medulloblastoma, also has led to greatly improved survival.[1,2]

Radiation therapy has played a major role in control and eradication of pediatric brain tumors.[3] All brain tumors, though, are not equally sensitive to radiation therapy. Large doses of radiation have been associated with intellectual compromise, growth retardation, and various other endocrine abnormalities. The intellectual compromise is most evident in children less than three years of age. These children have to be placed in special education classes or require supplemental educational services.[1]

Chemotherapy was initially used to treat recurrent tumors. Good responses were seen in medulloblastomas, germinomas, and anaplastic astrocytomas.[1,2] Another advantage of chemotherapy may be its ability to shrink very large tumors, thereby providing a chance for second-look surgery. In children less than three years of age, chemotherapy could potentially be used to delay the onset of radiation therapy, thereby minimizing its toxicity in these young children. Some chemotherapy agents can also act as radiation sensitizers and thus enhance the effect of radiation.[1,2] This potential role of chemotherapy in enhancing cures and reducing radiation-associated neuropsychological toxicity prompted investigators to further explore its potential in an adjuvant fashion with surgery and radiation therapy in treating brain tumors.

BARRIERS TO CHEMOTHERAPY RESPONSE

Chemotherapy response in brain tumors is limited by structural constraints to drug delivery, the most important of which is the blood–brain barrier (BBB).[1,4] The anatomical basis of the BBB is the presence of tight junctions between the capillary endothelial cells. Drugs that have high lipid solubility, weak protein bind-

TABLE 1. CNS Tumors

Tumor	Incidence (%)
Supratentorial	40
Gliomas	21.5
Ependymomas	2.7
Craniopharyngioma	12.4
Infratentorial	54.9
Medulloblastoma	30.4
Cerebellar astrocytoma	7.3
Ependymoma	5.3
Brain stem glioma	11.8
Others	4.4

ing, and low ionization are able to cross the BBB and attain adequate concentration in the central nervous system (CNS).[5] Capillaries within the brain tumor itself can be normal or abnormal. The abnormalities include thickened capillary walls, fenestrations, altered basement membrane, increased pinocytic vesicles, and surface projections. Abnormal capillaries form a blood–tumor barrier.[6] Existence of this barrier leads to varied uptake of chemotherapy drugs in different parts of the tumor. In general, drug uptake is least in the central necrotic area of the tumor and the normal brain surrounding the tumor. This area has finger-like projections of tumor that escape the chemotherapy effect and thus lead to tumor recurrence. The density of capillaries is decreased in the tumor area. This leads to decreased blood flow and perfusion, leading to lower drug concentration in the tumor cells. Uptake of water-soluble drugs in the brain is mainly limited by the BBB, whereas that of lipophilic drugs is limited by decreased blood flow. White matter has decreased amounts of capillaries and decreased blood flow.[7,8] Because the brain stem is primarily composed of white matter, it might explain why brain stem tumors are not extremely responsive to chemotherapy or radiation therapy as both of these modalities require a good blood supply.

Another reason for the poor response to chemotherapy may be intrinsic development of resistance to various drugs.[9] Capillary endothelium and brain tissue possess enzymes that may inactivate the chemotherapy drugs.[10] There may be rapid repair of DNA or overexpression of a multidrug-resistant gene (MDR).[11] Possible inactivation of commonly used drugs for brain tumors, such as nitrosoureas, vincristine, epipodophyllins, and platinum compounds; and the possible modulation of their drug resistance is outlined in TABLE 2.

TREATMENT OF SPECIFIC BRAIN TUMORS

Medulloblastoma or Primitive Neuroectodermal Tumor

Medulloblastoma or primitive neuroectodermal tumor (PNET), an embryonal tumor of ectodermal origin, has a tendency to seed along the cerebrospinal axis and invade cerebrospinal fluid (CSF).[12] Standard treatment for this tumor is radiation therapy (RT). The usual dose is 3600 cGy to the whole craniospinal axis with a boost of an additional 1800 cGy to the tumor bed. Thus, the total dose delivered to the tumor bed is 5500 cGy.[13] Follow-up of these patients revealed a significant loss of intelligence.[1,14] In an effort to reduce this sequelae of radiation, the Children's

Cancer Group (CCG) and Pediatric Oncology Group (POG) conducted a study comparing standard-dose RT to reduced-dose RT (2400 cGy) to craniospinal axis. There was an unacceptably high incidence of neuraxis failure with reduced-dose radiation compared to standard RT (31% vs. 15%), and the overall survival of the low-dose radiation group was also poor when compared to standard RT (52% vs. 77%).[15]

Medulloblastomas, which present as very large tumors at the time of initial surgery, residual tumors of greater than 1.5 cm post surgery, or tumors that have seeded along the cerebrospinal axis, have the worst prognosis when treated with surgery and radiation therapy only.[4,12] Improved response to adjuvant chemotherapy in these tumors was first demonstrated by simultaneous studies carried out by the CCG[16] and the International Society of Pediatric Oncology (SIOP).[17] Chemotherapy agents used were lomustine (CCNU) and vincristine following radiation therapy of 5500 cGy to the tumor bed and 3600 cGy to the craniospinal axis. Statistically significant improved survival was seen only in high-risk tumors with adjuvant chemo-

TABLE 2. Mechanism and Modulation of Resistance of Chemotherapy Drugs in Brain Tumors

Drugs	Mechanism of Resistance	Modulation of Resistance
BCNU	• ↑ O^6—methyl guanine—DNA methyl transferase • ↑ Glutathione S-transferase • ↑ Expression of metallothionein mRNA • Impaired drug transport due to decreased uptake and increased efflux at cellular level	• Reducing levels of O^6—methyl guanine DNA methyl transferase by treatment with methylating drugs • Pretreatment with glutathione transferase inhibitors such as ethacrynic acid • Treatment with streptozotocin • Pretreatment with ornithine decarboxylase inhibitors (DFMSO), Misonidozole combination of 6 thioguanine, dibromodulcitol, or procarbazine • Hyperthermia
Platinum compounds	• Impaired transport due to alteration of cell membrane • Increased intracellular levels of sulfhydryl proteins, metallothionein and glutathione • Amplification of oncogene related to cell proliferation	• Increase intracellular concentration by intracarotid route or continuous infusion • Inhibit DNA repair by using caffeine, ARA-C, aphidicolin • Hyperthermia • Decrease glutathione by using buthionine sulfoximine
Vincristine	• Decreased binding • Multidrug resistance phenotype increased	• Verapamil
Epipodophyllotoxins	• ↑ Multidrug resistance (MDR) gene expression • Rapid DNA repair • Altered topoisomerase activity • Prior radiation therapy	• Use preradiation or with radiation therapy • MDR modulation such as verapamil, cyclosporine • Combining topoisomerase I and II together

therapy when compared to RT alone (59% vs. 50%). Survival in localized tumors was better than disseminated tumors (57% vs. 38%). In an effort to intensify chemotherapy further, the CCG used a combination of eight drugs: vincristine, hydroxyurea, prednisone, lomustine, cisplatin, procarbazine, cytosine arabinoside, and cyclophosphamide (8 in 1) given in one day to treat high-risk tumors in an adjuvant fashion. This regimen had about a 50% response.[18] When this regimen was compared to CCNU and vincristine in a randomized study, no clear advantage was seen for 8 in 1 over CCNU and vincristine. Toxicity was greater with the 8 in 1 regimen.[19] Packer *et al.* conducted an adjuvant chemotherapy trial with CCNU, vincristine, and cisplatin following a conventional dose of craniospinal RT. They reported a long-term disease-free survival of 90% at four years for localized tumors.[20] A significant reduction in IQ was seen in Packer's original study. In an attempt to reduce radiation sequelae, these investigators carried out another study with decreased radiation dose to craniospinal axis (1800 cGy) in children less than five years of age. The chemotherapy regimen was kept the same. In their pilot study, 10 children were treated. The actuarial survival was 69%. Three children relapsed, one in the primary site and two outside the primary site. Interestingly, the mean IQ of this group remained unchanged, and the overall survival was also unchanged at four years.[21]

Some investigators believe that chemotherapy given after RT may have limited access to the tumor bed due to microvascular changes and healing of the BBB following RT.[4] Thus, chemotherapy given in a preradiation setting may have better access to the tumor cells and may lead to decreased radiation toxicity. In these clinical trials, patients responding well to preradiation chemotherapy also had a better response with postradiation adjuvant chemotherapy.[22-26] A summary of all significant clinical trials in medulloblastoma is presented in TABLE 3.

TABLE 3. Chemotherapy Trials in Medulloblastoma

Author/Ref.	Chemotherapy	Patient No.	Response
Evans *et al.*[16] (CCG)	CCNU, VCR, PRED	233	59% RT + CT 50% RT only 48% vs. 0% RT + CT vs RT alone in high-risk patients
SIOP[17]	CCNU, VCR	286	45% overall 54% RT + CT 23% RT only T3 & T4—38% RT + CT 19% RT alone
Packer[20]	VCR, CCNU, Cisplatin	26	90%
Allen[22]	High-dose Cytoxan ± 1.T Ara-C Hydroxyurea	8	37%
Pendergrass[19] (CCG)	8 in 1	15	60%
Mosijczyk[26] (POG)	VCR, Cisplatin High-dose Cytoxan	32	47%
Kovnar[25] (St Jude)	Cisplatin VP-16	8	87.5%
Freeman[4]	Carboplatin, BCNU Streptorotocin	8	87.5%

Malignant Gliomas

In various chemotherapy trials, anaplastic astrocytomas and glioblastoma multiforme were grouped together. This is unwise, because response in anaplastic astrocytomas is much better than in glioblastoma multiforme.[1,4] An adjuvant chemotherapy study conducted by the CCG randomized patients to receive RT alone vs. RT plus prednisone, CCNU, and vincristine (PCV). Five-year disease-free survival was 18% with RT alone vs. 48% for the PCV in anaplastic astrocytomas. In glioblastoma multiforme, survival was 6% with RT alone and 42% with PCV.[27] In an effort to intensify treatment, CCG compared the PCV regimen with their 8 drugs in one day regimen. No difference was seen in disease-free survival in the two groups (33% vs. 36%).[28] When toxicity was compared in the two regimens, it was much greater in the 8 drugs in one day regimen. The National Cancer Institute Canada conducted an adjuvant chemotherapy trial in patients with anaplastic oligodendroglioma using intensified procarbazine, CCNU, and vincristine. A total of 24 patients were treated with an overall complete response (CR) and partial response (PR) of 65% (7CR + 10PR).[29] Currently, both the CCG and the POG are looking at different drug pairs in a preradiation setting. The CCG is looking at pairing VP16 with cyclophosphamide, ifosfamide, or carboplatin for four cycles followed by standard RT. In the POG study, VP 16 will be combined with BCNU/cisplatin or BCNU/cyclophosphamide before hyperfractionated RT.

Ependymomas

Adjuvant chemotherapy trials have failed to demonstrate any advantage over standard RT in ependymomas.[1,4] Preradiation chemotherapy used in infants yielded about a 40 to 45% overall response (CR + PR).[30] Most of the failures occurred in the primary site.

Pineal Tumors

These account for 1–2% of all childhood brain tumors. Histological subtypes include pineal parenchymal tumors and germ cell tumors.[1,31] Chemotherapy has not proven to be of any benefit for the pineal parenchymal tumors. Germ-cell tumors present as germinomas or teratomas. These tumors have the propensity to involve CSF. They behave very much like the gonadal germ cell tumors. They respond well to combinations of cisplatin or carboplatin, VP16, and bleomycin. Usual practice is to give three to six cycles of combination chemotherapy utilizing these agents followed by involved field and craniospinal irradiation.[4] Finlay *et al.*[32] treated primary CNS germ cell tumors with a combination of carboplatin, VP 16, and bleomycin. They got a complete response in all 14 patients entered in the study. These patients did not receive radiation and have done well.[32] This study seemed to indicate that radiation therapy may be omitted in patients with germ cell tumors who achieve complete response with chemotherapy. A large pool of patients and a longer follow-up, though, is needed to validate this data.

Infant Brain Tumors

Nearly 15 to 20% of brain tumors occur below the age of three years.[1,2] The most common histological types seen include medulloblastomas (23%), low-grade supratentorial astrocytomas (21%), ependymomas (18%), and high-grade supratentorial astrocytomas (11%). Overall survival in these children with surgery and RT is less than 20%.[33] Radiation is associated with severe psychological sequelae, as well as neurological deficits.[34] Preradiation chemotherapy has been successfully utilized to delay the start of RT and thus minimize neuropsychological toxicity. The most effective agents used to date include vincristine, cyclophosphamide, VP16, and cisplatin. In nearly two-thirds of the patients, RT was successfully delayed without any evidence of tumor progression.[1,4] A study conducted by Bramm *et al.* at M. D. Anderson Hospital used MOPP (nitrogen mustard, vincristine, prednisone, procarbazine) in infants with brain tumors and found an extremely good response.[35] Some of these infants attained a complete response and didn't require radiation therapy. Duffner *et al.*[30] carried out a study in 102 infants under three years of age using vincristine, cyclophosphamide, cisplatin, and VP16. They found objective responses (CR + PR) in medulloblastomas (48%), ependymomas (48%), malignant gliomas (60%), and poorly differentiated embryonal tumors (29%).[30] Incomplete removal of tumor, tumor dissemination, and location in brain stem area were poor prognostic markers. Strauss *et al.* got an 86% objective response in 8 children treated with cisplatin and VP16.[36] The CCG used the 8 in 1 regimen in infants younger than 18 months of age. Progression-free survival of 25% was seen in ependymomas and PNET tumors and 33% with malignant gliomas, excluding brain stem gliomas.[37] Interestingly, survival was identical whether or not these children received radiation therapy. In an effort to further improve these results, the CCG is embarking on a randomized trial for induction and a standard maintenance in infant brain tumors. The induction regimen would be a randomization between vincristine, VP 16, cisplatin, and cyclophosphamide (Regimen A) vincristine, VP 16, carboplatin, ifosfamide, and mesna (Regimen B). Five cycles of this induction treatment will be followed by eight cycles of maintenance using vincristine, cyclophosphamide, VP 16, and carboplatin. At the end of treatment, patients will be evaluated. Patients with no residual tumor will be followed with observation only, while the ones with residual tumor will be treated with conventional RT.

NEWER APPROACHES IN TREATMENT OF BRAIN TUMORS

Due to limited success observed with chemotherapy, newer strategies are being explored for treating high-risk tumors, especially gliomas and recurrent tumors. These approaches include the following:

- Intraarterial chemotherapy,
- Intratumor chemotherapy, and
- High-dose chemotherapy with autologous bone marrow transplantation.

The major effort in these approaches is to intensify the drug dosage.[1,38]

Intraarterial Chemotherapy

Intraarterial chemotherapy has been used in an effort to achieve high concentrations of drugs in the brain tumor tissue. The agents used have included BCNU, ACNU, cisplatin, carboplatin, VP 16, and methotrexate. Chemotherapy has been delivered by selective catheterization of anterior, middle, or posterior cerebral artery. These drugs have been given with or without disruption of blood–brain barrier using mannitol.[4] Most of the studies have been done in high-grade gliomas and recurrent tumors in adults. Some responses, both CR and PR, have been seen in malignant gliomas. Complication of intraarterial chemotherapy are many and include difficulty in canulization, eye damage, blindness, leukoencephalopathy, and various thromboembolic phenomena.[1,4] At this point, the safety and efficacy of this particular approach in children is still unproven.

Intratumor Chemotherapy

Brain tumors remain confined to the cranial cavity. Metastases outside the brain are extremely rare. This type of malignancy should be an ideal one to eradicate with multimodality treatment. This unfortunately has not been accomplished in brain tumors due to the dose-limiting toxicity of radiation therapy to normal brain tissue and the systemic toxicity of chemotherapy. If chemotherapy drugs could be delivered into the tumor tissue in brain, they would achieve higher concentrations locally, have minimal systemic toxicity from diffusion, and could be used in an adjuvant fashion with RT. Before these novel approaches can be incorporated in a clinical setting, however, safety and efficacy as well as undue toxicities need to be checked in a preclinical model. Such a model is the intracerebrally implanted 9L rat brain tumor.[39] This tumor is induced in CD Fisher rats by weekly i.v. injections of nitrosomethylurea. The tumor then is implanted intracerebrally (i.c.). The tumor displays the histological characteristics of a mixed glioblastoma multiforme and a gliosarcoma. This model was established in our laboratory and used to study the effectiveness of single-agent or combination chemotherapy with and without RT.[39] The first drug we tested was bleomycin. This drug has definite activity against 9L glioma cells *in vitro*. Its large molecular weight restricts its transport across the BBB. Even large systemic doses failed to show an appreciable effect in rats burdened with 9L tumor. However, when the drug was delivered i.c. even in much smaller doses, significant improvement in survival was seen.[40] In subsequent experiments, this drug potentiated radiation therapy and was additive in prolonging the survival when used with other chemotherapy drugs.[39] Prompted by this response, we conducted a clinical phase I trial and demonstrated that bleomycin can be safely and repetitively delivered via an ommaya reservoir into the tumor bed.[41] To this date, combination of standard or fractionated RT and direct intratumor administration of bleomycin has not been evaluated clinically.

Other drugs that we have tested included diazoquinone (AZQ), platinum compounds, BCNU, and dibromodulcital (DBD).[39] This model can thus provide an ideal setting to test intratumor administration of various chemotherapy drugs singularly or in combination plus RT for various high-grade gliomas.

High-Dose Chemotherapy and Autologous Transplantation

Standard-dose chemotherapy has failed to show any distinct advantage in malignant gliomas, brain stem tumors, and recurrent resistant tumors. In the treatment of many solid tumors, there is evidence that escalation of the dose of a particular chemotherapeutic agent may permit improvement in drug response.[38,42] Alkylating agents demonstrate a linear dose–response curve, and the intrinsic resistance to these agents can also be overcome by increasing the concentration of the drug. The ability to increase the dose of chemotherapy drugs is hampered by life-threatening myelosuppression. However, this dose-limiting toxicity can be overcome by administration of bone marrow or peripheral stem cells following chemotherapy to ensure hematopoietic reconstitution. In brain tumor patients, marrow involvement is extremely rare, so the stem cells collected are not mixed with tumor cells. Thus, this approach seems ideal to test in high-grade brain tumor patients, particularly those with malignant gliomas. BCNU at a single dose of 600 to 1400 mg/m² has shown improved response in patients who recurred after conventional dose treatment with nitrosoureas.[43] Etoposide at a dose of 600 to 800 mg/m² per day for three days (1800–2400 mg/m²) similarly demonstrated good clinical response in high-grade brain tumor.[44] Other agents that have shown improved clinical activity with intensified dosage include CCNU, carboplatin, cisplatin, thiotepa, and cyclophosphamide.[38] These drugs in recent trials have even been combined with a local RT boost to the tumor area in the brain.

To date, most of these trials in children have been directed only against malignant astrocytomas, but they have the potential for being used for other high-grade brain tumors. Finlay *et al.* treated 10 patients under the age of 21 with malignant astrocytoma. Eight of ten patients had recurrent tumor, and two patients had prior treatment with high-dose adjuvant chemotherapy. Five patients were treated with thiotepa and etoposide, and the remaining five patients received BCNU, thiotepa, and etoposide. There were four complete responses measured at day 28 following stem cell infusion. Two of these complete responses had recurrence of disease 300 days post transplant, but the remaining two patients were disease-free at more than 440 days after infusion.[45] Khalifa *et al.*[46] treated 20 children with a median age of six. The histological subtypes in this group of patients were mixed, and included medulloblastoma, ependymomas, brain stem gliomas, and primitive neuroectodermal tumor. These patients were pretreated with busulfan for four days followed by thiotepa for three days. The overall response rate in this study was 26%, including three partial responses.[46] Heideman *et al.* treated 13 children with malignant glioma, including eleven newly diagnosed patients and two patients with recurrent disease. Most patients had significant residual tumor. These patients received thiotepa and cyclophosphamide as their preconditioning regimen. Following marrow infusion, these patients also received radiation therapy to the tumor bed. The response rate in this study was 31%. There was one complete response, two partial responses, and seven stable disease. Overall median survival was 14 months, and progression-free survival was nine months.[47] Significant toxicity was hematopoietic and neurological; the major neurological toxicity was drowsiness. In all three series, there was only one toxic death due to hemorrhage. Current studies are in progress both by the POG and the CCG. The POG is looking at cyclophosphamide and melphalan as the conditioning regimen and the CCG is looking at thiotepa, BCNU, and etoposide for preconditioning and radiation therapy to tumor bed before transplantation.

High-dose chemotherapy with bone marrow transplantation is an approach with considerable morbidity, and when deciding whether this is a reasonable treatment option, the toxicity of such an approach must be taken into consideration. A major toxicity to monitor with this procedure would be neurological, primarily the toxic encephalopathy, which is a late complication, and intracranial hemorrhage. With improved supportive care, the toxicity of this procedure is continually decreasing. Utilization of the peripheral blood progenitor cells decreases the need for a painful bone marrow harvest and is associated with hastened engraftment of neutrophils and platelets when compared with the use of autologous bone marrow. Hematopoietic growth factors can also decrease the transplant-related mortality and morbidity. With decrease in morbidity, this procedure may have a place in treatment of patients with primary progressive disease, recurrent disease, or failure of front-line therapy.

Future Directions in Treatment of Brain Tumors

Chemotherapy, along with surgery and radiation therapy, has now become an important part of treatment for brain tumors. Medulloblastomas and germinomas have shown tremendous improvement with the use of adjuvant chemotherapy. With other tumors, such as gliomas and brain stem tumors, new approaches to chemotherapy need to be looked at. These newer approaches may include high-dose chemotherapy with stem cell or bone marrow rescue; modulation of the intrinsic resistance of chemotherapy drugs; and innovative methods of drug delivery, such as binding drug to liposomes, degradable wafers, or other carriers such as dehydro-pyridine, intraarterial therapy, or intratumor therapy. Intratumor therapy holds a great potential because it would deliver a high concentration of anticancer drug to the tumor cells with much smaller than conventional intravenous dosage. This approach will keep systemic toxicity at its minimum and may be looked at in an adjuvant fashion with radiation therapy once the toxicity and efficacy of this combination can be defined. Some other novel approaches to treatment of brain tumor include gene therapy and antisense technology. Antisense technology is a technique in which single-strand RNA or DNA interacts with complementary nucleic acid sequences to block a known biological effect. These single-stranded RNA or DNA sequences may act as downregulators of chemoresistant genes such as multidrug resistance gene or the metallothionine gene and thus enhance chemosensitivity of the tumor. Newer approaches need to be developed where there is sparing of normal tissue from the harmful effect of the treatment itself. This would make the future treatments more curative while also providing improved quality of life.

REFERENCES

1. HEIDEMAN, R. L. 1993. Tumors of the central nervous system. *In* Principles and Practice of Pediatric Oncology. P. A. Pizzo & D. G. Poplack, Eds.: 633–681. J. B. Lippincott Publishers. Philadelphia, PA.
2. COHEN, M. E. & P. K. DUFFNER. 1984. Trends in treatment, referral patterns and survival statistics of children with brain tumors in the United States. *In* Brain Tumors in Children: Principles of Diagnosis and Treatment. M. E. Cohen & P. K. Duffner, Eds.: 349–352. Raven Press. New York.
3. BLOOM, H. J. G. 1982. Intracranial tumors: Response and resistance to therapeutic endeavors, 1970–1980. Int. J. Radiat. Oncol. Biol. Phys. **8:** 1083–1113.
4. FREEMAN, A. I. 1994. Chemotherapy of pediatric brain tumor. *In* Brain Tumors, a Comprehensive Text. R. A. Morantz & J. W. Walsh, Eds.: 745–762. Marcel Dekker, Inc. Publishers. New York.
5. RALL, D. P., *et al.* 1962. Extracellular space of brain as determined by diffusion of insulin from the ventricular system. Life Sci. **2:** 43.
6. BLASBERG, R. G., *et al.* 1986. Chemotherapy of brain tumors: Physiological and pharmacokinetic considerations. Sem. Oncol. **13:** 70.
7. SAKURADA, O., *et al.* 1978. Measurement of local cerebral blood flow with iodo[^{14}C]antipyrine. Am. J. Physiol. **234:** H59.
8. LEVIN, V. A. 1986. Pharmacokinetics and CNS chemotherapy. *In* Fundamentals of Cancer Chemotherapy. K. Hellman & S. K. Carter, Eds.: 28–40. McGraw-Hill Publishers. New York.
9. FEUN, L. C., *et al.* 1994. Drug resistance in brain tumors. J. Neuro. Oncol. **20:** 165–176.
10. SCANLON, K. J., *et al.* 1989. Molecular basis of cisplatin resistance in human carcinomas, model systems. Anticancer Res. **9:** 1301–1312.
11. MELTZER, P. S., *et al.* 1989. Drug resistance in multiple myeloma and non-Hodgkin's lymphoma: Detection of *p*-glycoprotein and potential circumvention by addition of verapamil to chemotherapy. J. Clin. Oncol. **7:** 415–424.

12. KOPELSON, G., *et al.* 1983. Medulloblastoma: The identification of prognostic subgroups and implications for multimodal management. Cancer **51:** 312–319.

13. PARK, T. S., *et al.* 1983. Medulloblastoma: clinical presentation and management. J. Neurosurg. **58:** 543–552.

14. MULHERN, R. K., *et al.* 1992. Neuropsychological status of children treated for brain tumors: A critical review and integrative analysis. Med. Pediatr. Oncol. **20:** 181–191.

15. KUN, L. E., *et al.* 1990. Medulloblastoma—caution regarding new treatment approaches. Int. Radiat. Oncol. Biol. Phys. **20:** 897–899.

16. EVANS, A. E., *et al.* 1990. The treatment of medulloblastoma. Results of prospective randomized trial of radiation therapy with and without CCNU, vincristine, and prednisone. J. Neurosurg. **72:** 572–582.

17. TAIT, D. M., *et al.* 1990. Adjuvant chemotherapy for medulloblastoma: The first multicentre control trial of the International Society of Pediatric Oncology (SIOP). Eur. J. Cancer **26:** 464–469.

18. PENDERGRASS, T. W., *et al.* 1987. Eight drugs in one day chemotherapy for brain tumors: Experience in 107 children and rationale for preradiation chemotherapy. J. Clin. Oncol. **5:** 1221–1231.

19. ZELTZER, P. M. 1995. Radiation and 8 drugs in one-day chemotherapy (8 in 1) (regimen B) or (regimen A) for high-risk intracranial primitive neuroectodermal tumors of childhood (PNET): Preliminary results from the Children's Cancer Study Group. Med. Pediatr. Oncol. **23:** 238.

20. PACKER, R. J., *et al.* 1994. Outcome for children with medulloblastoma treated with radiation and cisplatin, CCNU and vincristine. Chemother. J. Neurosurg. **81**(5): 690–698.

21. GOLDWEIN, J. W., *et al.* 1993. Results of a pilot study of low dose craniospinal radiation therapy plus chemotherapy for children younger than 5 years with primitive neuroectodermal tumors. Cancer **71**(8): 2647–2652.

22. ALLEN, J. C., *et al.* 1983. Pre-radiation chemotherapy for newly diagnosed childhood brain tumors. Cancer **52:** 2001–2006.

23. LOEFFLER, J. S., *et al.* 1988. Preradiation chemotherapy for infants and poor prognosis children with medulloblastoma. J. Radiat. Oncol. Biol. Phys. **15:** 177–181.

24. FOSSATI-BELLANI, F. 1989. Preradiation chemotherapy for childhood and poor prognosis medulloblastoma [abstract]. International Symposium on Pediatric Neuro-oncology. June 1–3, 1989. Seattle, WA.

25. KOVNAR, E. H., *et al.* 1990. Pre-irradiation cisplatin and etoposide in the treatment of high-risk medulloblastoma and other malignant embryonal tumors of the central nervous system: A phase II study. J. Clin. Oncol. **8:** 330–336.

26. MOSIJCZUK, A. D., *et al.* 1997. Preradiation chemotherapy in advanced medulloblastoma. Pediatric Oncology Group Study. Cancer **12:** 2755–2762.

27. SPOSTO, R., *et al.* 1989. The effectiveness of chemotherapy for treatment of high grade astrocytoma in children: Results of a randomized trial. J. Neuro. Oncol. **7:** 165–177.

28. FINLAY, J. L., *et al.* 1995. Randomized phase III trial in childhood high grade astrocytoma comparing vincristine, lomustine and prednisone with the eight drugs in one day regimen (CCG study). J. Clin. Oncol. **13**(1): 112–123.

29. CRAINCROSS, G., *et al.* 1997. Chemotherapy for anaplastic oligodendroglioma. J. Clin. Oncol. In press.

30. DUFFNER, P. K., *et al.* 1993. Postoperative chemotherapy and delayed radiation in children less than three years of age with malignant brain tumors. N. Engl. J. Med. **328:** 1724–1731.

31. PACKER, R. J., *et al.* 1984. Pineal region tumors of childhood. Pediatrics **74:** 97–103.

32. FINLAY, J., *et al.* 1992. Chemotherapy without irradiation (XRT) for primary central nervous system (CNS) germ cell tumors (GCT): Report of an international study. Proc. Am. Soc. Clin. Oncol. **420:** Abstr.

33. DUFFNER, P. K., *et al.* 1986. Survival of children with brain tumors: SEER program 1973–1980. Neurology **36:** 597.

34. MULHERN, R. K., *et al.* 1992. Neuropsychological status of children treated for brain tumors: A critical review and integrative analysis. Med. Pediatr. Oncol. **20:** 181–191.

35. BARAM, T. Z., *et al.* 1987. Survival and neurological outcome of infants with medulloblastoma treated with surgery and MOPP chemotherapy. Cancer **60:** 173–177.
36. STRAUSS, L. C., *et al.* 1991. Efficacy of postoperative chemotherapy using cisplatin plus etoposide in young children with brain tumors. Med. Pediatr. Oncol. **19:** 16–21.
37. GEYER, J. R., *et al.* 1994. Survival of infants with primitive neuroectodermal tumors of malignant ependymomas of the CNS treated with eight drugs in 1 day: A report from the Childrens Cancer Group. J. Clin. Oncol. **12:** 1607–1615.
38. PETERSDORF, S. H., *et al.* 1994. High dose chemotherapy for the treatment of malignant brain tumors. J. Neuro. Oncol. **20:** 155–163.
39. KIMLER, B. F., *et al.* 1994. The 9L rat brain tumor model for preclinical investigation of radiation–chemotherapy interactions. J. Neuro. Oncol. **20:** 103–109.
40. VATS, T. S., *et al.* 1979. Study of effectiveness of bleomycin in rat brain tumor model intravenously and intracerebrally. Int. J. Radiat. Oncol. Biol. Phys. **5:** 1527–1529.
41. MORANTZ, R. A., *et al.* 1983. Bleomycin and brain tumors: A review. J. Neuro. Oncol. **1:** 249–255.
42. CHESON, B. D., *et al.* 1989. Autologous bone marrow transplantation. Current status and future directions. Ann. Intern. Med. **110:** 51–65.
43. TAKVORIAN, T., *et al.* 1983. Autologous bone marrow transplantation: Host effects of high-dose BCNU. J. Clin. Oncol. **1:** 610–620.
44. GIANNONE, L., *et al.* 1987. Phase II treatment of central nervous system gliomas with high dose etoposide and autologous bone marrow transplantation. Cancer Treat. Rep. **71:** 759–761.
45. FINLAY, J. L., *et al.* 1990. High dose multiagent chemotherapy followed by bone marrow "rescue" for malignant astrocytomas of childhood and adolescence. Neurol. Oncol. **9:** 239–248.
46. KHALIFA, C., *et al.* 1992. High dose busulfan and thiotepa with autologous bone marrow transplantation in childhood malignant brain tumors: A phase II study. Bone Marrow Transplant. **9:** 227–233.
47. HEIDEMAN, R. L., *et al.* 1993. High dose chemotherapy and autologous bone marrow rescue followed by interstitial and external beam radiotherapy in newly diagnosed pediatric malignant gliomas. J. Clin. Oncol. **11:** 1458–1465.

Challenges in the Management of
Bone Tumors—1996

ALAN W. CRAFT[a]

Department of Child Health
Royal Victoria Infirmary
Newcastle upon Tyne NE1 4LP, United Kingdom

Before the advent of chemotherapy the prognosis for the two common bone tumors that occur in young people, Ewing's sarcoma and osteogenic sarcoma, had been relatively unchanged since the era when surgery was the only treatment available. For osteosarcoma surgery alone could cure about 20% of patients, with most of the remainder dying within two years with pulmonary metastases. The only role of radiotherapy was to delay amputation in patients who did not develop pulmonary metastases within six months, the so-called Cade principle. For Ewing's tumor the prognosis for surgery alone was very poor, with less than 5% of patients being long-term survivors; and, although the addition of radiotherapy to treat the primary tumor could prolong survival, the overall outcome remained very poor. The 1970s saw the beginning of the chemotherapy revolution. Ewing's tumor in particular seemed especially responsive to chemotherapy, with dramatic shrinkage of primary tumors; and the prognosis of osteosarcoma also seemed to improve dramatically. However, in spite of 25 years of chemotherapy, almost half of all patients with these two tumors will still die of their disease. The past 10 years have seen refinement of therapies, but there has been little improvement in overall survival. There remain many challenges, the most important of which must be to find new methods of treatment that will improve survival. Even obtaining a consensus on what is the current best chemotherapy remains to be defined for osteosarcoma, although there is general agreement about Ewing's sarcoma. The past 15 years have seen an increasing use of endoprosthetic replacement surgery with limb conservation; and even in axial skeleton tumors there is now scarcely a bone in the body that surgeons will not attempt to remove. There is a need to define the role of surgery and to determine the optimum methods to give the best long-term functional outcome.

Finally, if we are to move forward in our knowledge, there is a need to enter patients into clinical trials and studies. There is good evidence in childhood cancer that treatment at specialist centers and entry into trials does improve the outcome[1]; and the only way to assess new, and not so new, treatments is by means of well-designed clinical trials. In order to have an 80% chance of detecting a 15% difference in survival, a two-arm trial needs to have 200 patients in each arm. Given the rarity of bone tumors in young people, and the need for reasonably rapid conclusion of trials, the only solution is multicenter, and probably multinational, trials. In the U.K. the proportion of patients treated at one of the 24 U.K. Children's Cancer Study Group (UKCCSG) centers has increased steadily over the past 20 years, and the majority of these are now entered into clinical trials (TABLE 1). Some of the current challenges pertaining to these two bone tumors of children and young adults are reviewed.

[a] Telephone: 0191 2023009; fax: 0191 2023022; e-mail: A.W.Craft@Newcastle.ac.uk

TABLE 1. Entry of Patients into Clinical Trials and Referral to UKCCSG Centers

	Osteosarcoma	Ewing's Sarcoma
Entered into trials		
1986	36	62
1990	64	81
1994	66	84
Registered with UKCCSG		
1977–1980	35	55
1981–1984	50	72
1985–1988	54	78
1989–1991	59	87
1992–1994	86	89

OSTEOSARCOMA

The overall survival of patients in the U.K. improved substantially after 1982, as evidenced by figures from the national registry of childhood cancer at Oxford; but over the past 10 years there has been little improvement (FIG. 1).[2] As can be seen from TABLE 1, an increasing number of patients have been referred to pediatric oncology centers; FIGURE 2 shows the survival of these patients. Treatment in the U.K. has involved entry of patients into a series of studies run under the auspices of the European Osteosarcoma Intergroup (EOI). In Germany patients have been entered into the Cooperative Osteosarcoma Studies (COSS). Although the treat-

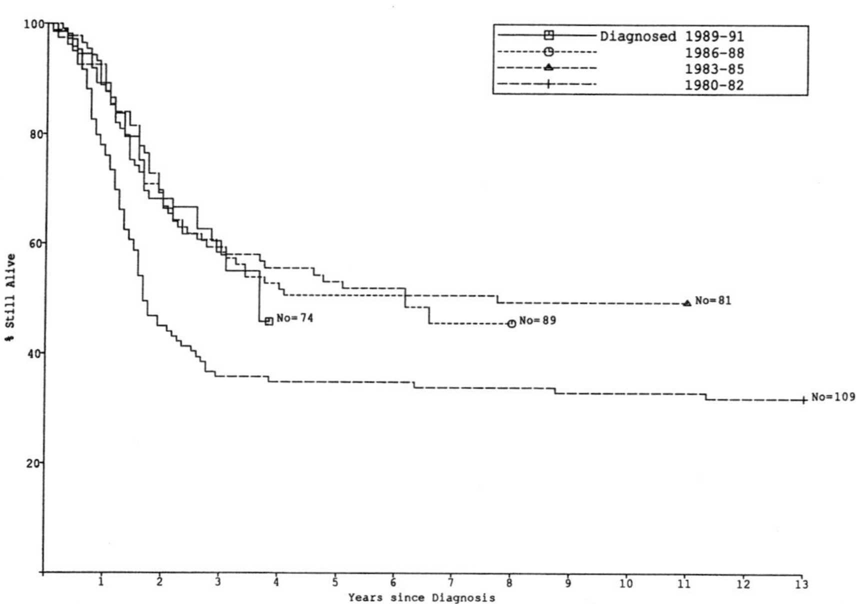

FIGURE 1. Childhood osteosarcoma survival in the U.K., 1980–1991.

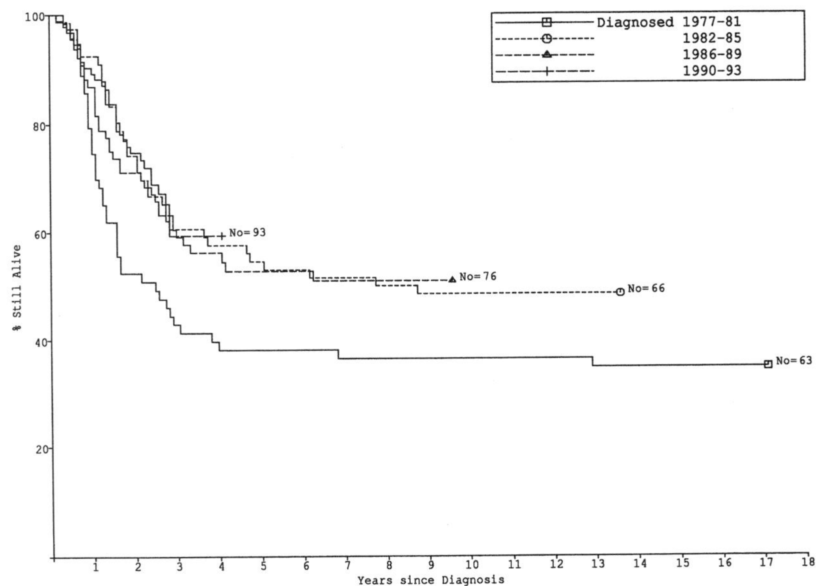

FIGURE 2. Survival of osteosarcoma patients referred to UKCCSG centers, 1977–1993.

ment strategies have been very different, the ultimate outcomes, in terms of population-based survival as opposed to trial results, are remarkably similar (Fig. 3).[3] Comparison of different treatment strategies can be done only in the context of a randomized control trial; but the remarkably similar population survival from the U.K. and Germany suggests that neither treatment strategy is better than the other,

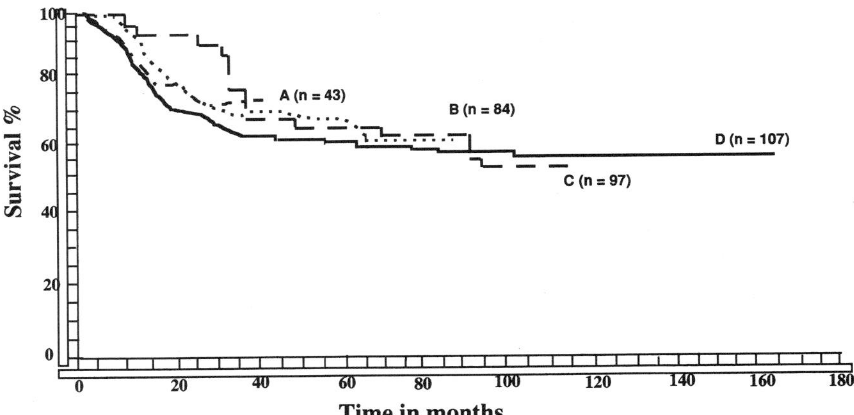

FIGURE 3. Survival of German osteosarcoma patients diagnosed in the years indicated. (Key: – – – A (1989–1991), - - - - - B (1986–1988), – – – – – C (1983–1985), ——— D (1980–1982).

and that both are unsatisfactory, given that at least 35% of the patients will still die of their tumor. What then are the key issues that need to be considered?

Is Methotrexate an Essential Drug?

There is no doubt that methotrexate as a single agent is an effective drug in the management of osteosarcoma. It has been difficult to show this, as it has usually been given in the context of multidrug protocols; but Meyers, in analyzing the Memorial Sloan-Kettering experience, has produced convincing evidence of a 19% histologically proved response rate.[4] This, however, is not substantially different from that reported for the other major agents, as reviewed by Bramwell.[5] No randomized trial of a chemotherapy, with or without methotrexate, has ever been undertaken in which the question asked was, Is methotrexate an essential drug or not? The EOI undertook a study of cisplatin and doxorubicin ± high-dose methotrexate.[6] The study was designed as a feasibility study to test two different chemotherapy regimes, and showed a significant benefit for survival of the two-drug arm.[5] This cannot, however, be interpreted as indicating that there is no place for methotrexate. The two-drug arm gave the doxorubicin and cisplatin every 21 days, and these two drugs were given six times. In the three-drug arm, cisplatin and doxorubicin were given only four times at 31-day intervals, and this was interspersed with methotrexate. The benefit for the two-drug arm can probably be best interpreted as a benefit for more, and more dose-intense, cisplatin and doxorubicin. The subsequent EOI study[7] compared a multidrug regime, including high-dose methotrexate, with the same two-drug cisplatin/doxorubicin regime used in the first EOI study. The preliminary results were presented at the ASCO (American Society of Clinical Oncology) meeting in Philadelphia in 1996 and showed a clear equivalence between the two arms. Once again, this does not prove that methotrexate is of no value, but merely that a non–methotrexate-containing regime gives a similar outcome to one that does include this drug. It is indeed very difficult to design a study that could test the hypothesis that methotrexate is an essential drug. There is increasing evidence that dose intensity of all drugs is important, and although methotrexate is relatively nonmyelotoxic, its addition to a multidrug regime will always compromise the dose intensity of the other drugs. Decisions on its use therefore need to be made on other grounds—e.g., economic and social. The second EOI study demonstrated that an 18-week treatment without methotrexate was equivalent to a 44-week multidrug regime with methotrexate. In terms of patient preference the two-drug arm is clearly superior. From the economic point of view the cost of the cytotoxic drugs in the two-drug arm, at 1996 prices, was $3,000 and in the multidrug arm $15,000, an excess of $12,000. In addition, the extra cost of 16 nights in hospital needs to be considered, along with the cost of methotrexate assays. Methotrexate has become one of the "staple" drugs for osteosarcoma, and it is going to be very difficult to drop it from regimes.

If methotrexate is to be given, then it is important that it be given optimally. Even the optimum dose level has proved difficult to establish, although it now seems clear that higher doses in the range 8–12 gm/m^2 are superior to lower doses of 6 gm/m^2.[8,9] High-dose methotrexate can be given with safety only if serum monitoring of levels is undertaken and folinic acid rescue adjusted accordingly. This has meant that a vast amount of data has been accumulated on the pharmacokinetics of the drug, and using retrospective analysis it has been possible to determine an optimum threshold value above which response seems to be improved. Delepine, analyzing data from Paris, found a significant association between a critical serum level of 1000 μmol/l at the end of the methotrexate infusion and tumor control.[10] In using pharmacokinetic data in a clinical setting it is important to know whether it is possible to predict serum levels in a given situation and whether or not prospective dose adaptation will give optimum levels. Delepine and his colleagues were able to use individual dose adaptation to obtain the minimum serum level and showed improved clinical outcome.[11] In order to try and confirm these findings, Graf and his colleagues retrospectively analyzed data from the COSS studies.[12] They confirmed in the COSS 80 study that a mean peak threshold of >1000 μmol/l gave a highly significant improvement in survival (64% v. 18% disease-free survival at 10 years, $p < 0.0001$). In addition they were able to suggest that, in any individual patient, six courses of methotrexate with serum level above the threshold were enough to show improvement in survival. Although the peak level correlated with the area under the curve, it was the former that was the better predictor of response. In subsequent COSS studies, many of which included the use of cisplatin, it was not possible to correlate methotrexate pharmacokinetics with outcome. They concluded that individual dose adaptation of methotrexate is not necessary if a fixed dose, 12 mg/m^2, of the drug is given; hydration is restricted to 3 l/m^2/24 hours; and cisplatin is also contained in the treatment regime.

Intravenous or Intraarterial Therapy

In view of the risk of local relapse in osteosarcoma, the possibility has been suggested that intraarterial delivery of cytotoxic drugs directly into the tumor might improve survival. Cisplatin is one of the most active agents against osteosarcoma, and it is this drug that has been most widely used in this context. The delivery of intraarterial therapy is more difficult, with the need for surgical intervention for arterial access. There is also a significant risk of vascular complications. The possibility that giving the drug arterially might compromise the systemic effect must also be considered, for the risk of metastatic relapse is much greater, and more difficult to deal with, than that of local relapse. There are theoretical advantages to the use of intraarterial therapy, particularly with regard to a possible greater uptake of drug by the tumor because of higher serum concentrations. Indeed, there is some evidence to suggest that this might be so in other types of tumor.[13] A number of investigators have shown that the technique is feasible.[14,15] The COSS 86 study included a randomization to compare IV or IA cisplatin. The randomization was very difficult in the context of the other important trial factors, but by stratification of risk factors it was possible to obtain a balanced randomization with two reasonably comparable groups. The end point of the study was the proportion of patients with >90% tumor necrosis following the IV or IA therapy. The proportion of patients with such a favorable result was 69% versus 68% in the IV and IA groups.[16] Intraarterial therapy therefore does not appear to confer any therapeutic benefit; and, in view of its potential for complicating treatment, it is no longer a treatment that is widely recommended.

Is Dose Intensity Important?

There is increasing evidence, from a variety of different tumors in both children and adults, that the dose intensity of treatment is important.[17] There are many methodological difficulties when one assesses this issue retrospectively.[18] One of the most important factors is that patients who are *able* to receive the most dose-intense treatment are likely to be those who would have the best prognosis in any case. They are generally fitter, in adult practice often younger, and may have less burden of disease. Nevertheless, a number of studies have shown in bone tumors that dose intensity is important. Smith *et al.* reviewed the evidence available in 1990[19] by looking at the doses recommended to be given in a number of published studies. They used patients with >90% tumor necrosis as the end point, and this is reasonable given the strong association between this and eventual outcome. Utilizing logistic regression analysis, they were able to assess the contribution of the various drugs in the treatment regimes assessed. The relative dose intensity of doxorubicin was the single most important factor influencing outcome. They also found a positive effect for cisplatin and high-dose methotrexate; and there was a suggestion that this might also be true for ifosfamide, but at the time of analysis this had not been extensively reported in osteosarcoma. This is a theoretical analysis based on recommended doses, which of course is not what patients necessarily receive. In both the COSS (Bielack, personal communication) and EOI studies (unpublished data), there is evidence that those patients receiving a higher dose of chemotherapy have a more favorable outcome. As already outlined, the first EOI study[6] showed a clear difference in survival for the two-drug arm, which is probably attributable to more intensive chemotherapy.

Dose intensity is therefore almost certainly a critical factor determining outcome in osteosarcoma; a pilot study to determine whether the dose intensity could be increased was recently reported by Ornadel *et al.*[20] Twenty-four patients were given cisplatin/doxorubicin at a target of two- rather than three-week intervals with the addition of granulocyte colony stimulating factor (GCSF) to speed marrow recovery. While it was not always possible to achieve the desired increased frequency, especially for the postoperative courses, there was overall an increased dose intensity over that previously achieved with the three-weekly regime. Toxicity was increased but manageable, although there was a clinically significant increase in the amount of perioperative bleeding.

The current EOI study is investigating the possible benefit of increasing dose intensity. The standard, three-weekly, cisplatin/doxorubicin regime is being compared with the same drugs given at a target time of two weeks (FIG. 4). The target relative dose intensity of the accelerated arm is 1.5.

Are There Any New Drugs Available?

In 1987 Bramwell[5] reviewed the drugs available for the management of osteosarcoma. Most of those assessed and found to be active were already included in standard protocols. Ifosfamide was considered, but very little data was available, much of it unpublished. However an overall response rate of 24% in 78 evaluable patients made it one of the most active drugs when used in a single-agent setting. Most of the patients, however, were treated following relapse. Harris *et al.*[21] recently reviewed the activity of ifosfamide when given in a front line, "window," setting or when given after multiple other chemotherapies. The response rate was 27% for the former compared to only 10% for the latter. Ifosfamide therefore does appear to be an active agent in osteosarcoma, and it has been incorporated into the major studies of both COSS 95 and in the United States by the Children's Cancer Group (CCG), CCG7921 (POG-9351).

In 1989 it was reported that in dogs with osteosarcoma a significant improvement in outcome could be achieved by the use of liposome-encapsulated muramyl tripep-

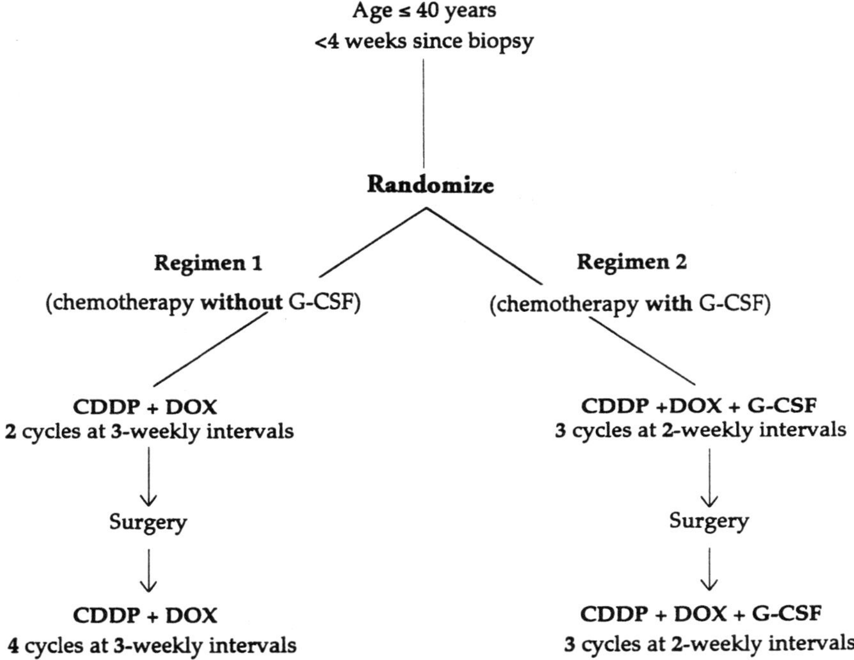

FIGURE 4. Design of EOI 80931 trial involving patients with untreated, nonmetastatic, biopsy-proven osteosarcoma of long bones of an extremity.

tide-phosphatidylethanolamine (MTP-PE). This is a synthetic molecule that resembles fragments of the wall of mycobacteria and probably acts by activation of macrophages.[22] The EORTC (European Organisation for Research and Treatment in Cancer) undertook a phase II study[23] and found no response in patients with metastatic soft-tissue sarcoma. However, it has been incorporated as a randomized variable in the current CCG-7921 study.

Topotecan, a water-soluble camptothecin analogue, which acts as a topoisomerase 1 inhibitor, has shown significant activity against a number of murine tumors and in human xenografts of osteosarcoma.[24] A phase I study in children has now been completed,[25] and it is currently being investigated in a phase II window study by CCG in their CCG-7943 study.

What Are the Important Prognostic Variables?

In a tumor that has a 65% survival and in which dose intensity of chemotherapy seems to be important, it is of paramount importance to determine the prognostic variables that predict for outcome. The aim of this determination is to be able to identify patients who may be at higher or lower risk so that this treatment can be either intensified or reduced accordingly. Ideally, what is needed is a "front-end" prognostic factor that can be determined at the time of diagnosis so that treatment can be planned accordingly. Davis *et al.*[26] recently reviewed the literature in nonmetastatic patients and evaluated age, sex, tumor site and size, and histological response to preoperative chemotherapy. Both tumor size and histologic response were significant factors in a univariate analysis, but tumor size was no longer significant when the other factors were allowed for. The most important factor therefore is the histological response to initial therapy. This, of course, is not a factor that can be determined before the institution of treatment. The current EOI and CCG studies are not using any stratification according to prognostic factors, but in the COSS 95 study all patients receive standard "induction" therapy and are then stratified into three risk groups according to tumor volume and histological response. The low-risk group are then given a very short, nonintensive, postoperative therapy, while therapy is sequentially increased for the other two groups (Bielack, personal communication).

Osteosarcoma—Conspectus

At present 60–65% of patients presenting with nonmetastatic osteosarcoma can probably be cured. Improvements in survival will probably come about with intensification of existing treatments, as there are no highly promising new drugs available. Identification of risk factors will be important so that patients with a very good outlook can be spared the undoubted increased toxicity of the new intensified regimes.

Multinational trials are necessary to evaluate new treatments.

EWING'S SARCOMA

The outcome for Ewing's sarcoma has steadily improved over the past 20 years. FIGURE 5 shows the data for the U.K. Patients diagnosed in 1980 had a less-than-30% chance of being cured compared to the more recently diagnosed cohorts, who have a substantially increased chance of long-term survival. However, it can also be seen from this figure that there is a substantial risk of late relapse and death in Ewing's tumor. Relapses are still being seen at 12 years and more from diagnosis, although the majority do occur within the first 6 years.

Unlike the situation with osteosarcoma, there has been a remarkable uniformity of treatment across the world in Ewing's sarcoma. The major study groups of UKCCSG,[27] CESS (Co-operative Ewing's Sarcoma Study),[28] and IESS (Intergroup Ewing's Sarcoma Study)[29] all utilized the four drugs vincristine, cyclophosphamide, doxorubicin, and actinomycin D in their first studies in the early 1980s. All then switched from cyclophosphamide to ifosfamide for the majority of patients in their subsequent studies. Consequently, there is less controversy in the management of Ewing's sarcoma, when compared to osteosarcoma; but there remain substantial challenges. Up to 40% of patients still cannot be cured, and the optimum method of local treatment has yet to be defined. Some of the current topics of debate are as follows.

Ifosfamide or Cyclophosphamide?

Ifosfamide was introduced in the 1980s as an alkylating agent that could be given in higher dosage and greater safety then cyclophosphamide, especially if it

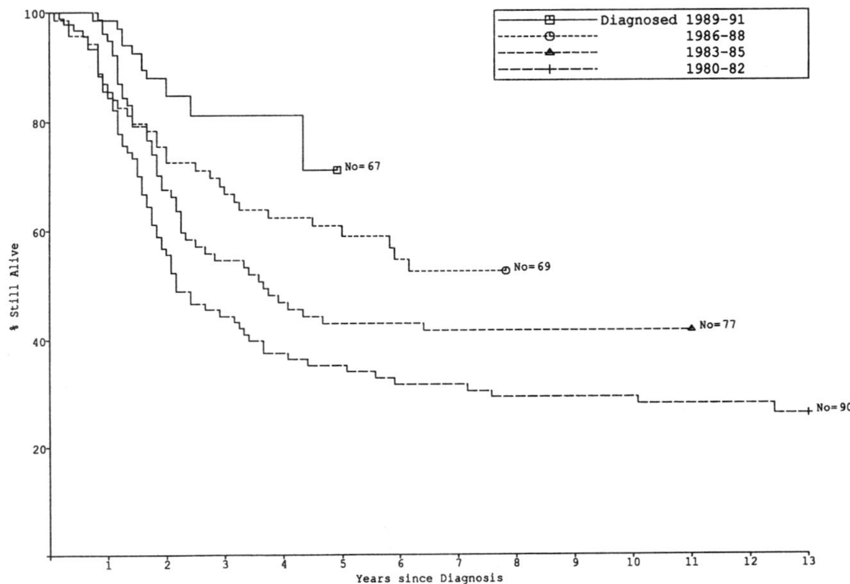

FIGURE 5. Childhood Ewing's sarcoma survival in the U.K., 1980–1991.

is used with the uroprotective agent mesna. The UKCCSG showed an improvement in survival of some 20% when the ET-1 and ET-2 studies were compared,[27] the only difference being the switch from cyclophosphamide to ifosfamide. A similar improvement in survival has been shown in the CESS 86 study.[30] However, Oberlin et al.,[31] reporting the results of the SFOP (Societé Francais Oncologie Pediatrie) study in France, could find no benefit for ifosfamide. In addition, their patients experienced a worrying incidence of severe cardiotoxicity, with two fatalities. They attributed this to the possible cumulative toxicity of doxorubicin and ifosfamide. An editorial in the *Journal of Clinical Oncology* in 1995 entitled "Ifosfamide: Should the honeymoon be over?" reviewed the evidence for the efficacy of this drug over its unfashionable cousin cyclophosphamide.[32] They pointed out that truly comparable studies had not been done, and this is certainly true for Ewing's sarcoma. Both the UKCCSG and CESS studies used retrospective comparisons with earlier studies that had used cyclophosphamide in a far-from-maximal dose. The current availability of mesna and other methods of supportive care, should enable much higher doses of cyclophosphamide to be given, and it is quite possible that these might be equally efficacious. However, no randomized study has yet been undertaken in Ewing's sarcoma comparing equidose cyclophosphamide and ifosfamide. Given the significant incidence of persistent renal toxicity in patients treated with ifosfamide,[33] it would seem not unreasonable that such a study should be considered.

Is Dose Intensity Important?

The general problems of determining the importance of dose intensity have been alluded to above regarding osteosarcoma. In their review of dose intensity,

Smith *et al.*[19] also looked at reported studies in Ewing's sarcoma. Once again they found that the dose intensity of doxorubicin was the single most important factor. Increasing dose intensity of actinomycin D was an important negative factor that seemed to reduce the chances of survival. However, in most studies the increased dose of actinomycin was at the expense of a decreased intensity of doxorubicin. They concluded that in the design of future trials "careful attention should be given to optimising the doxorubicin dose intensity." Both the UKCCSG (unpublished data) and the CESS (Jurgens, personal communication) have shown in their studies that the amount of drug received by the patients is an important prognostic factor correlating strongly with survival. However, once again, the pitfalls of such analyses must be emphasized and the results interpreted with caution. Nevertheless, it is likely that dose intensity is as important in Ewing's sarcoma as it seems to be in osteosarcoma.

How Do We Treat Metastatic Patients?

The presence or otherwise of metastases at diagnosis is the single most important prognostic factor in Ewing's sarcoma. The common sites of metastases are the lung and/or bone. For those patients with lung-only disease, there would appear to be a 20% chance of survival with contemporary, ifosfamide-based chemotherapy. However, bone metastases confer a much more sinister outlook, with only, at best, 5–10% of patients who can be cured. Attempts have therefore been made to try and intensify therapy in such metastatic patients. Burdach and his colleagues in Germany[34] have reported on their initial experience of myeloablative radiochemotherapy with hematopoietic stem-cell rescue in such poor-prognosis Ewing's tumor patients. They studied 17 patients who received increasingly intensive treatment in a dose-escalation study. They compared them with 41 historic controls matched for the important prognostic factors. They reported a probability of relapse-free survival for the study patients of 45% ($\pm$12%) at six years compared to only 2% ($\pm$2%) for the historic controls. This improvement in survival has been maintained with prolonged follow-up (Burdach, personal communication).

The combined UKCCSG and CESS groups (EICESS) are currently planning a study in metastatic patients based on the pioneering work of Burdach.

What Is the Optimum Local Treatment?

Unlike osteosarcoma, for which the only possible local treatment is surgery, there are therapeutic options in Ewing's sarcoma. Radiotherapy was the preferred treatment until the early 1980s, since when there has been an increasing use of surgery with or without radiotherapy. The issues that need to be taken into account when deciding on local treatment are the age of the patient, the site of the tumor, and the possible late effects. The late effects can be categorized as either functional—i.e., degree of limitation of physical activity—or pathological. In the latter the problems of failure of growth of long bones, radiation necrosis with pathological fracture, and the subsequent development of second primary tumors in the radiation field all need to be considered. There has been an increasing use of surgery over the past 10 years, but outside the context of a randomized trial of surgery it is not possible to determine whether surgery itself has any influence on survival. In most studies patients who have surgery have a more favorable outcome than those who have radiotherapy, but the issue is confounded by the fact that those patients in whom surgery is possible have tumors that have an inherently better prognosis—i.e., they are usually small and in the limb, compared to large pelvic tumors. Optimum management therefore is defined by patient factors along with local experience and practice.

CONCLUSION

There remain many challenges in the management of bone tumors in children and young adults. Well over half of all patients can be cured, but the challenges for the next few years must be to identify "front-end" prognostic factors, develop better and probably more intensive treatments for those with a poor chance of cure, and optimize—perhaps reduce—treatment for those who have an excellent chance of cure. The optimum modality of local treatment remains to be defined in Ewing's sarcoma, but in both types of tumor the best surgical methods are not clear. The striking differences in type of surgery across Europe clearly indicate that further studies are needed. Rotationplasty is commonly undertaken on the continent of Europe, but rarely in the U.K. However, "growing" endoprostheses are often used in even very small children in the U.K., whereas on the mainland similar children would be subjected to either amputation or rotationplasty. The correct answer will probably come only from late functional outcome studies, but it may be 10–20 years before the results of such investigations will be known.

We have come through two exciting decades in the management of bone tumors. There remain significant challenges for the next decade.

ACKNOWLEDGMENTS

Mr. C. A. Stiller and Professor J. Michaelis are thanked for providing national survival data and Professor Herbert Jurgens and Dr. Stefàn Bielack for data on current European studies.

REFERENCES

1. STILLER, C. A. 1988. Centralisation of treatment and survival rates for cancer. Arch. Dis. Child. **63:** 23–30.
2. STILLER, C. A. 1994. Population based survival rates for childhood cancer in Britain. Br. Med. J. **309:** 1612–1616.
3. KALETSCH, U., P. KAATSCH & J. MICHAELIS. Jahresbericht 1995 des Deutschen Kinderkrebsregisters, Mainz, 1996.
4. MEYERS, P. A. *et al.* 1992. Chemotherapy for nonmetastic osteogenic sarcoma: The Memorial Sloan-Kettering Experience. J. Clin. Oncol. **10:** 5–15.
5. BRAMWELL, V. H. C. 1987. Chemotherapy of operable osteosarcoma. Bailliere's Clin. Oncol. **1:** 175–195.
6. BRAMWELL, V. H. C. *et al.* 1992. A comparison of two short intensive adjuvant chemotherapy regimens in operable osteosarcoma of limbs in children and young adults: The first study of the European Osteosarcoma Intergroup. J. Clin. Oncol. **10:** 1579–1591.
7. SOUHAMI, R. L. *et al.* 1996. A randomised trial of two regimens of chemotherapy in operable osteosarcoma. A study of the European Osteosarcoma Intergroup. Proc. Am. Soc. Clin. Oncol. **15:** 520.
8. ROSEN, G. *et al.* 1981. Osteogenic sarcoma: Eighty-percent three-year disease free survival with combination chemotherapy (T-7). Nat. Cancer Inst. Monogr. **56:** 213–220.

9. WINKLER, K. *et al.* 1984. Neoadjuvant chemotherapy for osteogenic sarcoma: Results of a German/Austrian study. J. Clin. Oncol. **2:** 617–624.

10. DELEPINE, H. *et al.* 1989. Individualisation de posologie de Methotrexate haut dose par le dosage des concentrations plasmatique interet therapeutique dans le sarcome osteogenique. Bull Cancer (Paris) **76:** 913–918.

11. DELEPINE, N. *et al.* 1989. High value of pharmacokinetics (PK) adaptation in success of treatment of osteosarcomas (OS) by protocols based on high dose methotrexate (HDMTX). Proc. Am. Soc. Clin. Oncol. **8:** 319. (Abstr.)

12. GRAF, K. *et al.* 1994. Methotrexate pharmacokinetics and prognosis in osteosarcoma. J. Clin. Oncol. **12:** 1443–1451.

13. GOUGETTE, A., A. APCHIN, M. FOKA & J. M. RICHARDS. 1986. Pharmacokinetics of intra-arterial and intravenous cisplatin in head and neck cancer patients. Eur. J. Cancer Clin. Oncol. **22:** 257–263.

14. CHUANG, V. P. *et al.* 1981. The therapy of osteosarcoma by intra-arterial cis-platinum and limb preservation. Cardiovasc. Intervent. Radiol. **4:** 229–235.

15. JAFFE, N. *et al.* 1985. Comparison of intra-arterial cis-diamminedichloroplatinum II with high-dose methotrexate and citrovorum factor rescue in the treatment of primary osteosarcoma. J. Clin. Oncol. **3:** 1101–1104.

16. WINKLER, K. *et al.* 1990. Effect of intraarterial versus intravenous cisplatin in addition to systemic doxorubicin, high-dose methotrexate, and ifosfamide on histologic tumor response in osteosarcoma (Study COSS-86). Cancer **66:** 1703–1710.

17. DODWELL, D. J., H. GURNEY & N. THATCHER. 1990. Dose intensity in cancer chemotherapy. Br. J. Cancer **61:** 789–794.

18. REDMOND, C., B. FISHER & W. H. SAMUEL. 1983. The methodologic dilemma in retrospectively correlating the amount of chemotherapy received in adjuvant therapy protocols with disease-free survival. Cancer Treat. Rep. **67:** 519.

19. SMITH, M. A. *et al.* 1991. Influence of doxorubicin dose intensity on response and outcome for patients with osteogenic sarcoma and Ewing's sarcoma. J. Natl. Cancer Inst. **83:** 1460–1470.

20. ORNADEL, D. *et al.* 1994. Doxorubicin and cisplatin with granulocyte colony-stimulating factor as adjuvant chemotherapy for osteosarcoma: Phase II trial of the European Osteosarcoma Intergroup. J. Clin. Oncol. **12:** 1842–1848.

21. HARRIS, M. B. *et al.* 1995. Treatment of osteosarcoma with ifosfamide: Comparison of response in pediatric patients with recurrent disease versus patients previously untreated: A Pediatric Oncology Group Study. Med. Pediatr. Oncol. **24:** 87–92.

22. MACEWEN, E. G. *et al.* 1989. Therapy of osteosarcoma in dogs with intravenous injection of liposome-encapsulated muramyl-tripeptide. J. Natl. Cancer Inst. **81:** 935–938.

23. VERWEIJ, J. *et al.* 1994. Phase II study of liposomal muramyl tripeptide phosphatidylethanolamine (MTP/PE) in advanced soft tissue sarcoma of the adult. An EORTC Soft Tissue and Bone Sarcoma Group Study. Eur. J. Cancer **30A:** 842–843.

24. HOUGHTON, P. *et al.* 1992. Evaluation of 9-dimethylaminomethyl-10-hydroxy camptothecin against xenografts derived from adult and childhood tumors. Cancer Chemother. Pharmacol. **31:** 229–239.

25. TUBERGEN, D. *et al.* 1994. Phase I study of topotecan in children with refractory solid tumors: A pediatric oncology group study. Proc. Am. Soc. Clin. Oncol. **13:** 463.

26. DAVIS, A. M., R. S. BELL & P. J. GOODWIN. 1994. Prognostic factors in osteosarcoma: A critical review. J. Clin. Oncol. **12:** 423–431.

27. CRAFT, A. W., S. COTTERILL & J. IMESON. 1993. Improvement in survival for Ewing's sarcoma by substitution of ifosfamide for cyclophosphamide. Am. J. Pediatr. Hematol. Oncol. **15**(Suppl. A): S31–S35.

28. JURGENS, H. *et al.* 1988. Multidisciplinary treatment of Ewing's sarcoma of bone. A 6-year experience of a European Cooperative Trial. Cancer **61:** 23–32.

29. NESBIT, M. E. *et al.* 1990. Multimodal therapy of primary nonmetastatic Ewing's sarcoma of bone: A long term follow-up of the first intergroup study. J. Clin. Oncol. **8:** 1664–1674.

30. JURGENS, H. *et al.* 1988. The GPO Cooperative Ewing's Sarcoma Studies CESS 81/86 Report after 6½ years. Klin. Paediatr. **200:** 243–252.

31. OBERLIN, O. *et al.* 1992. No benefit of ifosfamide in Ewing's sarcoma: A nonrandomized study of the French Society of Pediatric Oncology. J. Clin. Oncol. **10:** 1407–1412.

32. KAMEN, B. A. E. FRENKEL & O. M. COLVIN. 1995. Ifosfamide: Should the honeymoon be over? J. Clin. Oncol. **13:** 307–309.

33. SKINNER, R. *et al.* 1996. Risk factors for ifosfamide nephrotoxicity in children. Lancet **348:** 578–580.

34. BURDACH, S. *et al.* 1993. Myeloablative radiochemotherapy and hematopoietic stem-cell rescue in poor-prognosis Ewing's sarcoma. J. Clin. Oncol. **11:** 1482–1488.

Long-Term Survival in Childhood and Adolescent Cancer

Five-Center Study—U.S.A.

GRACE E. FOEGE HOLMES

Department of Pediatrics and Preventive Medicine
4004 Robinson Hall
University of Kansas Medical Center
3901 Rainbow Boulevard
Kansas City, Kansas 66160-7313

INTRODUCTION

The survival rate among patients diagnosed with childhood or adolescent cancer has shown marked improvement, and many of these individuals have reached adulthood. However, cancer treatment often includes both ionizing radiation and potent chemotherapeutic agents that are potential teratogens and mutagens. Uncertainty about the long-term effects of these treatments may cause anxiety in survivors of childhood cancer regarding not only their general health but also fertility, plans for marriage, general academic ability, and health of offspring. Physicians often share in this uncertainty and, as a result, may disseminate incomplete and even erroneous information to survivors regarding prognosis.

In order to address these and other long-term health issues, the National Cancer Institute (NCI) in Bethesda, Maryland, in 1980 embarked on a large collaborative effort with five institutions to investigate late effects of cancer and its treatment and to evaluate the physical and psychosocial morbidity of childhood and adolescent cancer survivors and their offspring. The institutions, which included the California Department of Health Services, the University of Iowa, the University of Kansas, M. D. Anderson Hospital with the University of Texas, and Yale University, participated with the Clinical Epidemiology and Biometry branches of the NCI to study survivors and their sibling controls. This paper will review and summarize their findings related to marriage, fertility, pregnancies, educational attainment, and offspring of these groups.[1–6]

METHODS

Subjects/Data Collection and Analyses

Survivors of childhood or adolescent cancer were diagnosed between 1945 and 1975, and for study selection must have survived more than five years after the histological diagnosis of their first malignancy or of any brain or CNS tumor, and to have reached the age of at least 21 years by 12/31/79. These survivors belonged to a population who were diagnosed up to the time when childhood cancer became a curable disease and who had reached their reproductive years. A control subject was eligible if born before 1961, if a sibling of survivor (full-blood relation preferred), same sex if possible, and closest age possible. Up to two controls were sought per survivor.

Interviewers for the five sites received special training. Fieldwork was carried out between August 1980 and April 1983. Interviews in person or by telephone lasted approximately one hour. If the survivor was deceased or incompetent, a suitable proxy was interviewed (usually parent or spouse).

Questions were asked on many subjects regarding health, education, employment, psychosocial issues, and offspring. Siblings were identified, contacted, and interviewed as with the survivors. Neither cases nor controls were aware of the hypotheses linking childhood cancer with subsequent problems.

Data Analyses

Descriptions of data analyses for each subject discussed here are published elsewhere.[1,2,4–6] For the convenience of the reader, these are reproduced with permission in an APPENDIX.

RESULTS

For all five sites, 2498 survivors were identified who fit selection criteria; and, after survivor interviews, 3604 appropriate controls were selected. The interview response of both groups was 91% (case survivors $n = 2283$, controls $n = 3270$), and there was no significant difference between survivors and controls with respect to center, sex, or race. Average age of survivors at the time of study was 31.1 years and of controls 32.5 years. Proxy interviewees were required for 10% of cases and 4% of controls. The largest proportion of survivors had Hodgkin's disease, central nervous system (CNS) tumors, and soft tissue sarcoma.

Marital Status[1]

Question 1: Compared to their siblings, were survivors of childhood and adolescent cancer less likely to marry?

The probability of marriage in 2170 eligible survivors relative to their 3138 control siblings was 0.86. Thus, overall 14% of survivors studied were less likely to marry than were the controls. The risk ratio (R.R.) or the risk of marrying for all male cases was 0.87 (99% confidence interval 0.76–0.99) and for female cases was 0.86 (99% confidence interval 0.76–0.97).

Most of the marriage deficit was for survivors of CNS tumors, especially in males, and this was true for all ages. The subgroup of men with CNS tumors had a relative risk of marrying of only 0.48 with a decrease in the R.R. to 0.34 if diagnosis was made before 10 years of age. If diagnosis was made at 10 years or older, the R.R. was 0.60. Women with CNS tumors had a R.R. of marrying of 0.73.

Treatment for medulloblastoma was associated with a pronounced negative effect in marrying. This was true regardless of gender.

Fertility and Pregnancy[2,3]

Question 2: Compared to their siblings, were survivors of childhood and adolescent cancer less likely to become pregnant?

TABLE 1. Adjusted Relative Fertility and 95% Confidence Intervals of Survivors of Cancer, According to Type of Treatment Received in Subjects Presumed to Be at Risk for Pregnancy[2]

Treatment	All	Males	Females
No radiation or chemotherapy	0.91 (0.82–1.00)	0.81 (0.70–0.94)	0.98 (0.86–1.11)
Radiation above diaphragm	0.88 (0.75–1.02)	0.89 (0.71–1.12)	0.85 (0.69–1.05)
Radiation below diaphragm	0.76 (0.63–0.92)	0.74 (0.57–0.96)	0.78 (0.59–1.03)
Alkylating agents (A.A.)	0.67 (0.52–0.87)	0.42 (0.27–0.62)	1.02 (0.74–1.42)
Radiation above diaphragm +A.A.	0.65 (0.49–0.86)	0.46 (0.30–0.71)	0.89 (0.62–1.28)
Radiation below diaphragm +A.A.	0.57 (0.43–0.75)	0.38 (0.25–0.58)	0.81 (0.57–1.17)

Infertility was defined as present if there was a history of unsuccessful attempts for pregnancy, if case or control had consulted a physician because of not achieving a pregnancy, or if medication was being taken to promote fertility. Subjects presumed not to be at risk for pregnancy were not included in the analysis. After exclusions for various reasons, there were 1232 of the 2283 cases and 2253 of 3270 controls available for analysis.

Overall, 15% of cases who were at least potentially fertile were less likely to report a pregnancy than were controls. Married cancer survivors with no known cause of infertility had 85% of the risk of fertility of their sibling controls or adjusted R.R. = 0.85 (95% confidence interval, 0.78–0.92), with a greater fertility deficit among male survivors.

The adjusted R.R. for fertility for male cases was 0.76 and for females 0.93. Relative fertility was not significantly affected by age, income, race, educational attainment, medical center, type of interview, or other variables that were considered.[2]

By the end of the study 16% of the male controls and 12% of the female controls had reported no pregnancy. This compares closely with results of a national U.S. data study from 1982[7] that showed that 14% of married women in the U.S. who were potentially fertile had never reported a pregnancy.

When considering the effect of treatment on fertility, both sexes showed impairment. With treatment by surgery alone, there was a 9% reduction in fertility. The relative risk for fertility varied with treatment and with gender. Fertility estimates were markedly decreased among male survivors, particularly if treatment had included alkylating agents as shown in TABLE 1.

There were no differences in the rates of spontaneous abortion among cases and controls (12.6% vs. 12.5%, respectively), stillbirths, or ectopic pregnancies. In preliminary analyses,[3] there was a significant difference ($p < 0.02$) in number of pregnancies in 1268 cases (mean of 2.4 pregnancies) and in 2272 controls (mean of 2.6 pregnancies).

In spite of reduced fertility among cases, however, there appeared to be only a few difficulties carrying a pregnancy to term once it was initiated.

Educational Achievement[4]

Question 3: Compared to their siblings, did survivors of childhood and adolescent cancer have the same educational achievement?

Educational data of 2283 cases and 3261 controls were compared regarding the level of attainment. No attempt was made to address the quality of academic performance. In general, non-CNS tumor survivors were not significantly different from controls in their educational level achievement.

There were educational deficits among CNS tumor survivors, but, in spite of this, 77% were high school graduates. Of the high school graduates, 49% entered college. Once in college, there was no significant difference between cases and controls in the percentage of those who completed four years.

Offspring (Cancer and Birth Defects)[5,6]

Question 4: Compared to the offspring of their sibling controls, were there higher incidences of cancer or birth defects among the offspring of survivors of childhood and adolescent cancer?

There were seven cases of cancer among 2308 offspring of 2283 case survivors (0.30%) and 11 cases among 4719 offspring (0.23%) of 3604 controls. The observed number was not significantly different from that expected in the general population.

No statistically significant excess of therapy-induced cancers was present in children of cases versus controls during the follow-up period. Familial aggregation of known single-gene traits associated with cancer was observed.

Records were sought to document offspring with major genetic disease, defined as cytogenetic or single-gene disorder or one of 15 isolated birth defects monitored by the Centers for Disease Control. Genetic disease occurred in 3.4% of 2198 offspring of survivors and in 3.1% of 4544 offspring of sibling controls ($p = 0.33$, N.S.).

DISCUSSION

Marriage

In most societies throughout the world, marriage is considered an important indicator of the quality of life. Most of the marriage deficit among case survivors was found in CNS tumor survivors, especially in males. Perhaps this signals biological differences in gender response to the tumor and/or the treatment. Society's expectations in a spouse may also have been a factor. CNS tumor survivors also married at slightly older ages than did controls.

Treatment for medulloblastoma in both sexes was associated with a negative effect in marrying. Presumably, this was related to a decrease in cognitive ability.

For the most part, these studies showed that survivors of childhood and adolescent cancer treated through the mid-1970s could and do look forward to the future regarding marriage with as much optimism as their siblings.

Infertility

Infertility is a complex mix of psychosocial, sexual, and biological factors. In childhood and adolescent cancer survivors, this mix is particularly complex. Citing one general estimate of infertility conceals a possible positive prognosis for many patients.

The risk of side effects on fertility must take several factors into account, and not merely the therapeutic agents. The risk also depends on the sex of the patient, the type of cancer, and the type and regimen of therapy.

Physicians have frequently been guilty of providing erroneous and negative information to survivors of cancer regarding family planning. Our findings indicated that, in spite of reduced fertility among survivors, there were only a few difficulties carrying a pregnancy to term once it was initiated.

Education

As with marriage deficits, educational deficits in career survivors versus their sibling controls appeared to be directly related mostly to CNS tumors. Quality of education as well as the learning process, though important, were not studied.

Offspring of Survivors

Just as with fertility, there appears to be realistic optimism regarding offspring of survivors in terms of occurrence of cancer or genetic disease. No significant increase with either problem was demonstrated among offspring of survivors versus those of controls. Unless DNA molecular studies are sensitive enough and able to detect problems among offspring, it appears that with our present knowlege, the future for normal offspring among cancer survivors may be the expected outcome.

CONCLUSION

A large, collaborative group investigated late effects of cancer and its treatment and evaluated the physical and psychosocial morbidity of childhood and adolescent cancer survivors and their offspring. Cancer survivors were diagnosed between 1945 and 1975 and were identified according to detailed eligibility criteria. Siblings of survivors were selected as controls. We identified 2498 survivors and 3604 appropriate controls, with a 91% response rate for each group. They were interviewed in person or by telephone on many subjects regarding health, education, employment, psychosocial issues, and offspring. Marriage and educational deficits among survivors were directly proportional to the diagnosis of CNS tumors; for survivors of non-CNS tumors, the deficits were not great. Though there was reduced fertility among survivors, there were few difficulties carrying a pregnancy to term once initiated. There was no significant increase in occurrence of cancer or birth defects among offspring of survivors versus controls. Our findings justify more overall optimism for the future of long-term survivors of childhood and adolescent cancer than was previously assumed.

REFERENCES

1. BYRNE, J., T. R. FEARS , S. C. STEINHORN, J. J. MULVIHILL, R. R. CONNELLY, D. F. AUSTIN, G. F. HOLMES, F. F. HOLMES, H. B. LATOURETTE, M. J. TETA, L. C. STRONG & M. H. MYERS. 1989. Marriage and divorce after childhood and adolescent cancer. JAMA **262:** 2693–2699.
2. BYRNE, J. , J. J. MULVIHILL, M. H. MYERS, R. R. CONNELLY, M. D. NAUGHTON, M. R. KRAUSS, S. C. STEINHORN, D. D. HASSINGER, D. F. AUSTIN, K. BRAGG, G. F. HOLMES, F. F. HOLMES, H. B. LATOURETTE, P. J. WEYER, J. W. MEIGS, M. J. TETA, J. W. COOK &

L. C. STRONG. 1987. Effects of treatment on fertility in long-term survivors of childhood or adolescent cancer. N. Engl. J. Med. **317:** 1315–1321.

3. BYRNE, J., J. J. MULVIHILL, M. H. MYERS, S. C. ABBOTT, R. R. CONNELLY, M. R. HANSON, D. D. HASSINGER, M. D. NAUGHTON, D. F. AUSTIN, V. A. GURGIN, F. F. HOLMES, G. F. HOLMES, H. B. LATOURETTE, P. J. WEYER, M. J. TETA, L. C. STRONG & J. A. COOK. 1985. Risk of infertility among survivors of childhood and adolescent cancer. Am. J. Hum. Genet. **37**(4): A24, (066).

4. KELAGHAN, J., M. H. MYERS, J. J. MULVIHILL, J. BYRNE, R. R. CONNELLY, D. F. AUSTIN, L. C. STRONG, J. W. MEIGS, H. B. LATOURETTE, G. F. HOLMES & F. F. HOLMES. 1988. Educational achievement of long-term survivors of childhood and adolescent cancer. Med. Pediatr. Oncol. **16:** 320–326.

5. MULVIHILL, J. J., R. R. CONNELLY, D. F. AUSTIN, J. W. COOK, F. F. HOLMES, M. R. KRAUSS, J. W. MEIGS, S. C. STEINHORN, M. J. TETA, M. H. MYERS, J. BYRNE, K. BRAGG, D. D. HASSINGER, G. F. HOLMES, H. B. LATOURETTE, M. D. NAUGHTON, L. C. STRONG & P. J. WEYER. 1987. Cancer in offspring of long-term survivors of childhood and adolescent cancer. Lancet **8563**(ii): 813–817.

6. BYRNE, J., S. A. RASMUSSEN, S. C. STEINHORN, R. R. CONNELLY, M. H. MYERS, C. E. LYNCH, J. FLANNERY, D. F. AUSTIN, F. F. HOLMES, G. E. HOLMES, L. C. STRONG & J. J. MULVIHILL. 1996. Genetic disease in offspring of long-term survivors of childhood and adolescent cancer. N. Engl. J. Med. Submitted.

7. MOSHER, W. D. & W. F. PRATT. 1985. Fecundity and infertility in the United States, 1965–1982. *In* Advance Data From Vital and Health Statistics; no. 104. (DHHS publication no. PHS 85-1250.) United States Public Health Service. Hyattsville, MD.

APPENDIX

Data Analyses

Marriage[1]

[*Reprinted from Byrne et al.,*[1] *which was written by employees of the federal government and thus in the public domain.*]

The duration of marriage was counted in months from the start of the marriage or live-in relationship to its end. Rate ratios for marriage in survivors were stratified by various factors and compared with the entire group of controls, except for analyses by sex, where survivors were compared with controls of like sex. The proportion of subjects who were divorced at follow-up was defined as those currently divorced or separated divided by the total of those currently divorced or separated plus those currently married. Subjects who were widowed and those in a homosexual relationship were excluded from analysis of divorce. Analysis was by single and multiple cross-tabulations and by Cox proportional hazards models. Fit of the proportional hazards model was evaluated by permitting the proportionality assumption to vary with time and with risk factor level. The results supported the use of the proportional hazards model as appropriate. The final proportional hazards models included an indicator variable for each of the individual cancer sites as well as terms for income, age at follow-up, and vital status; education was included as a stratified variable (grades 0 through 10, 11 through 15, 16+). Confidence intervals were set at 99% to account for multiple comparisons. Differences between survivors and control subjects in average duration of first marriages for subjects whose marriages had ended and average age at first marriage were compared using a t test.

To account for different lengths of time to marriage, the probability of marriage in survivors relative to control subjects (rate ratio [RR]) was computed using the rate (percentage) per person-years at risk derived from Cox proportional hazards models.[A1,2] Because the earliest marriage to a subject in our study occurred at age 14 years (and the latest at age 50 years), we counted the length of time from age 13 years to the date of the first marriage and censored on date of interview or death. Divorce rates were calculated as the number of first marriages or cohabiting relationships that had ended before the end of follow-up. Breakdown of first marriage was calculated as the time elapsed from the beginning of the first marriage or relationship to its end, censoring on interview or death of the respondent. Kaplan-Meier[A3] methods were used to plot actuarial life-table estimates of time to first marriage and time to first divorce.

Matched proportional hazards analyses were carried out in SAS, with family as the blocking factor.[A4] The inferences drawn from the matched analyses were the same as those from analyses using pooled control subjects. We report the results of analyses of pooled control subjects, because the smaller numbers of like-sexed sibling pairs or triads made detailed analyses impossible.

Infertility[2,3]

Two groups of study subjects were defined with use of a series of exclusions (Table 1). Only 3% of pregnancies occurred among unmarried or non-cohabiting subjects; therefore, only married or cohabiting survivors were included in the analysis. Nine controls who had had a childhood or adolescent cancer were excluded; 113 survivors who married before their cancer was diagnosed were also excluded because of the inverted sequence of events. A number of women who became pregnant or men who fathered a child before their first marriage were excluded because their interval between marriage and pregnancy was "negative" and therefore they could not be included in the proportional-hazards analysis.

Finally, we excluded other subjects because of known sterility or voluntary infertility. Among the 62 who were known to be sterile before their first marriage were 7 women who had never menstruated, 21 women with secondary amenorrhea, and 34 people who had had sterilizing surgery. The results presented in Tables 2, 3, and 4 and in Figures 1 to 3 apply to the subjects remaining after all the exclusions—i.e., those presumed to be at risk of pregnancy. Table 4 also shows the relative fertility associated with treatment before the exclusion of subjects who were at little or no risk of pregnancy.

The age at diagnosis was dichotomized at 15 years because of small numbers of subjects; for instance, of 27 males treated with alkylating agents, only 6 received their treatment before they were 10 years old.

The numbers of male subjects treated before 10 years of age in the other treatment groups were also small. "Pregnancy status" refers to a first pregnancy, regardless of the age of the subject at the time, and was dichotomized as "ever or never pregnant." The measure of association between survivor and control status and fertility is called "relative fertility" and is equivalent to a relative risk. "Pregnancy rate" refers to the proportion of subjects who had ever been pregnant or who had fathered a child. Data analysis was done with crude and multiple contingency tables, using the Mantel-Haenszel method.[A5] Proportional-hazards models[A1] were used to estimate the fertility of survivors relative to that of their siblings while controlling for the effects of covariates. We obtained estimates and 95% confidence limits for relative fertility after adjustment for age and calendar year of marriage (as continuous variables), sex, and varying duration of the risk of pregnancy after first marriage in the presence of censoring due to death, end of follow-up, or menopause. Survival curves were calculated with the Kaplan-Meier method.[A3]

The analyses reported here used quasi-matched controls—i.e., survivors in each treatment and site category were compared with their own sibling controls. Results obtained from exactly matched analyses—i.e., proportional-hazards analysis stratified according to family—were closely similar but were not reported because of the 80% loss of subjects, due partly to the number of survivors who had no siblings and partly to the lack of sibling controls of the same sex.

Educational Achievement[4]

[*Reprinted from Kelaghan* et al. *with permission of Editor and by permission of Wiley-Liss, Inc., a subsidiary of John Wiley & Sons, Inc.*]

Rate ratios were calculated as the percentage of survivors achieving a given grade level (eighth, twelfth, etc.) to the percentage of controls achieving that level. Approximate 95% confidence intervals for the ratio were calculated using test-based methods.[A6] The null hypothesis that the percentage achieving a given grade level was the same for survivors and controls was rejected if the confidence interval did not include 1.0. To test for linear trend in a series of grade-level ratios, the method of Breslow and Day was used.[A7] In addition, matched analyses (taking family into account) were performed using logistic regression (PROC PHGLM).[A8] The matched and unmatched analyses produced very similar results, so only the unmatched results are presented.

Initial examination of educational achievement as measured by completing 12 years of school (high school) was based on all interviewed subjects. For subsequent analyses, we defined four levels of educational achievement (completion of eighth grade, completion of high school, entrance to college, and completion of 4 years of college) and made the completion of the last three levels conditional on completion of the previous levels. For instance, all subjects (with exceptions explained below) were eligible for the analysis of completion of eighth grade, but only those who completed eighth grade were included in the analysis of graduation from high school.

The index cancer could have been diagnosed after some educational levels were passed. To ensure that any differences in educational achievement could be related to the history of cancer, we eliminated from each analysis those survivors (and their controls) who were diagnosed at an age later than the customary attainment of a given educational level. Therefore, analysis of completion of eighth grade was limited to those diagnosed before age 15 years, and analysis of completion of high school and entrance to college were limited to those diagnosed under age 19 years. Furthermore, the completion of 4 years of college usually requires the individual to be at least 22 years old, so we eliminated from the analysis of this variable those under 22 years of age at time of follow-up. Five survivors and four controls with missing educational data were not included in any calculations.

Cancer in Offspring[5]

[*Reprinted from Mulvihill* et al.[5] *with permission of* The Lancet.]

The exact version[A9] of the non-parametric Wilcoxon rank-sum test[A10] was used to compare the frequency of cancer in offspring of survivors with the frequency in like-sexed sibling-controls (1-tailed test, $p = 0.05$). To control for interfamilial differences, the exact version of the blocked Wilcoxon two-sample rank-sum test was also used, with families as the blocking factor. The blocked Wilcoxon test was further modified for number of offspring by matching each survivor to like-sexed sibling controls with regard to number of offspring. Observed numbers of cancers among offspring of survivors and controls were compared with expected numbers computed with life-table methods.[A11] Person-years at risk of development of cancer were calculated from date of birth of the offspring to date of cancer diagnosis, interview, or child's death, whichever occurred first. Expected numbers of cancers (except intracranial tumours and non-melanotic skin cancers) were calculated by multiplying the person-years at risk by age-specific, sex-specific, and calendar time-specific incidences, based on data from the Connecticut Tumor Registry. The ratio of observed to expected numbers of cancers (relative risk [RR]) was tested for statistical significance (1-tailed test, $p = 0.05$) by assuming that the observed numbers of cancers followed a Poisson distribution.

Birth Defects in Offspring[6]

At interview, subjects were asked specifically about each conception and live born child. One question inquired about "conditions sometimes present at or soon after birth" and referred the subject to two printed lists of conditions. The list entitled "Health Conditions at Birth" included the following: crossed eyes (strabismus), stomach blockage (pyloric stenosis), hole in roof of mouth (cleft palate), hare lip (cleft lip), rupture in groin (inguinal hernia), club foot, absent, fused, or extra fingers or toes, hole in the heart, hip displacement, diverted urinary stream (hypospadias), mongolism (Down syndrome), open spine (spina bifida), water on the brain (hydrocephalus), exposed brain (anencephaly), undescended testicle (cryptorchidism), prematurity, hyaline membrane disease, and other. The second list entitled "Other Conditions" included achondroplasia, acrocephalosyndactyly, aniridia, Apert's syndrome, cancer, dystrophia myotonia, Gardner's syndrome, Marfan's syndrome, multiple polyposis, neurofibromatosis, osteogenesis imperfecta, polycystic disease of the kidney, von Recklinghausen's disease, retinoblastoma, and Steinert's syndrome. Efforts to verify the reported condition through review of hospital records, death certificates and cancer treatment records were made for each reported event. All disorders were included in the analysis unless records specifically denied the report.

"Genetic disease" was defined as a syndrome of malformations known to have an associated cytogenetic abnormality (whether or not karyotypes were available), a single gene (Mendelian) disorder, or any one of 15 common simple birth defects. A condition was described as "sporadic" if no relative had the same or related genetic disease in the opinion of two reviewing physicians. All recessive disorders were called "familial" (nonsporadic), since both parents were gene carriers. If a child had both a familial and a sporadic condition, the child was classified as sporadic; when the number of genetic conditions was tabulated, only the condition determined to be sporadic was counted. Chi-square analysis and Fisher's exact test were used to test statistical significance.

APPENDIX REFERENCES

A1. Cox, D. R. 1972. Regression models and life tables (with discussion). JR Stat Soc Series B. **34:** 187–220.

A2. LEE, E. T. 1980. Statistical Methods for Survival Data Analysis. Lifetime Learning Publications. Belmont, CA.

A3. KAPLAN, E. L. & P. MEIER. 1958. Nonparametric estimation from incomplete observations. J. Am. Stat. Assoc. **53:** 457–481.

A4. HARRELL, F. E., JR., Ed. 1986. The PHGLM Procedure: SUGI Supplemental Library User's Guide. Version 5. SAS Institute Inc. Cary, NC.

A5. MANTEL, N. & W. HAENSZEL. 1959. Statistical aspects of the analysis of data from retrospective studies of disease. JNCI **22:** 719–748.

A6. ROTHMAN, K. J. & J. D. BOICE. 1979. "Epidemiologic Analysis With a Programmable Calculator." NIH Pub. 79-1649. Government Printing Office. Washington, DC. p. 11.

A7. BRESLOW, N. E. & N. E. DAY. 1980. Statistical Methods in Cancer Research: Volume 1. The Analysis of Case-Control Studies. IARC Pub. No. 32. Lyon: International Agency for Research on Cancer.

A8. SAS INSTITUTE INC. 1983. Supplemental Library Users Guide. SAS Institute Inc. Cary, NC.

A9. SIMON, R. 1980. Statistical methods for evaluating pregnancy outcomes in patients with Hodgkin's disease. Cancer **45:** 2890–2892.

A10. LEHMANN, E. L. 1975. Nonparametrics: Statistical methods based on ranks. Holden-Day. San Francisco.

A11. MONSON, R. R. 1974. Analysis of relative survival and proportional mortality. Comput. Biomed. Res. **7:** 325–332.

Quality Assurance of Chemotherapy

SÁNDOR ECKHARDT

National Institute of Oncology
Budapest POB 21.H-1525, Hungary

Recently, not only evidence-based medicine but also quality of life aspects became topics of paramount importance in clinical research. Among quality of life issues, quality assurance (QA) is one of the most controversial subjects. QA procedures are required for all medical activities. Cost-effectiveness—including benefit/risk calculations for the patient—may only be performed if the procedure is based on a generally accepted, optimalized, standard activity.

What is the definition of QA? Optimalization of health care activities according to internationally agreed standards. In the clinical management of malignancies, major QA segments exist that are separate categories for the care of both children and adults.[1] These are the following: screening, early detection, diagnosis, staging, prognosis, treatment, rehabilitation, and terminal care.

Currently, the treatment of malignancies consists of several modalities often referred to as "complex" therapy. This means that the combination of surgery, radiotherapy, chemotherapy, and different methods of biotherapy associated with psychological assistance and the necessary supportive measures are applied with success to the patient. The goal of this review is to survey QA issues of major importance in cancer chemotherapy.

The most important aims of any QA procedures can be summarized as follows: (a) optimalization of professional standards in protocols, (b) adequate treatment management, (c) protection of patient, and (d) protection of health care personnel. From QA aspects of cancer chemotherapy, the following main criteria have to be fulfilled: the drug has to be stable and possess a tolerable level of toxicity; its dosage and regimen known; phase I, II, and III trial information available; and the statistical evaluation of its activity published. Whenever conflicting results are achieved, they should be widely discussed, and a consensus must be reached—preferably on an international basis. Thereafter, the agreed recommendations may be summarized in a protocol.

The protocol always means a state-of-the-art document valid for a certain period. If any new treatment method (new drug, new regimen, etc.) becomes available, it has to be clinically investigated. In cases in which there is statistical evidence for its greater effectiveness in comparison to the method described in the protocol, the original protocol has to be replaced by a new one.

Thus, a constant protocol revision committee is needed to monitor progress and development. This is especially important for patients with a long-run follow-up, sometimes also called phase IV, for whom not only recurrence but also late toxicity might occur, necessitating establishment of a new protocol.[2] A good example is tamoxifen, an almost harmless hormonal drug used for more than 20 years. Its dosage has been revisited in the past years because of its recently revealed late toxicity. In this case a protocol revision is obviously needed.

Protocols are also professional education guidelines. They can be effectively used in universities and teaching hospitals for disseminating standard, accepted, and agreed-upon practice.

190

Which criteria have to be fulfilled in a chemotherapy protocol? Basically, rules of good clinical practice (GCP) have to be considered: for example, the drug must be chemically defined and stable and its solubility known. The substance should be active and its toxicity tolerable; its dosage and regimen published. Pharmacokinetic data—including potential drug interactions—have to be available. Dose reduction schedules as well as the list of the necessary protective measures against toxicity should also be listed. The question of what drug(s) to give and how long to administer them to the patient should be correctly answered in detail.

A further question awaits reply in the protocol: to whom should chemotherapy be delivered? The answer is that only patients with chemosensitive tumors should be treated by cytostatic drugs, since they are potentially curable. Which are the main groups of chemosensitive tumors? The great majority of pediatric solid tumors, leukemias and lymphomas, testicular and ovarian malignancies, trophoblastic tumors, osteogenic sarcomas, and breast and prostatic cancer. There are also carcinomas in which chemotherapy might be beneficial by inducing complete response associated with a prolongation of life but without a cure (colorectal, head and neck, small-cell lung cancer, multiple myeloma, renal and bladder cancer, apudomas, melanoma, etc.) The treatment results have justified the establishment of protocols for these diseases as well. All other chemotherapeutic trials with new drug regimens remain investigational until the value of their effectiveness has been statistically proven. Accordingly, these methods are only candidates for being incorporated into protocols.

The exact delivery of any chemotherapeutic treatment according to the prescriptions of a protocol depends on the availability of adequate manpower. A sufficient number of physicians and the corresponding paramedical personel is essential for the success of therapy, along with the requisite skill and expertise. QA also means that the protocol prescribes the quantitative and qualitative minimum requirements. The same applies to the resources. Cytostatics, supportive drugs, drug delivery instruments for administration, and any kind of tools necessary for the given procedure should be also listed in a protocol. By this means any mismanagement of the patient—including overtreatment or undertherapy—should be avoided. At the same time, care providers, when following the guidelines of a protocol, may find themselves under the protective umbrella of the agreed standard procedures. An evaluation of the cost-effectiveness, including risk assessment, may direct health insurance policy in a realistic direction. The patient's satisfaction analysis may also assist in the reaching of this goal.

The patient protection segment of the QA includes the following aspects: (a) the affected person should have access to full information related to his disease, including prognostic possibilities and therapeutic options with their potential outcome. If he is well and correctly informed, then (b) he is able to make a treatment choice together with his physician, (c) he must give his consent to the therapy chosen by the common patient/doctor decision, (d) the patient's active participation in the therapy is essential; and (e) his satisfaction analysis may assist in the evaluation of the whole procedure. All these aspects could be formulated in the QA guidelines. It may be difficult, however, to treat patients with limited intellectual capacity (infants, children, incapacitated persons, very old individuals, etc.) according to these principles. Nevertheless, in such cases a proper substitute decision-maker for the patient has to be found (parents, social worker etc.).

The health personnel must be protected not only against discontent on the part of the patient, as mentioned above, but also against conflicting or fraudulent activities of the other care providers. Shelter is sometimes required against media attacks as well. Health insurance companies may also respect QA guidelines. All in all, a protocol describing therapeutic procedures could serve as a basis for the calculations needed by insurers.

The QA guidelines in chemotherapy are always valid for a standard treatment. All other therapeutic activities are investigational. It is very difficult to distinguish between "standard" and "investigational" therapy. What is standard today may be outdated tomorrow. Conversely, any investigational method, if based on sufficient evidence, may be standard therapy in the future (e.g., gene therapy).[3]

REFERENCES

1. POLLOCK, B. H. 1994. Quality assurance for interventions in clinical trials: Multicenter monitoring, data management, and analysis. Cancer **74:** 2647–2652.
2. CASTIGLIONE-GERTSCH, M., *et al.* 1994. The international (Ludwig) Breast Cancer Study Group trials I–IV: 15-year follow-up. Ann. Oncol. **5:** 717–724.
3. DEISSEROTH, A. B. 1993. Current trends and future directions in the genetic therapy of human neoplastic diseases. Cancer **72:** 2069–2074.

The Problem of Pediatric Malignancies in the Developing World

H. P. WAGNER[a] AND V. ANTIC[b]

[a]Swiss Institute for Applied Cancer Research
Effingerstrasse 40
3008 Berne, Switzerland

[b]Division of Clinical and Experimental Research
University of Berne
Tiefenaustrasse 112
3004 Berne, Switzerland

INTRODUCTION

According to Magrath,[1] a major benefit of the problem of pediatric malignancies in developing countries is the pressure to rethink the essentials of pediatric oncology and the opportunity to design clinical trials that pay particular attention to simplicity, economy, short-duration therapy, and safety. In addition, pediatric oncology in developing countries offers the opportunity to gain information about the relevance of the environment to the development of pediatric neoplasia.

In industrialized countries, within four decades and despite the rarity of pediatric cancers, the survival of affected children increased more than threefold, to about 75%. This development was characterized, essentially, by a concentration of specialized personnel in pediatric cancer units (PCU), providing optimum treatment and training, and by the establishment of multidisciplinary, multicenter, cooperative trial groups, testing new therapies. Adequate referral, supportive parents, and good insurance are additional requirements for successful coping with pediatric cancer. Unfortunately, worldwide, fewer than 20% of all children with cancer get adequate treatment.

DEMOGRAPHIC DEVELOPMENT AND PEDIATRIC ONCOLOGY

Pediatric Population and Cancer Incidence

As shown in TABLE 1, approximately 1.45 billion or 84% of all children less than 15 years old lived in developing countries in 1990.[2] Assuming an average cancer incidence of 105 and $125/10^6$ children per year in developing and in industrialized countries, respectively, the absolute number of new pediatric cancers can be estimated.[3] Around the world the cancer incidence in boys is approximately 10% above and in girls 10% below the overall average incidence. The absolute number of pediatric cancers is increasing in the developing, not in the industrialized countries (TABLE 1). In the latter 3 to 4% of all cancer patients are children, but only 0.5% of the deaths due to cancer occur in childhood. In developing countries, the respective figures are 7% and 4%. In industrialized countries, 20% of the population are killed by cancer; in developing countries only 6%.[1]

TABLE 1. Population ($\times 10^6$) and Number of New Pediatric Cancers 1990 (2000)[2,3]

Type of Countries	Population ($\times 10^6$)		Absolute Number of New Cancers
	All Ages	<15 yrs	
Industrialized	1,221 (1,286)	267 (265)	33,000 (33,000)
Developing	4,071 (4,965)	1,446 (1,692)	152,000 (178,000)

Factors Hindering the Development of Pediatric Oncology

The most important hindering factors are poverty, poor hygiene, and lack of education. In 1989 the mean gross national product (GNP) and the mean health expense per capita differed by a factor of 55 and 72, respectively, between low- and high-income countries[4,5] (TABLE 2). Often combined with poverty are poor hygiene, malnutrition, a high infant mortality (10–20%, compared to <0.1% in industrialized countries) and a high mortality of children less than 5 years old (up to 26%).[1] In 1986 it was estimated that in the 36 poorest countries only 36% of the population had access, within 15 minutes walking distance, to clean water.[6] The consequences are diarrheal diseases, a high parasitic infestation, and diminished resistance to infections. If the mean health expense per capita is low, there are fewer physicians (1 per 1,000–6,000 inhabitants compared to 1 per 500 in industrialized countries) and fewer nurses (1 per 1,000 to 2,000 compared to 1 per 150), and both are residing preferentially in urban areas. This means that people in rural areas have to travel far to find a doctor, and the few who can pay for traveling often arrive too late, with a child unable to withstand an anticancer treatment.

Poverty is also related to a lack of education and an inability to earn, particularly in rural areas, where the illiteracy rate is high, in women higher than in men. The consequences are not only a high fertility rate, but also an inability to recognize or understand diseases, to comply with therapy, and to react to complications.

Health Budgets and Pediatric Oncology

In 1989 the average health budget of 16 developing countries accounted for 7.75% of the total budget, but varied from 0.8 to 27.2%.[4] In 1991 WHO and UNICEF

TABLE 2. Mean Gross National Product (GNP) per Capita, Mean Health Expense, Life Expectancy, and Literacy Rate of Adults[4,5]

Type of Countries (as defined, 1989)	Mean GNP (1989, US $)	Mean Health Expense (early 80ies, US $)	Life Expectancy at Birth (1978; years)	Adult Literacy Rate (1978; %)
Low income (GNP < 580 US $)	330	9	50	38
Middle income (GNP 580–6,000 US $)	2,040	31	61	71
High income (GNP > 6,000 US $)	18,330	650	74	99

TABLE 3. Death Rates/100,000 According to Age and Disease in China 1989[7]

	Years		
Disease	0	1–4	5–14
Infections and parasites	61	14	4
Respiratory	458	34	5
Cancer	7	6	4

estimated that with 2 to 3 US $/year per child for oral rehydration, 13 US $/child for vaccines, and a few additional cheap measures 43% of the 14.6 million children lost each year due to diarrheal diseases and infections could be saved.[7] In very poor countries, pediatric oncology has, therefore, no priority. The situation changes, however, if the socioeconomic status of a country improves. This has been the case in Latin America as well as in India, Malaysia, Thailand, and China, where infant mortality has decreased dramatically (over 40%) and where the mortality rate of children below 5 years has fallen to less than 10%.[1] In these countries cancer has become a leading cause of death for children between 5 and 15 years[7] (TABLE 3). It was hoped that by the year 2000, worldwide, 90% of all children born alive would reach the age of 5, but part of the progress has been annihilated by the HIV pandemic.[8] In black Africa 16 million people, 3% of the population, are HIV positive. In some cities, for example, Abidjan, Kinshasa, Lusaka, Kampala, and Blantyre, up to 35% of the pregnant women are infected.[9] If the mother is HIV positive, there is about a 25% chance that the baby will be infected too.[10] If this is the case, the mortality rate during the first 5 years will be approximately 30% higher than for uninfected babies. It has been estimated that by the year 2000 the mortality during the first 5 years of life will have gone up again from 15.9 to 18.9% in 10 Central and East African countries.[11] In Asian countries the HIV infection rate is going up too, not only among homosexuals and sexual workers, but also in the population at large.[8]

GEOGRAPHY OF PEDIATRIC CANCER

Worldwide, acute lymphoblastic leukemia (ALL) is the most common neoplasia. In industrialized countries the incidence of ALL is $>25/10^6$ children below age 14 or 15,[12,13] and the incidence curve shows a peak between 2 and 5 years due to a high incidence of the common (CD10-positive) ALL subtype.[14] In developing countries the incidence of ALL is $<25/10^6$ (in low-income countries $<10/10^6$),[15,16] the percentage of T ALL is higher,[17,18] and there is no peak incidence between 2 and 5 years. Besides, ALL in developing countries is more often associated with lymphadenopathy and splenomegaly. These observations suggest that the incidence of the common ALL, but not of T ALL, is related to socioeconomic status.

Acute nonlymphoblastic leukemia (ANLL) occurs at an incidence rate of 4–10/10^6 and is not related to socioeconomic status.[12,13,19] In developing countries chloromas are found in up to 30% of all cases with ANLL.[19,20]

In Latin America and the Middle East, the M. Hodgkin is more frequent in children than adults and may even be found in children below age 2.[21,22] The mixed-cellularity and lymphocyte-depleted types predominate; and in a high percentage of patients EBV DNA, RNA (EBER), and latent membrane protein (LMP1) are localized in Hodgkin's cells, Reed-Sternberg cells, and their variants (see Refs. 12 and 24 for review). In North America and Europe the nodular-sclerosis type is more common. The incidence rate varies between 4 and $8/10^6$ children in industrialized and between <1 and $10-11/10^6$ children in developing countries.

Among the non-Hodgkin's lymphomas, it is mainly the incidence, the symptomatology, and the EBV association of Burkitt lymphomas that vary.[1,25] In equatorial Africa up to $75/10^6$ children may develop a Burkitt's lymphoma each year.[1,16,26] In Europe and North America the incidence rate is $<10/10^6$ children.[12-14] According to the localization of the break-points on the chromosomes involved in the typical translocations, subtypes of Burkitt's lymphoma can be differentiated.[27] Burkitt's lymphoma illustrates the role of environmental factors in the development of a cancer of the immune system.

Tumors of the central nervous system account for approximately 18% of pediatric cancers in industrialized countries[28] (incidence rates for white children: $27-30/$ million[12,13]), but are less common in the developing world (see Ref. 1 for a review), although frequencies as high as in Western countries have been reported.[29] Of great interest is the high incidence (frequencies up to 13%[30] instead of about 3%[28]) of retinoblastoma in Latin America,[31] Africa,[30,32-34] and Asia,[29,33] but only hospital-based frequency and no population-based incidence rates are available. It is not clear whether these high frequency rates are mainly due to an increased incidence of nonfamilial cases or not.[34,35]

Neuroblastoma ($8-10/10^6$ children in Europe[12] and North America[13] develop neuroblastoma each year) occurs less often in developing countries,[33,36] but the true incidence of this tumor is difficult to determine.[37]

In Asia the incidence of nephroblastoma is lower than in the rest of the world (<4 instead of $8/10^6$ children (see Ref. 1 for a review)).

Due to endemic hepatitis B and C and vertical virus transmission,[38] hepatocellular carcinoma is common in sub-Saharan Africa and New Guinea.[33] Poor hygiene and malnutrition diminish the resistance to infection and promote carcinogenesis associated with chronic viral disease.[39]

In the Sudan and in North Africa nasopharyngeal carcinoma and in the state of São Paulo in Brazil adrenocortical carcinomas occur more frequently than elsewhere.[33] Finally, it is noteworthy that in North America the incidence of Ewing's sarcoma is higher in whites than in blacks.[13] The epidemiology of childhood cancer in Africa has recently been reviewed by Wessels and Hesseling.[40]

THE STATUS OF PEDIATRIC ONCOLOGY IN REPRESENTATIVE DEVELOPING COUNTRIES

Africa

Sub-Saharan countries are among the poorest in the world. There are only a few pediatric oncologists; and children with cancer, if treated, are either in the hands of pediatricians or of surgeons, hematologists, or other doctors interested in one or another form of pediatric cancer. Very few referal centers exist. Examples are the pediatric oncology unit at the Kenyatta National Hospital in Nairobi, Kenya, established in 1986, or the Treichville Hospital in Abidjan, Ivory Coast. Between 1970 and 1995, 1253 children with cancer were sent to Treichville Hospital, corre-

TABLE 4. Overall Survival at 2 Years of 900 Children with Cancer Treated at Treichville Hospital, Abidjan, Ivory Coast, 1970–1995[40]

Type of Cancer	NHL	Hodgkin's	Acute Leukemia	Nephroblastoma	Others
Total treated	455	36	171	128	110
2-yr survival (%)	18	<2	<1	4	<1

sponding to about 10% of the expected number at the Ivory Coast (average of 5 million children, expected pediatric cancers per year around 525). Only 900 of the 1253 children were treated.[41] The 2-year survival was dismal (TABLE 4); 100 died while on chemotherapy.

In Nairobi 67 children with nephroblastoma were operated on from 1987 to 1993.[42] For 32 no records were available; of the remaining 35 (among them 18 stage III, 18 stage IV, and 2 stage V) 12 had no follow-up and 6 a follow-up for less than 3 years. Of the remaining 17 adequately evaluable patients, seven survived 3+ years. In approximately one-third of the patients the tumor ruptured.

At the Queen Elizabeth Central Hospital in Blantyre, Malawi (9 million, 50% less than 15 years old, GNP per capita 230 US $), only 25% of 158 children with NHL completed therapy and only 7% returned within the first two years after treatment for a control.[43]

At the King Edward Hospital in Durban, Natal, South Africa, with the same treatment, the 5-year survival of Indian children with ALL increased from 30 to 60% (78 patients total) but remained at 10% for 115 black children with the same disease.[44]

In North Africa there are a number of pediatric oncology units treating 100 to over 300 new pediatric cancers per year. Compared to sub-Saharan Africa, more pediatric oncologists are available, and the diagnostic and treatment facilities are better; but referals are often late, and drugs as well as blood products are not always available as needed.

In Egypt (60 million, 40% children) at the National Cancer Institute in Cairo, approximately 10% of all histopathologic examinations concern children. Of 594 solid pediatric tumors seen from 1985 to 1989, 36% were lymphomas, 18% bone and 16% soft tissue sarcomas, 7% CNS tumors, and 4% nephro- and 3% neuroblastomas.[45] Of 1277 children with cancer seen from 1970 to 1996 at the Department of Clinical Oncology and Radiotherapy of the University Hospital in Alexandria, there were 34% leukemias, 19% CNS tumors, 15% lymphomas, 12% nephro- and 9% neuroblastomas, 6% bone, and 3% soft tissue sarcomas.[46] The problems in the management of ALL in Egypt are well illustrated by a report on 104 children with ALL treated at the Cairo University Kasr El Eint Center of Oncology.[47] In 19 out of 104 children, no therapy could be given due to a very poor general status at entry. Fourteen died during induction, and in 23 the treatment could not be properly administered and the patients relapsed or disappeared. Of the remaining 48, 45 achieved complete remission and of these 36 remained in complete remission after a minimum follow-up of 2 years. More satisfactory are the treatment results of 166 children with Morbus Hodgkins at the National Cancer Institute in Cairo: Of 152 evaluable patients, 88% survived relapse-free at 4 years, and the overall survival was 92%.[48] Frequent infestation with *Schistosoma mansoni* as well as hepatitis B and C often complicate the therapeutic efforts.

Asia

India has a population of 882 million. Thirty-six percent are children. Around 30,000 children get cancer every year, but fewer than 3,000 survive. There are big cancer centers with pediatric divisions in Bombay (Tata Memorial), New Delhi (All India Insitute of Medical Science), Madras (Cancer Institute at Adyar), Bengalore (Katway Memorial Cancer Insitute), Trivandrum, Vellore, Ahmedabad, Calcutta, and Chandighar (Post Graduate Medical Institute). Each of these centers admits several hundred new pediatric cancer patients each year, but only about one out of three children can be treated because of late referal, poor conditions, lack of drugs and blood products, and lack of parental compliance (see Ref. 49 for review). In 1992 the Tata Memorial admitted 745 new children with cancer (60% with leukemia or lymphoma), accounting for 5% of all admissions. There are few specialists devoted to pediatric oncology only. Good diagnostic and radiation facilities are available at the big cancer centers, but access to these centers is limited. Multicenter cooperative groups have been established to test antileukemic therapy.[1,49]

China has a population of 1150 million, and one-third to one-fourth are children. The GNP per capita has doubled within 10 years. Cancer is becoming one of the leading causes of death for children 5 to 15 years old (TABLE 3); 30 to 35,000 new pediatric cancers are expected annually. There are structures for pediatric oncology in pediatric clinics of university hospitals, cancer centers, research institutes, and military hospitals, for example, in Beijing (Beijing Children's Hospital, Beijing University Hospitals, Beijing Institute for Tumor Treatment), Shanghai (Shanghai Children's Hospital, Pediatric Hospital of the Shanghai Medical University, Xinhua Hospital of the Shanghai Second Medical University, Shanghai Institute for Pediatric Research), Tianjin, Shenyang, Haerbing and Jilin, in Nanjing, Suzhou, Hefei, Jinan, Zengzhou, Xian, in Fuzhou and Nanchang, in Guangzhou, Chongqing, and Nanning, and in Xining and Urumchi. In 1992 the Pediatric Oncology Committee of the Chinese Anti-Cancer Association comprised 78 oncologists, 34 surgeons, 8 radiooncologists, 12 pathologists, and 23 epidemiologists and researchers, a total of 155 individuals.[50] The majority of these came from Shanghai, Beijing, Guangzhou, Tianjing, Shenyang, and Nanjing, but all of the aforementioned cities were represented. In Shanghai and Beijing and a few other cities, a small fraction of children with cancer are treated like children in industrialized countries, but the vast majority of pediatric cancer patients do not receive adequate treatment. Only very recently efforts were made to establish cooperative trial groups. A specific aspect of Chinese medicine is the coexistence of traditional and Western medicine, the latter being used for acute, the former for more chronic ailments.[51] The main problems in China are the lack of specialists, of drugs, of supportive care, and late referals.

In the newly industrialized nations of Southeast Asia (Singapore, Hong Kong, Taiwan, South Korea) pediatric oncology is similar to that in North America or Europe.

Latin America

The Latin Amercian countries belong to the more advanced developing countries with a GNP per capita between US $600 and 2,700 (TABLE 5). For pediatric oncology this means that the treatment results obtained at big pediatric cancer units are comparable to those in industrialized countries. This is certainly true for Chile, Argentina, and Brazil.

TABLE 5. Demography (1988–1992) of South American Countries Listed According to Infant Mortality[a]

Country	Population $\times 10^6$	Percent Living in Urban Areas	Infant Mortality (per 1000 live births)	Life Expectancy (yr)		Literacy Rate over Age 15		Gross National Product per Capita (US $)
Chile	14	85	17	68	75	94	93	1940
Uruguay	3	86	20	69	75	95	95	2560
Argentina	33	86	31	68	74	96	95	2370
Colombia	33	67	37	66	72	88	86	1240
Paraguay	5	48	39	65	70	92	88	1110
Ecuador	11	55	63	63	68	92	88	960
Brazil	151	75	68	64	69	82	81	2680
Peru	23	71	81	62	65	96	83	1160
Bolivia	8	51	102	51	55	85	71	620

[a] From *Encyclopedia Britannica.*

In 1991 Argentina had a population of 33 million. Only 14% lived in rural areas. The infant mortality was 3.1%, and the mean life expectancy was 68 years for men and 74 years for women. Approximately 850 to 900 children develop a new cancer each year. Most of the pediatric oncology specialists live in Buenos Aires, but there are some in Rosario, Cordoba, and Mendoza. In Buenos Aires several hospitals take care of pediatric cancer patients, for example, the Hospital de Pediatria "JP Garrhan" or the Hospital de Niños "R. Gutierrez." These hospitals are well equipped and treat practically all patients according to GATLA (the Argentinian Group for the Treatment of Acute Leukemia) or GLATHEM (the Latin American Group for the Treatment of Hematopoietic Malignancies) protocols. Even children in the rural areas have access to a pediatric cancer unit, and the follow-up is continously improved.

In 1992 Brazil had a population of 151 million, and 25% were living in rural areas. The infant mortality rate was 6.8%. In 1991 the mean life expectancy was 64 years for men and 69 for women. It has been estimated that approximately 4,000 children develop a new cancer each year. The access to medical facilities varies and is better in the south than in the north. The most important pediatric cancer units are situated in São Paulo, Rio de Janeiro, Campinas, Curitiba, Porto Allegre, Salvador, and Recife. One of the three centers for pediatric oncology in São Paulo, the A.C. Camargo Hospital, treats over 400 new cancer patients every year. Standardization of treatment (protocols for nephroblastoma, non-Hodgkins lymphomas, leukemias) and advances in supportive care have improved treatment results to close to those obtained in industrialized countries.

STRATEGIES FOR THE IMPROVEMENT OF PEDIATRIC CANCER TREATMENT IN DEVELOPING COUNTRIES

The definition of minimal requirements to guarantee effective treatment and the enhancement of the percentage of children with cancer having access to treatment are the main goals. The establishment, at university hospitals or cancer centers, of pediatric cancer units (PCU) is the most efficient way to concentrate specialists; increase interdisciplinary cooperation; make an optimum use of resources; develop supportive care; and assure training, data management, and follow-up.[1,52]

In developing countries PCUs can function efficiently without sophisticated imaging facilities, intensive-care units, and hematopoietic precursor cell transplantation, provided they have a good hematology, microbiology, and histopathology laboratory; good surgery and radiation facilities; the necessary drugs and blood products; and minimum facilities for data management and follow-up.

One important requirement is that immunosuppressed patients are separated from patients with infectious diseases and that a sufficient number of rooms exists for isolation, infusion of drugs and blood products, and outpatient services. Very important are housing facilities for accompanying family members and availability of support for transportation for the follow-up.

Diagnoses and staging should be accurate enough to permit adequate registration and stratification for treatment. Standardization, adjusted to the local conditions, of initial investigations, treatment, and evaluation of toxicities and treatment results are a *conditio sine qua non* in order to make progress. Malnutrition, parasitic infestation, preexisting infections (e.g., tuberculosis, HIV), parental failure to support therapy, and limited supportive care have to be kept in mind when decisions regarding the aggressiveness, the complexity, and/or the duration of treatment are made. In addition, the importance of an adequate infection prophylaxis for febrile, neutropenic patients cannot be overemphasized. Raje *et al.*[53] and Kurkure *et al.*[54] reported from India that between 1986 and 1992 mortality due to infection after introduction of prophylactic therapy decreased from 20 to less than 5%, while disease-free survival increased from 40 to 60%.

As mentioned before, PCUs are essential for the formation of pediatric oncology specialists at all levels. By exchanging staff with peripheral hospitals and dispenseries and by establishing contacts with practitioners, PCUs can contribute to an earlier detection and referal of cancer patients. In addition, by collaborating with other PCUs in the same or in other countries, they can participate in multicenter trials or take part in international research projects, for example, through the mechanism of "twinning," that is, long-lasting cooperation between a center in a developing and a center in an industrialized country.

Careful planning with a clear definition of priorities and goals is required for adequate development of pediatric oncology in a country. This planning can be done by PCUs, cancer organizations, or governmental authorities and may concern the localization of PCUs, the development of networks for efficient detection and referral, and the institution of cooperative groups such as national associations and/ or parents' groups as well as educational programs providing standardized training and degrees or diplomas for specific functions.

International societies such as the International Society of Pediatric Oncology and subcommittes of international nongovernmental or governmental organizations may support conferences, educational exchange programs, and scholarships and may provide assistance in planning or negotiations with governmental authorities. Last but not least, pharmaceutical companies or agencies like the World Bank could promote the development of medical industries in developing countries covering the local needs more specifically and providing drugs at a fair price. By promoting interdisciplinary cooperation and supportive care, pediatric oncology could there-fore contribute to the establishment of modern and efficient treatment facilities, serving not only pediatric cancer patients but also the population at large.

SUMMARY

In industralized countries, 70 to 80% of children with cancer can be cured by expensive interdisciplinary teamwork and by cooperation on the national or international level. As a result of poor socio-economic conditions, worldwide, less than 20% of the approximately 185,000 children developing cancer each year get adequate treatment. This is also true for 14 to 15 million children dying each year of diarrhea and infections, but while this number is decreasing, the number of children with cancer is increasing. In "middle income" developing countries, cancer is now a leading cause of death for children between 5 and 15 years of age.

The "geography" of pediatric cancer reveals complex interactions between environment, lifestyle, and carcinogenesis. The "mapping" of pediatric cancer is far from complete, and the investigation of carcinogenetic interactions has barely started.

A great challenge is the planning of pediatric oncology in developing countries. The goals are to improve the access to treatment and treatment results. Even if pediatric oncology has a low priority, the institution, in each country or large province, of at least one pediatric cancer unit may improve not only cancer treatment but medical care in general. By promoting education, organizing meetings, and setting minimum standards for training and care, international organizations can contribute to the development of pediatric oncology worldwide.

REFERENCES

1. MAGRATH, I., *et al.* 1993. Pediatric oncology in less developed countries. *In* Principles and Practice of Pediatric Oncology. P. A. Pizzo & D. G. Poplack, Eds.: 1225–1251. Lippincott. Philadelphia, PA.
2. DEMOGRAPHIC YEARBOOK 1988. United Nations. New York.
3. PERKIN, M., *et al.* 1988. International Incidence of Childhood Cancer. IARC publication No. 87. International Agency for Research on Cancer. Lyon, France.
4. WORLD DEVELOPMENT REPORT 1991 (published for the World Bank). 1991. Oxford University. Oxford, England.
5. POVERTY AND HUMAN DEVELOPMENT (published for the World Bank). 1980. Oxford University, Oxford.
6. WORLD HEALTH STATISTICS ANNUAL, 1986. 1987. World Health Organization. Geneva.
7. WORLD HEALTH STATISTICS ANNUAL, 1990. 1991. World Health Organization. Geneva.
8. QUINN, T. C. 1996. Global burden of the HIV infection. Lancet **348:** 99–106.
9. MIOTTI, P. G., *et al.* 1990. HIV-1 and pregnant women: Associated factors, prevalence, estimate of incidence and role in fetal wastage in central Africa. AIDS **4:** 733–736.
10. PECKHAM, C. & D. GIBB. 1995. Mother-to-child transmission of the human immunodeficiency virus. N. Engl. J. Med. **333:** 298–302.
11. CHIN, H. J. 1990. Current and future dimensions of the HIV/AIDS pandemic in women and children. Lancet **336:** 221–224.
12. STILLER, C. A., *et al.* 1995. Childhood cancer in Britain: The national registry of childhood tumours and incidence rates 1978–1987. Eur. J. Cancer **31A:** 2028–2034.
13. GURNEY, J. G., *et al.* 1996. Trends in cancer incidence among children in the U.S. Cancer **78:** 532–541.
14. GURNEY, J. G., *et al.* 1995. Incidence of cancer in children in the United States. Cancer **75:** 2186–2195.
15. WESSELS, G. 1995. A survey of childhood cancer in Namibia 1983–1988. *In* Proceedings of the First Continental Meeting of the International Society of Paediatric Oncology in Africa. Stellenbosch, April 6–9, 1994. P. B. Hesseling & G. Wessels, Eds.: 9–11. University of Stellenbosch. Stellenbosch, RSA.
16. MUKIIBI, J. M., *et al.* 1995. Pattern of the most common childhood malignancies in Malawi. In Proceedings of the First Continental Meeting of the Internatonal Society of Paediatric Oncology in Africa. Stellenbosch, April 6–9, 1994. P. B. Hesseling & G. Wessels, Eds.: 12–13. University of Stellenbosch, Stellenbosch, RSA.
17. KAMAT, D. M., *et al.* 1985. Pattern of subtypes of acute lymphoblastic leukemia in India. Leuk. Res. **9:** 927–934.
18. KAMEL, A., *et al.* 1990. Phenotypic analysis of T-cell acute lymphoblastic leukemia in Egypt. Leuk. Res. **14:** 602–609.
19. MAC DOUGALL, L. G., *et al.* 1986. Acute childhood leukemia in Johannesburg. Am. J. Pediatr. Hematol. Oncol. **8:** 43–51.
20. SHOME, D. K., *et al.* 1992. Orbital granulocytic sarcomas (myeloid sarcomas) in acute nonlymphocytic leukemia. Cancer **70:** 2298–2301.
21. WACHTEL, A., *et al.* 1995. Childhood Hodgkin's disease (HD) in Peru: Clinical presentation and treatment (abstract). Med. Pediatr. Oncol. **25:** 330.
22. ROGUIN, A., M. WEYL BEN-ARUSH & J. DALE. 1995. Incidence of childhood lymphoma in Northern Israel. Pediatr. Hematol Oncol. **12:** 447–454.
23. AMBINDER, R. F., *et al.* 1993. Epstein-Barr virus and childhood Hodgkin's disease in Honduras and the United States. Blood **81:** 462–467.
24. HUH, J., *et al*, 1996. A pathologic study of Hodgkin's disease in Korea and its association with the Epstein-Barr virus infection. Cancer **77:** 949–955.
25. MAGRATH, I. T. 1991. African Burkitt's lymphoma. Am. J. Pediatr. Hematol. Oncol. **13:** 222–246.
26. PLO, K. J., *et al.* 1995. Childhood malignancies—an 18-year experience in Abidjan, Côte d'Ivoire, West Africa. *In* Proceedings of the First Continental Meeting of the International Society of Paediatric Oncology in Africa. Stellenbosch, April 6–9, 1994. P. B. Hesseling & G. Wessels, Eds. : 17. University of Stellenbosch. Stellenbosch, RSA.

27. MAGRATH, I. 1993. Malignant non-Hodgkin's lymphomas in children. *In* Principles and Practice of Pediatric Oncology. P. A. Pizzo & D. G. Poplack, Eds.: 537–575. Lippincott. Philadelphia, PA.
28. MILLER, R. W., J. L. YOUNG, JR. & B. NOVAKOVIC. 1994. Childhood cancer. Cancer **75:** 395–405.
29. KUSUMAKUMARY, P., *et al.* 1991. Childhood cancer in Kerala, India (letter). Lancet **338:** 455–456.
30. MASSABI, M., B. K. MUAKA & N. TAMBA. 1989. Epidemiology of childhood cancer in Zaire (letter). Lancet **ii:** 501.
31. PEREZ, C., *et al.* 1995. Advanced retinoblastoma (RB): Peruvian experience (abstract). Med. Pediatr. Oncol. **25:** 261.
32. ABIOSE, A., *et al.* 1985. Childhood malignancies of the eye and orbit in northern Nigeria. Cancer **55:** 2889–2893.
33. PERKIN, D. M., *et al.* 1988. The international incidence of childhood cancer. Int. J. Cancer **42:** 511–520.
34. DRAPER, G. J. 1990. Childhood cancer in Zaire (letter). Lancet **335:** 553–554.
35. ANTONELI, C. B. G., *et al.* 1996. Retinoblastoma (RB) experience in a developing country (abstract). Med. Pediatr. Oncol. **27:** 267.
36. MILLER, R. W. 1989. No neuroblastoma in Zaire (letter). Lancet **ii:** 978–979.
37. LUCAS, S. B. & P. R. FISCHER. 1990. No neuroblastoma in Zaire (letter). Lancet **335:** 115.
38. ALTER, M. J. 1994. Transmission of hepatitis C virus—route, dose and titer (editorial). N. Engl. J. Med. **330:** 784–786.
39. MORRIS, J. D. H., A. L. W. F. EDDLESTON & T. CROOK. 1995. Viral infection and cancer (review). Lancet **346:** 754–758.
40. WESSELS, G. & P. B. HESSELING. 1995. Epidemiology of childhood cancer in Africa. Int. J. Pediatr. Hematol. Oncol. **2:** 263–268.
41. PLO, K. J., *et al.* 1996. Chemotherapy in pediatric oncology: Past, present and future in the Ivory Coast. Abstracts of the Second Continental Meeting of the International Society of Pediatric Oncology. Cairo, March 6–9, 1996: 12–13.
42. SAFWAT, A. S. & W. M. MACHARIA. 1995. Surgery in advanced Wilms' tumour. *In* Proceedings of the First Continental Meeting of the International Society of Paediatric Oncology in Africa. Stellenbosch, April 6–9, 1994. P. B. Hesseling & G. Wessels, Eds.: 134–136. University of Stellenbosch. Stellenbosch, RSA.
43. VAN HASSELT, E. J. & R. BROADHEAD. 1995. Burkitt's lymphoma: A case file study of 160 patients treated in Queen Elizabeth central hospital from 1988 to 1992. Pediatr. Hematol. Oncol **12:** 277–281.
44. THEJPAL, R. &. J. A. NAIDOO. 1995. Poor outcome in African ALL. *In* Proceedings of the First Continental Meeting of the International Society of Paediatric Oncology in Africa. Stellenbosch, April 6–9, 1994. P. B. Hesseling & G. Wessels, Eds.: 163–164. University of Stellenbosch. Stellenbosch, RSA.
45. TAWFIK, H. N. 1996. Spectrum of pediatric oncology at the pathology department, NCI, Cairo. Abstracts of the Second Continental Meeting of the International Society of Pediatric Oncology. Cairo, March 6–9, 1996: 5–6.
46. MAHMOUD, L. M., *et al.* 1996. Paediatric malignant tumours in Alexandria: An overview from January 1970 to December 1996. Abstracts of the Second Continental Meeting of the International Society of Pediatric Oncology. Cairo, March 6–9, 1996: 90.
47. NAYEL, H. 1996. Problems in the management of pediatric ALL in Egypt: Clinico-pathologic characteristics and treatment results of 104 cases (POCH, 1991–1993). Abstracts of the Second Continental Meeting of the International Society of Pediatric Oncology. Cairo, March 6–9, 1996: 25–26.
48. EL BADAWY, S. A., *et al.* 1996. Pediatric Hodgkin's disease, treatment results of risk adopted combined chemotherapy and radiotherapy. Abstracts of the Second Continental Meeting of the International Society of Pediatric Oncology. Cairo, March 6–9, 1996: 66.
49. CHANDY, M. 1995. Childhood acute lymphoblastic leukemia in India: An approach to management in a three-tier society. Med. Pediatr. Oncol. **25:** 197–203.
50. WANG, Y. Personal communication.

51. FAZEL, M. 1995. Now show me your tongue: A taste of medicine in China. Lancet **346:** 1687–1688.
52. SIOP COMMITTEE ON STANDARDS OF CARE AND TRAINING IN PAEDIATRIC ONCOLOGY. 1991. Requirements for the Training of a Paediatric Haematologist/Oncologist and Recommendations for the Organisation of a Paediatric Cancer Unit (PCU).
53. RAJE, N. S., *et al.* 1994. Infection analysis in acute lymphoblastic leukemia: A report of 499 consecutive episodes in India. Pediatr. Hematol. Oncol. **11:** 271–280.
54. KURKURE, P., *et al.* 1995. Impact of improved supportive care on treatment outcome in acute lymphoblastic leukemia—an Indian experience (abstract). Med. Pediatr. Oncol. **25:** 261.

Nutrition and Pediatric Cancer

HEDVIG E. BODÁNSZKY[a]

Second Department of Pediatrics
Semmelweis University Medical School
Tüzoltó u 7-9
Budapest, Hungary

Nutrition is one of the most important factors in treating patients with malignant diseases, especially while they are receiving chemotherapy. Patients with malignant diseases who are able to sustain their weight appear better able to tolerate their illness and the treatment program. Good nutritional status is especially important in childhood, because the nourishment consumed serves not only the daily energy requirements, but also the development of the child's body.

There are many reasons that patients with malignant diseases lose weight. Important among these is the disease itself, the chemotherapy received, and the environment as well as psychological and emotional factors (FIG. 1).

Complications are more common during chemotherapy, because the metabolic alterations often affect the patient's sense of taste and of smell—even the composition of the saliva may undergo changes. Also during chemotherapy, some organs may be damaged, depending on the drugs used (TABLE 1). One of the most sensitive tissues in this respect is the small intestinal mucosa, which is involved with all drugs.

In order to determine the condition of the small intestinal mucosa, we can measure the enzyme lactase, which is the most sensitive enzyme of the mucosa because it is located in the highest layer of the brush border. In our hemato-oncology department, we have been studying patients with acute lymphoid leukemia during chemotherapy and examining the extent of the intestine's mucosal damage; the duration of the damage; and, after the treatment, the time needed for healing. To determine the condition of the small intestinal mucosa, we measured the activity of the enzyme lactase. For this, we used the expiratory H_2 breath test, one of the most informative noninvasive methods for diagnosing functional disorders of the small intestine.[1]

We have studied 25 children with acute lymphoid leukemia: 10 girls and 15 boys between 2.5 and 17.5 years with a mean age of 8.8 years. The risk factor was considered "medium" in 4 out of 10 girls and "low" in 6 out of 10. In the group of boys, one was high risk, one was medium, and the others had low risk factors (TABLE 2).

These children have been treated by means of the European Leukemia Study Group (SIOP) protocols according to the risk factors. We determined the degree of the small intestinal mucosal damage and followed the degree of the damage. The duration of the damage was measured after the conclusion of the therapies.

We found that the damage usually lasts for 6 to 8 weeks. It persisted for the longest time after methotrexate administration. In the overall clinical picture of

[a] Address for correspondence: Hedvig E. Bodánszky, M.D., Ph.D., Kelenhegyi ut 81, 1118 Budapest, Hungary.

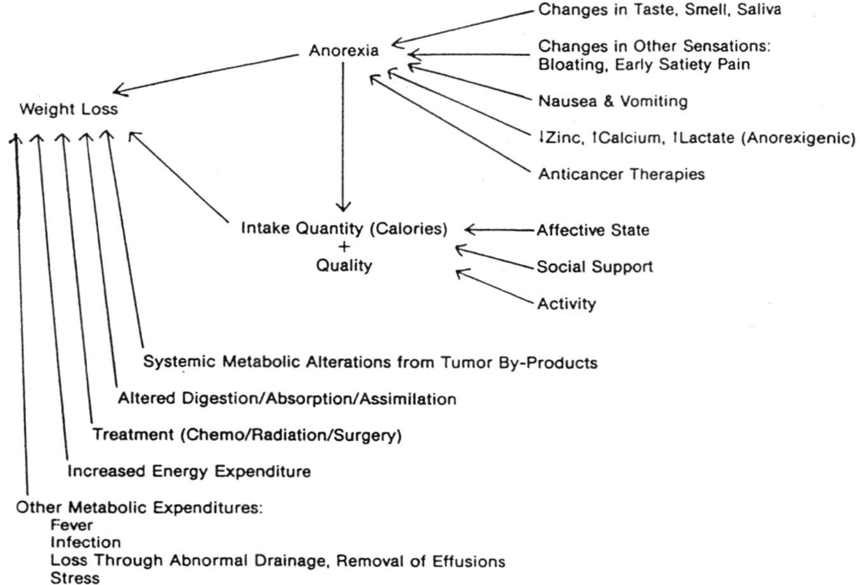

FIGURE 1. Factors influencing weight loss in cachexia. (From Lindsey A. M. 1986. Cancer cachexia: Effects of the disease and its treatment. Sem. Oncol. Nursing **2:** 1929. Reprinted with permission of Grune & Stratton, Inc.)

these children, the most dangerous condition is cachexia, which develops secondary to the progressive growth of the malignancy and has grave implications for the quality of life, length of survival, and decreased tolerance of the treatment.

The etiology of nutritional disorders is complex.[2] Nausea, vomiting, and anorexia

TABLE 1. Gastrointestinal Side Effects of Chemotherapeutic Agents

Chemotherapeutic Agents	Gastrointestinal Side Effects
Adriamycin	Anorexia, nausea, vomiting
Cisplatinium	
Cyclophosphamid	
Procarbazine	
Actinomycin D	Mucositis
Adriamycin	
Methotrexate	
Vinblastin	
Actinomycin D	Diarrhea
Methotrexate	
Vinblastin	Constipation
Vincristin	Adynamical ileus
Methotrexate	Organ damage, liver
Asparaginase	Pancreas

TABLE 2. Risk Factors of 25 Children with ALL

Sex	BFM	Age (Mean)
10 girls	19 low (6 girls, 13 boys)	2.5–17.5 years (8.86 years)
15 boys	5 medium (4 girls, 1 boy)	
	1 high (1 boy)	

induced by chemotherapy or radiotherapy usually result in a greater degree of anorexia than that caused by the cancerous process itself (FIG. 2).

Feeding children with malignant diseases is not easy, because we have to take into consideration that an ill child usually has no appetite, feels full very quickly, and a change in metabolism also develops as a result of a change in the functions of the hypothalamus. A change in taste and the ability to distinguish between different tastes occurs.[8]

In our department, we studied the alteration in taste of children with acute lymphoid leukemia, and we found significant changes. According to the literature,[3] we found the threshold for sweet-tasting food is higher and the threshold for bitter-tasting food is lower among children. Therefore, we had to calculate with aversion to certain foods taken into consideration.

All this leads to a situation in which the patient loses his or her ability to sustain normal weight because of an inability to eat enough proper food. Other factors also contribute to the loss of weight. With a drastic change in the metabolism, the

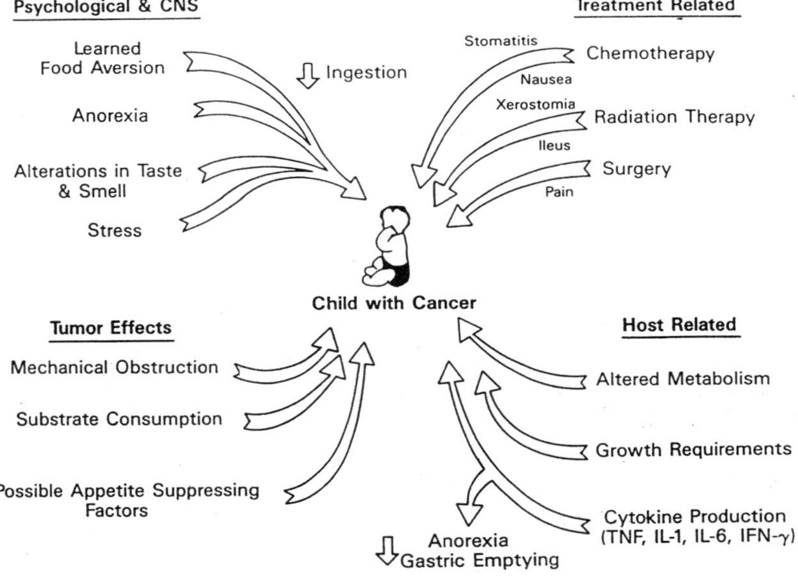

FIGURE 2. Factors contributing to the origin and progression of cachexia in a young patient with cancer.

balance of salt and water in the body is overturned, with progressive diminishing of the vital functions.

PROTEIN

The most dangerous effect of insufficient nutrition is protein-energy malnutrition. The human body has no protein reserves; therefore, if protein intake is not sufficient the body begins to consume itself, beginning with muscle fibers, including heart muscle.[4] We speak of protein loss if the weight loss is more than 5%, if the weight/length ratio is smaller than 5%, or if the value of serum albumin is less than 33 g/l.

Animal experiments indicate that the problem is even more complex than it first appears, because a high-protein diet can also help the growth of the tumor tissue. The literature of the last few years shows that certain amino acids are responsible for promoting tumor growth in certain cases in which the development and avoidance of protein malnutrition was monitored. It is possible that by extracting certain amino acids from the diet we can slow down or even stop the growth of the tumor.[5]

FAT

Fat is the most important source of energy. At the same time, it has several other important biological functions: It is a structural part of cell membranes, a major component in metabolism, and also has a role in immune function. Well-absorbed medium- and short-chain fatty acids are an important part of the diet.[9]

CARBOHYDRATES

The most sensitive part of chemotherapy is its effect on the absorption of carbohydrates, because of the damage to the small intestinal mucosa. Therefore, the avoidance of milk and milk products during chemotherapy is recommended.[10]

Naturally, the role that vitamins and minerals as well as trace elements play is more important in these patients than under normal circumstances. For example, because of their antioxidant effect, vitamin E and selenium are more important for the chemotherapy patient than for a healthy person.

We should not neglect the role of the bacterial flora in the intestine, because chemotherapy also affects bacterial microflora.[4] The pathogenic microbes increase because of the damage to the mucosa, and this also has a negative effect on the absorption of nutrients. Therefore, a feeding program has to be established (FIG. 3).

SUMMARY

Good nutritional status is very important in patients with malignant diseases, especially during chemotherapy. It is most important to avoid cachexia: One should avoid factors that may cause it or feed the patient by tube or intravenously if not possible orally, even if it is necessary to give total parenteral nutrition. The target

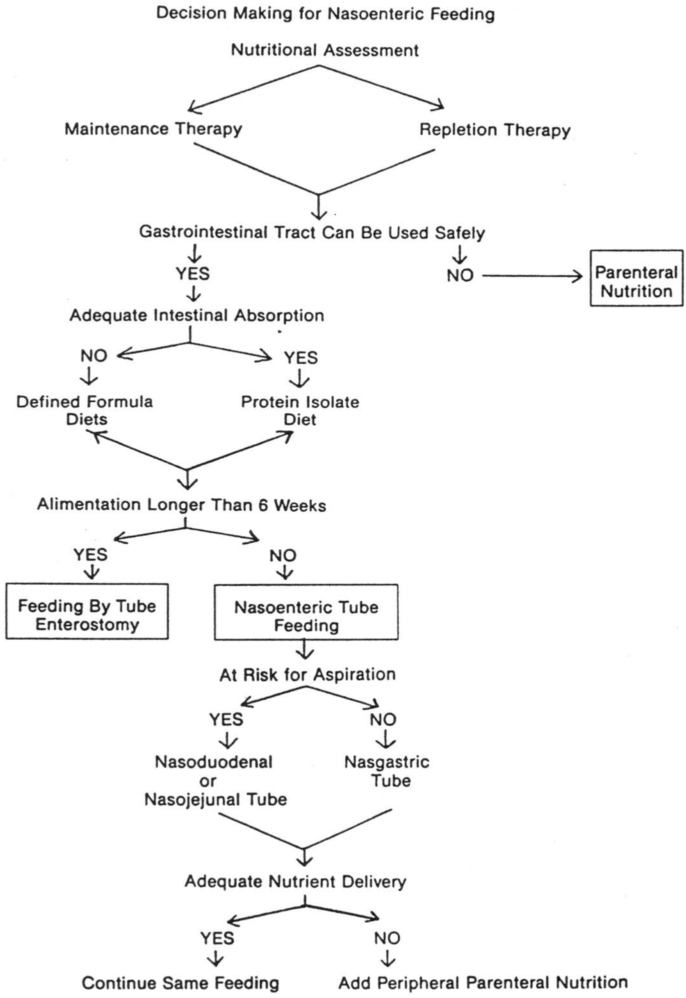

FIGURE 3. Patient selection for nasoenteric feeding. (From Rombeau, J. & M. Caldwell, Eds. 1983. Enteral and Tube Feeding. W. B. Saunders. Philadelphia, PA. Reprinted with permission of publisher.)

is to ensure that the infants and children with malignant diseases during chemotherapy will retain optimal nutritional condition, which helps the children's growth development and helps them tolerate the treatment better. As the patient's age changes, so the need for certain nutritive materials is modified, and the usage is changed also.

The Psychosocial Program for Childhood Leukemia in Monza, Italy

GIUSEPPE MASERA,[a,b] MOMCILO JANKOVIC,[b]
LUIGIA ADAMOLI,[c] ADELE CORBETTA,[c]
DONATELLA FRASCHINI,[c] ROMANA LIA,[c]
LAURA COLLINO,[c] ANNA LOCATI,[c] SILVIA PERTICI,[c]
ROSANNA BISSI,[c] MARIO BERTOLINI,[d]
GIOVANNI VERGA,[c] AND JOHN J. SPINETTA[c,e]

*[b]Clinica Pediatrica
[c]Comitato M. L. Verga
Per lo Studio e la Cura delia Leucemia del Bambino
[d]Cattedra di Neuropsichiatria Infantile
Università di Milano
Ospedale San Gerardo
Monza, Italy*

*[e]San Diego State University
Department of Psychology
San Diego, California 92182*

INTRODUCTION

Since the early 1970s, a cure for childhood leukemia has become a reality: Over 60% of cases are now cured. Yet, despite the relatively high cure rate, the diagnosis of leukemia continues to place a heavy burden on family functioning. The parents must walk the narrow line between focusing too much on the child's disease and treatment and maintaining a normal family life. Because cure is such a real possibility, the children must be prepared for a full and active participation in adult life, just like their peers. Proper discipline must be maintained within as normal a family life as possible. The child's continued attendance at school and participation in normal childhood activities is imperative in the child's preparation for adulthood. For all these reasons, psychosocial intervention has become a necessity in the treatment of the child, even for those children who eventually will die from the disease.

Although there is little disagreement that the ultimate goal of treatment for childhood leukemia is the total cure of the child—medical, educational, psychological, and social—the issue is how best to achieve this end.[1,2] The literature is filled with research-based conclusions on which type of psychosocial intervention is best,[3] including when and how one should communicate with the child about the diagnosis,[4,5] how to help the parents maintain some sense of normality in their family life,[6–8] how to help the child return to school,[9] how to keep the siblings informed,[10,11] how to start parent groups,[12] how to involve parents in medical decision-making,[12]

---

[a] Address for correspondence: Giuseppe Masera, M.D., Ospedale S. Gerardo, Centro di Ematologia Pediatrica, Via Donizett, 106, Monza, Italy 20052.

how to prepare for the terminal phase when it occurs for some children,[13,14] and how to continue to monitor long-term survivors.[15-18] Problems occur whenever the approach must be modified to meet the needs and cultural preparation and expectations of the children and their families. This is especially true when one tries to apply conclusions that are appropriate in one culture to other centers and to other cultures.

What works in one cultural setting may not work as well in others. How might a center apply programs from one country or setting to another? Not all hospitals can afford a psychosocial team. Not all cultures appreciate intervention by a psychologist or psychiatrist. What can pediatric hematologists do to modify their approach to the children and their families with maximal success, in a manner most appropriate to and respectful of the needs of the families within their own cultural setting? And above all, how can a center best monitor its intervention programs, to ensure that the needs of the children and their families are being met appropriately, in their best interests, and with greatest effectiveness and use of resources?

Communication of the diagnosis to the child, for example, becomes problematic in a culture not accustomed to such openness with children. In such a culture, both the child and the family must be prepared, taught the importance of such communication, and taught how to communicate at the child's level of development. Involving the parents in the decision-making process regarding their child's treatment, especially at the terminal phase, is very difficult in a culture that is used to a physician making all such decisions. Any movement in the direction of greater parental involvement must be tailored not only to cultural expectations, but also to the individual family's level of preparedness for such sharing. Sending a child back to school as soon as possible after initial treatment might not be acceptable to many parents or teachers, who must be trained to accept the value of an early return to school. Asking parents to discipline the child in a manner consistent with family history may become problematic for parents culturally expected to make fewer demands upon or give special treatment to a sick child.

No matter how valid a particular suggestion for intervention may be for the center doing the original research, a center with different family needs and expectations must apply the research findings in a reasonable and monitored manner. Continued research is critical in the application of psychosocial programs to different cultures, to determine if that particular intervention works in that culture and is appropriate for the children and families at that particular center. A continued monitoring of intervention programs and a modification of the programs based on research findings is critical.

With these issues in mind, we want to present a comprehensive program incorporating psychosocial support and research elements as an integral part of the total management of children with leukemia and their families. This program developed from more than 20 years of experience in this area. It is offered as a possible model for others attempting to manage the different and sometimes complex psychosocial issues deriving from care of the child with cancer. It must be kept in mind, of course, that in each part of the program the most important obligation for members of the health care team is to reserve time and space for meeting with parents and patients.

THE PROGRAM AT MONZA

The psychosocial program began in 1970, when medical advances in the treatment of childhood leukemia made cure increasingly possible in a large percentage of cases. Back then, it also became clear that the problems of children with cancer, as for any chronic illness, must be confronted in an environment of cooperation between physicians and parents. In 1974, the Milan Lombardy Association for the Hemopathic Child was founded, with the goal of gathering together parents of children of the Region of Lombardy (9 million people in northern Italy) with various hemopathies (leukemia, thalassemia, and hemophilia in particular).

The next major development came in 1978 with the birth of a parent association (Comitato Maria Letizia Verga) dedicated specifically to leukemia and to helping meet the needs of the health care team at Monza. At that time a very productive and enthusiastic "therapeutic alliance" began between the parents (as represented by the Comitato Verga) and the members of the health care team.

Since then, the program has developed with a basic understanding and conviction that by working together parents and health care team members can more effectively confront the disease and continue to offer the best possible and most comprehensive treatment, overcoming economic, organizational, and structural difficulties.

The health care team at Monza has always operated under the jurisdiction of the University of Milan, initially at the De Marchi Pediatric Clinic in Milan, and since 1984 at the Pediatric Clinic of the San Gerardo Hospital in Monza. In 1985, Monza's team added the collaboration of the division of child neuropsychiatry of the University of Milan and in 1987, the consultation of an American psychologist from San Diego, California, who in the 1970s had conducted critical studies on children with cancer.

Our approach has developed into a program by and for the parents: global; multidisciplinary; and with the goal, philosophy, and strategy of continual change and growth. During the last few years, psychosocial research has grown apace, as have training seminars for Italians from other cities and physicians from other nations. Our hematological center is part of the pediatric clinic of a general hospital, the San Gerardo Hospital. We diagnose about 50 new cases of leukemia per year. At present, we follow about 150 cases in treatment and 400 off therapy. A center for bone marrow transplantation is included: So far more than 120 children have received transplants.

Aim

To offer to every child and family a comprehensive program (medical, psychological, social, and educational) for guaranteeing to as many children as possible a true cure and to everyone the highest possible quality of life.

Method

1. The presence of a multidisciplinary health care team (pediatricians, nurses, psychologist, social worker, teachers, and volunteers) coordinated by a pediatric hematologist.
2. Definition of a common mission and promotion of a "therapeutic alliance" between families and health professionals.
3. Promotion of the most open possible level of communication at all levels, among child, family, physicians, and society to prevent the deleterious consequences of isolation.
4. Promotion, mobilization, and utilization of the family's and child's own resources.

5. Screening of families in greatest need of psychological and/or social support followed by individualized intervention.

The Team

The following members are included in the team: one chairman (GM) and three co-chairmen (JJS, MB, MJ); four pediatric oncologists (LA, DF, AC, AL) all employed by the parents' committee (MLVC, Maria Letizia Verga Committee); and nurses, who are all public service employees (hospital), but four of whom serve on the MLVC; one psychologist (LC), one neuropsychiatrist (RL), and one social worker (SP), all of whom are employed by the MLVC; five teachers employed in public service (school); one pedagogist (RB) employed by the MLVC; volunteers; and the parents' committee.

Components

Communication

Communication of the diagnosis to the family: This is the most important moment, crucial for a family that is in shock from the information that their child has a potentially fatal disease. Keeping in mind that communication is an ongoing effort, to be continued for many days and weeks to come, the following is our approach.

As soon as possible, when laboratory and clinical results are confirmed, the attending physician communicates the diagnosis to both parents together and tells them about our "program of care." At that point, a subsequent interview is set up with the director of the clinic. For this session we invite both parents, the family physician (so far we have obtained 50% attendance), any friend the parents may wish to have attend, and a nurse from the in-patient clinic where the child is being treated. The colloquium typically lasts about 45 minutes, in a private room. The program, the prognosis, and roles of each team member are carefully explained. The family is told how it might confront the difficult experience, and attention is paid to particular family needs, especially to the school and to other siblings.

For the past two years, with the consent of the parents, the colloquium is registered on an audiotape, and a copy of the tape is given to the parents. After a study at our center demonstrating the usefulness of this tape for the parents, the colloquia are now taped routinely (G. Masera *et al.*, unpublished data).

Communication of the diagnosis to the child: A specific physician (MJ) communicates the diagnosis of leukemia directly to the child without the parents being present, to avoid the child being distracted by the parents' presence. The dialogue lasts about 10 minutes and includes a set of 25 slides (mostly cartoons), where a simplified but scientific explanation for leukemia and chemotherapy is expressed. The word "leukemia" is used, unless parents explicitly forbid it (<5%). All of the children are approached at a level dependent on the age of the child. When possible, siblings are included.

The children are asked to explain their disease to their parents right after the physician's explanation, thus opening the door to family dialogue about the illness. The children are then asked to write a paragraph or to draw pictures about the disease and to discuss both with their parents. In this way parents and children can then speak more freely about the illness.

Communication following meetings: Having continuous informal and open communication during the course of therapy is crucial. Parents and children must plan, and be helped to plan, how to best use their own emotional and material resources and energies to deal with the disease and its treatment. In subsequent meetings information (and even suggestions) can be provided regarding daily activities such as: monitoring medication, maintaining a "normal" lifestyle for the family and the child, the child's continuing involvement in school and peer activities, dealing with hope or fear and realistic caution, family dynamics (e.g. spousal, division of roles, sibling concerns), the child's peer relations (awkwardness, teasing) and practical problems (hospital transportation, insurance, and finances).

The Global Approach

The core of the program is a well-integrated, multidisciplinary approach to total care coordinated by the pediatric hematologist, aimed primarily at helping the child and family confront the illness and its fallout with focused energies. The following are the program components:

- Screening for psychosocial and social problems that would likely place families at higher risk for eventual difficulties. A conjoint weekly discussion among the health care team members (in particular the attending physician, nurse, psychologist, and social worker) helps pinpoint families at greatest risk.
- The psychologist meets with the child and family, evaluates the family's psychological difficulties and concerns, and develops an intervention package specific to the psychological needs of that particular family.
- Similarly, the social worker meets with the family and determines if there are socio-economic needs (such as a need for monetary support for transportation or a need for a temporary residence for those families coming from a great distance). Thanks to the economic resources made possible by the parent group, Comitato M. L. Verga, the hospital is able to provide support for the most needy of the families.
- The nurses usually attend the meeting that physicians have every day in the ward. The meeting lasts nearly half an hour, and family needs can also be discussed. In fact, psychosocial problems form an integral part of the discussions.
- Nurses and parent volunteers from the Comitato M. L. Verga are always available to answer any questions the families or children may have.

The School

The educational program is a critical and essential part of the normal psychological development of any child. Depending on medical conditions, the child must be stimulated to return to full attendance at school as soon as possible. In-hospital teachers continue to follow the educational program of each child's school of origin. Depending on clinical conditions and the wishes of the child, channels of communication are opened between the hospital personnel and the teachers at the school of origin via a letter to the principal and informational booklets sent to the teachers. Telephone calls follow this preliminary contact, inviting the teacher to visit the center. Most recently, one of the doctors (LA) from the center has been going to the school and explaining the medical aspects of the leukemia to the teachers and to the child's classmates. The leukemic child attends the explanation, and usually he is actively involved.

A small manual for teachers (available in Italian upon request), including information about the disease and how to cope with the ill child, is systematically delivered to school personnel. If the child requires extra help, a personalized educational program is developed. To evaluate the specific educational competence and needs of each child, an educational instrument (DSBQ)[19] for both family and teachers is adopted. An annual meeting is held with all school personnel, including both in-hospital teachers and teachers from the child's school of origin.

Parents' Committee: Maria Letizia Verga

The Comitato M. L. Verga, formed in 1978, has developed over the years into a parent group intensely involved with the needs of the parents of children with leukemia. The parent group collaborates with the health care team, expanding the community's awareness of and sensitivity toward issues in childhood cancer. The parent group has raised funds for supplementing both the clinical and the research aspects of the program.
The Comitato M. L. Verga:

- Maintains an office with three staff members in the hospital.
- Twice annually publishes a professional-style bulletin (typically 32 or more pages) that is mailed to over 12,000 individuals (families of children with cancer, their friends, supporters).
- Coordinates over 100 parent-generated fund-raising activities every year all over the Region of Lombardy, with an overall participation of about 100,000 persons. These activities also sensitize the institutional structures, the governmental bodies, and public opinion to issues in pediatric oncology.
- Calls a general assembly of the parents every 4 months, chaired by the president of the association and the medical director of the department, to update parents on progress in treatments for leukemia and on the activities of the committee.
- Maintains ties with the administrations both of the hospital and of the region, working collaboratively toward solutions of organizational and space issues within the program.

Care of Long-term Survivors

The program of care for long-term survivors begins when the child or young adult goes off therapy. The physician, and usually also a nurse, meets with the family and with the patient, depending on age, to review the past and plan for the future. The family doctor is also involved at this point, receiving information about the clinical situation.

Our center offers a counseling service for serious medical and psychosocial problems and has developed a specialty clinic oriented toward the preventive medical and psychosocial care of long-term survivors. This clinic is managed by our pediatric hematology group, with a full range of adult and young adult specialists as consulting physicians (e.g., cardiologists, gynecologists, internists, etc.). This multidisciplinary follow-up clinic is able to pinpoint the needs of the individual subject based on a predefined scoring system and specific risk factors. A specifically tailored, computerized follow-up program (PPOA: Personalized Prevention Oriented Approach) (available upon request) is developed for each person. This computerized management system makes the problem easier to handle and more useful for the young adults, providing them with a written summary of their clinical history, a brief report of the received therapy, and associated toxicity.

Relapse

Relapse in leukemia brings with it a severe change in prognosis, depending on the type of relapse. The most critical and difficult decision facing parents and physicians at this point is the evaluation of the risk/benefit ratio to the child either of performing a transplant or of intensifying chemotherapy. The problem has become even more complex in recent years with the development of a variety of approaches to transplant when a fully matched family donor is not available: matched unrelated donor, aplo-identical family donor, cord blood stem cells completely or incompletely matched, autologous bone marrow, or peripheral stem cells. Given the lack of hard scientific data on the effectiveness of each of these alternate approaches to transplant, the difficulty and disorientation faced by the parents becomes obvious.

Given that even the physicians have a difficult time coming to a clear and precise decision on which approach to use, the director of the clinic meets with the parents and the physician in charge of the bone marrow transplant program (and, when possible, with the family physician) to discuss all of the options. If the family wishes, a second meeting is set up with the other members of the health care team (nurse, psychologist, social worker). This open-communication and shared-decision-making approach to the parents reduces preoccupations and anxiety in the face of this very difficult and controversial choice.

The Terminal Phase—A Palliative Approach

When all medical avenues have been exhausted and cure is no longer possible, with the consent of the parents we proceed to the phase of palliative therapy. The main goal is to obtain the best quality of life possible both for the child and for the family, through a supportive and palliative system of care based on the control of physical and psychic pain. Furthermore, we try as much as possible to avoid hospitalization and to give the child the possibility of continuing to attend school.

Our approach addresses these goals with:

a. A well-structured program coordinated by a pediatric hematologist (MJ);
b. Involvement of a few specific nurses;
c. Presence of another physician chosen according to the wishes or requests of the family;
d. Involvement of the family doctor from the beginning;
e. Periodic and continuous controls in the outpatient clinic to give the child and family the feeling of not being abandoned and to monitor the progress of the disease;
f. Phone calls often during the day to encourage the family in administering the drugs (anti-pain or sedatives) and in managing the difficult situation;
g. Dosage-adequate and easily applicable anti-pain medications that the child can

receive at home. Hospitalization is provided only if strictly necessary or requested by the child and/or family.

A meeting with parents after the death of their child is included as routine procedure: (a) for reviewing medical aspects of the illness the parents might wish to discuss or have explained and (b) for receiving feedback regarding possible suggestions for improving medical assistance during the terminal phase of the child.[20]

Research

It is critical to monitor psychosocial programs in order to evaluate the effectiveness of the program, both in whole and in part, to better the program over time. We have done this by: (a) Family Adjustment Scale (FAS)[21,22]; (b) Deasy-Spinetta Behavioral Questionnaire (DSBQ)[23]; and (c) evaluation of parents' satisfaction.

Monitoring is also important to verify and validate hypotheses derived from clinical experience (help to the family over time). We are doing this by: (a) studying the family over time to see if family members become stronger on basic human values, by use of the Current Adjustment Scale (CAS)[16,21]; (b) incorporating parents whose children have been cured or whose children have died to see if these experienced parents can be of help to the parents of these newly diagnosed children. The last point (b) is still being studied following the wishes expressed by parents.

Family Adjustment Scale (FAS): The FAS[21,22] was developed as a way of placing families onto a normal curve of adjustment and coping strengths, so that interventions might be geared toward specific family needs.

The FAS includes the following components:

1. A demographic form for collecting personal data regarding the patients, the family, and the treatment;
2. Four series of questions for evaluating how the family functions—the emotional needs of the patient (15 items), the siblings (12 items), the father (12 items), and the mother (12 items);
3. One set of questions for evaluating how the family copes with the medical needs of the patient (8 items) and one set for measuring the family's day-to-day functioning (7 items).

Each item is scored 0, 1, or 2. For every family, there is a separate score for each of the six sets of items, and a total score for the family (the total of the four family members). The FAS is filled out by a member of the research team and also by the physician and the nurse who take care of the child during hospitalization.

Some families adjust to the illness and its treatment much more adequately than others. Some families, often because of prediagnosis problems, might need extra support. One of the goals of the project was to help sort out families according to need, so that each might be helped in the way most appropriate and most helpful for that particular family. With this screening goal, our clinic uses an Italian translation of the FAS.

Having learned how to screen families from the use of the FAS over many years, the team screens the families more directly and more efficiently without it.

Deasy-Spinetta Behavioral Questionnaire (DSBQ): A second instrument in use in the Monza clinic is the DSBQ.[19,23] It is a 38-item questionnaire with dichotomous answers used to measure the child's current level of school functioning, both academic and social, from the teacher's perspective. Each child's classroom teacher receives two packets of material to complete, one regarding the patient and the other regarding a "typical child" in the same classroom who is the same age and sex as the patient and is representative of the general characteristics of the class as a whole. The completed forms are later processed. Children with leukemia differ significantly from controls both in overall school functioning and in the subordinate areas of learning, socialization, and emotional adjustment. Analysis of individual items demonstrates, however, that the areas of most concern are limited and that, in general, children with cancer do attend school willingly and regularly and do not exhibit apprehension regarding school. As a group, they are not at a disadvantage in pursuing a regular school program and are capable of a healthy integration into the school environment. Furthermore, only a small number of the leukemic children are responsible for the significant differences. Institutional help can aid the child in integrating with his or her peers. Even when the child's stay in the hospital is brief, the presence of an in-hospital teacher pursuing the regular academic program becomes for the child a symbol of reality and of the daily tasks that must be pursued despite the illness. School confirms that the child's life continues.

One of the primary uses of the DSBQ is in screening for this small number of children who score as most in need of preventive intervention. Our center is currently engaged in an ongoing prospective study of the efficacy of interventions for the child.

Evaluation of parental satisfaction: A questionnaire was administered in 1995 to 1996 to all of the parents of children treated between 1982 and 1992 at our clinic to measure their satisfaction with the various aspects of treatment. The results of this descriptive survey were discussed by the members of the health care team and presented to the parents at one of their general assemblies in order to obtain comments and feedback.[24] The survey supported our general treatment approach and reinforced our center's characteristically close working relationship between the members of the health care team as a group and the parents as a group (what we call a "therapeutic alliance"). We are currently engaged in an additional study of the satisfaction of the children with our approach to treatment.

The monitoring of the psychosocial intervention program results in a continuous updating of the program, with input from the families and children, resulting in a more effective, appropriate, and helpful use of resources.

Current Adjustment Scale (CAS): Families who have gone through the cancer experience, whether the child has died or has been cured, very often arrive at a point of greater strength and ability to confront life's difficulties because of what they have learned about themselves while confronting the cancer.[16,21] Since 1989, we have administered the 18-item CAS to families at various stages post-treatment in order to measure the family's current level of adjustment. We are evaluating the data from this study. We are attempting to understand this phenomenon in relation to other sociocultural values and family situations.

CONCLUSIONS

This program is feasible. Each team member has a role, but learns enough about the other disciplines (medicine, social work, psychology) to allow open communication regarding all aspects of the child's and family's needs, strengths, and weaknesses, so that the family can best be helped on the road to healthy coping. With such open communication, there is a unity of outlook and purpose, and parents are more willing to become active participants in the program.

The group is persisting; the parents are satisfied; the physicians are satisfied. We are writing this article in the hope that others may see this open approach as feasible in their own centers. We would like to (1) give support to the concerns of pediatric hematologists that, while psychosocial intervention programs are good, not all interventions good at one center are appropriate for other centers; (2) show that there are relatively clear and moderately simple ways of deciding which types of interventions might best fit a particular center's needs; (3) share instruments already in use at many centers, in different languages, that can help screen families for the most effective use of resources; (4) underscore the need for a continual monitoring of psychosocial intervention programs to insure that the particular interventions are in the best interests of the children and their families, and (5) to give an example of the possibility and the effectiveness of cooperative international psychosocial efforts.

One unavoidable issue is the scarcity of resources for carrying out so inclusive and vast a program. Health care workers are costly. We are aware that public entities, already overburdened by medical costs, are hard put to find the resources necessary for offering global assistance such as we have outlined, inclusive of education, psychosocial support, and research. We are confident, however, based on our 20 years of experience, that a close working alliance with parents both offers the possibility of collaboration and renders concrete this important and fundamental educational-psychosocial-research dimension of treatment.

We will be satisfied if this paper stimulates a dialogue among centers with different sociocultural conditions regarding a multidisciplinary approach appropriate to and feasible for each center. We underscore the critical therapeutic alliance that must take place at the most fundamental level at each center between the health care professionals as a group and the parents as a group.

ACKNOWLEDGMENTS

The authors would like to thank the "Fondazioni Tettamanti per lo Studio delle Leucemie ed Emopatie Infantili" for their support to this program and Mrs. Elena Berta for secretarial assistance.

REFERENCES

1. VAN DONGEN-MELMAN, J. E. W. M. & J. A. R. SANDERS-WOUSTRA. 1986. Psychosocial aspects of childhood cancer: A review of the literature. J. Child Psychol. Psychiat. **27:** 145–180.
2. VARNI, J. W. & E. R. KATZ. 1987. Psychological aspects of childhood cancer: A review of research. J. Psychos. Oncol. **5:** 93–110.
3. MASERA, G., J. J. SPINETTA, G. J. D'ANGIO, *et al.* 1993. Critical commentary: SIOP Working Committee on Psychosocial Issues in Pediatric Oncology. Med. Pediatr. Oncol. **21:** 627–628.
4. JANKOVIC, M., N. B. LOIACONO, J. J. SPINETTA, L. RIVA, V. CONTER & G. MASERA. 1991. Telling young children with leukemia their diagnosis: The flower garden as analogy. Pediatr. Hematol. Oncol. **11:** 75–81.
5. MASERA, G., M. CHESLER, M. JANKOVIC, *et al.* 1997. SIOP Working Committee on Psychosocial Issues in Pediatric Oncology: Guidelines for communication of the diagnosis. Med. Pediatr. Oncol. **28:** 6.
6. OVERHOLSER, J. C. & G. K. FRITZ. 1990. The impact of childhood cancer on the family. J. Psychos. Oncol. **8:** 71–85.

7. BIRENBAUM, L. K. 1991. Measurement of family coping. J. Pediatr. Oncol. Nurs. **8:** 39–42.
8. HORWITZ, W. A. & A. E. KAZAK. 1990. Family adaptation to childhood cancer: Sibling and family system variables. J. Clin. Child Psychol. **19:** 221–228.
9. MASERA, G., M. JANKOVIC, P. DEASY-SPINETTA, *et al.* 1995. SIOP Working Committee on Psychosocial Issues in Pediatric Oncology: Guidelines for school/education. Med. Pediatr. Oncol. **25:** 429–430.
10. CHESLER, M. A., J. ALLSWEDE & O. BARBARIN. 1991. Voices from the margin of the family: Siblings of children with cancer. J. Psychos. Oncol. **9:** 19–42.
11. MARTINSON, I. M., C. GILLIS, D. C. COLAIZZO, M. FREEMAN & E. BOSSERT. 1990. Impact of childhood cancer on healthy school-age siblings. Cancer Nurs. **13:** 183–190.
12. CHESLER, M. A. & O. A. BARBARIN. 1987. Childhood cancer and the family: Meeting the challenge of stress and support. Brunner/Mazel. New York.
13. SPINETTA, J. J., J. A. SWARNER & J. P. SHEPOSH. 1981. Effective parental coping following the death of a child from cancer. J. Pediatr. Psychol. **6:** 251–263.
14. SPINETTA, J. J. & P. DEASY-SPINETTA. 1981. Talking with children who have a life-threatening illness. *In* Living with Childhood Cancer. J. J. Spinetta & P. Deasy-Spinetta, Eds. C. V. Mosby. St. Louis, MO.
15. SPEECHLEY, N. & S. NOH. 1992. Surviving childhood cancer, social support, and parents' psychological adjustment. J. Pediatr. Psychol. **17:** 15–31.
16. SPINETTA, J. J., J. L. MURPHY, P. J. VIK, J. DAY & M. A. MOTT. 1989. Long-term adjustment in families of children with cancer. J. Psychos. Oncol. **14:** 179–191.
17. RAIT, D. S., J. S. OSTROFF, K. SMITH, D. F. CELLA, *et al.* 1992. Lives in a balance: Perceived family functioning and the psychosocial adjustment of adolescent cancer survivors. Family Process **31:** 383–397.
18. MASERA, G., M. CHESLER, M. JANKOVIC, *et al.* 1996. SIOP Working Committee on Psychosocial Issues in Pediatric Oncology: Guidelines for care of long-term survivors. Med. Pediatr. Oncol. **27:** 1–2.
19. ADAMOLI, L., P. DEASY-SPINETTA, A. CORBETTA, A. LOCATI, D. FRASCHINI, M. JANKOVIC, G. MASERA & J. J. SPINETTA. 1997. School functioning for the child with leukemia in continuous first remission: Screening high-risk children. Pediatr Hematol. Oncol. **16:** 121–131.
20. JANKOVIC, M., G. MASERA, C. UDERZO, V. CONTER, L. ADAMOLI & J. J. SPINETTA. 1989. Meetings with parents after the death of their child from leukemia. Pediatr. Hematol. Oncol. **6:** 155–160.
21. SPINETTA, J. J. 1991. Adjustment and adaptation in the child with cancer: A 3-year study. *In* Living with Childhood Cancer. J. J. Spinetta & P. Deasy-Spinetta, Eds.: 5–23. C. V. Mosby. St. Louis, MO.
22. ADAMOLI, L., D. FRASCHINI, A. CORBETTA, G. MASERA, J. J. SPINETTA, *et al.* 1996. A screening instrument for families at high risk for psychosocial problems in childhood leukemia: A multicenter study. Unpublished manuscript. University of Milan, Monza, Italy.
23. DEASY-SPINETTA, P. M. & J. J. SPINETTA. 1990. The child with cancer in school: Teachers' appraisal. Am. J. Pediatr. Hematol./Oncol. **2:** 89–94.
24. MASERA, G., G. TOGNONI, M. JANKOVIC, *et al.* 1996. La valutazione della soddisfazione delle famiglie in oncologia pediatrica. Riv. Inf. **15**(1): 5–13.

Anthracycline Cardiotoxicity and Its Prevention[a]

M. G. MOTT

University of Bristol
Institute of Child Health
Royal Hospital for Sick Children
St. Michael's Hill
Bristol, BS2 8BJ, United Kingdom

Most children treated for cancer in the modern era are cured of their disease and have become long-term survivors. The price these survivors may have to pay in terms of late effects of their treatment with surgery, radiation, and chemotherapy becomes increasingly important as alternative methods of treatment are developed; and a balance has to be struck between maximizing the chances of cure and minimizing the late effects for the long-term survivors.

A significant treatment component in the improved proportion of long-term survivors among childhood cancer patients has been the incorporation of the anthracycline drugs. The majority of long-term survivors will have been exposed to them during their treatment. The anthracycline antibiotics were isolated in the late 1950s and early 1960s. In 1957 a colony of *Streptomyces* that produced a red pigment was isolated and investigated at the Pharmitalia Research Laboratories in Milan, the product being named daunomycin after the pre-Roman tribe of Daunii from southern Italy.[1] A French group working independently in the Rhone Poulenc Laboratories isolated an antibiotic called Rubidomycin (Ruby) from *Streptomyces coeruleorubidis*. Rubidomycin and daunomycin were later found to be the same substance, and the name of the parent compound used in cancer trials was changed to daunorubicin.[2] Early clinical trials established the efficacy of daunorubicin and its C-14 hydroxy derivative, doxorubicin, in the control of pediatric malignancies, but as early as 1976 Gilladoga *et al.*[3,4] had already reported subsequent cardiotoxicity in pediatric patients. Since then, many attempts have been made to separate the antitumor effects from the cardiotoxic effects of anthracyclines. These have included the synthesis of a wide variety of analogues and the use of different infusion schedules, antioxidants such as alpha tocopherol (vitamin E) and *n*-acetyl cysteine, and metal-chelating drugs together with the development of carriers, in particular liposomes and conjugates with monoclonal antibodies, to improve targeting and attenuate toxicity.

Clinically, anthracycline-induced cardiotoxicity can be divided into acute, chronic, and late onset. Acute injury can occur immediately after treatment and includes transient arrhythmias, a pericarditis–myocarditis syndrome, or acute left ventricular failure. Chronic cardiomyopathy characteristically presents within one year of treatment, whereas late-onset cardiotoxicity manifests itself years to decades after treatment, often after a prolonged asymptomatic period. The pathogenesis of

[a] This work, undertaken in Bristol in association with my pediatric cardiology colleagues F. A. Bu'Lock and R. P. Martin, was funded by the Cancer Research Campaign and the CLIC Trust.

anthracycline cardiotoxicity is complex. The anthracycline antibiotics cause the selective inhibition of the expression of a wide variety of genes in cardiac myocytes. There is solid evidence of free radical–mediated myocardial damage, damage from calcium overload, disturbances in myocardial adrenergic function, release of vasoactive amines, and cellular toxicity from anthracycline metabolites as well as the production of pro-inflammatory cytokines. The increased oxygen radical activity is generated through the semiquinone moiety of the anthracycline molecule via an iron–anthracycline complex.[5] The irreversible loss of cardiac myocytes causes a dose-related cardiomyopathy, the effect of which is cumulative so that 18% of patients who have received a total dose of 700 mg/m^2 will develop congestive heart failure.[6] Most affected patients will respond to treatment with digoxin and diuretics in the short to medium term, but their long-term cardiac prognosis is not good.

There is a wide variation in individual sensitivity to anthracyclines. Known risk factors for the development of chronic anthracycline cardiotoxicity include higher rates of drug administration, mediastinal radiation, younger or older age, female sex, pre-existing heart disease, and hypertension. Lipshultz *et al.*[7] reported late abnormalities of afterload or contractility in 57% of 115 children who had been treated with doxorubicin for ALL one to fifteen years earlier, the cumulative dose of doxorubicin being the most significant predictor of abnormal cardiac function. Steinherz *et al.*[8] found that 23% of 201 children who had received anthracyclines had abnormal cardiac function at long-term follow-up, a percentage that rises to 38% of those followed more than 10 years and to 63% of those followed more than 10 years who had received a cumulative dose of 500 mg/m^2 or more. In our own studies of 226 survivors of childhood malignancy, more than 75% of those who had received more than 300 mg/m^2 of anthracycline had a left ventricular shortening fraction below the normal 25th percentile (less than 35%), and 23% had a frankly abnormal shortening fraction of less than 30%[9] (FIG. 1).

MONITORING AND PREDICTION

A real need exists for a safe and noninvasive test that can be repeated in children throughout treatment and used to monitor early toxicity as a guide to the prevention of chronic cardiomyopathy. Our experience with the monitoring of childhood patients over the last 20 years has led us and others[7,8] to adopt serial echocardiography as the optimal method to detect and quantitate subclinical myocardial damage. A wide variety of more or less complicated measures of both systolic and diastolic function can be followed by echocardiography, and normal values throughout childhood are now available for many of these[9] (FIG. 2). Our serial studies of cardiac function during treatment have shown that individual sensitivity to anthracyclines can be detected very early in treatment by simple functional parameters such as shortening fraction (SF). Those patients who go on to develop frankly abnormal SF values show a greater reduction of SF from their baseline (pretreatment) values even after the first doses of anthracycline when compared to those in whom SF is preserved (FIG. 3).[10] This should allow for selection of a cohort of patients who might benefit most from the introduction of effective cardioprotective measures.

CARDIOPROTECTION

The most likely cause of anthracycline-induced cardiac damage is the formation of an intracellular anthracycline–iron complex, which catalyzes the generation of

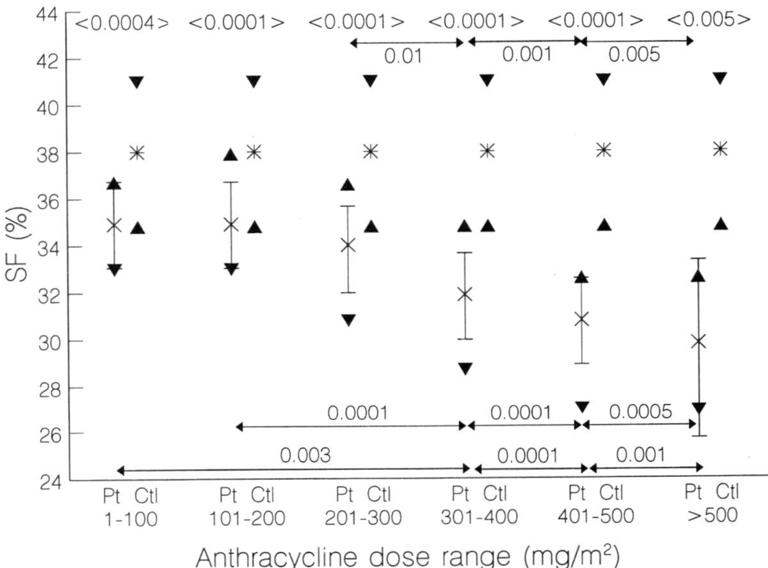

FIGURE 1. Shortening fraction (SF) percent by cumulative anthracycline dose. *Arrows* show interquartile range; *error bars*, the confidence limits of the mean.

reactive hydroxyl radicals which then react with membrane lipids and cause oxidative damage to the myocardial cells. Iron-chelating agents such as the metabolic product of ICRF187 appear to prevent the development of cardiac damage by this mechanism without antagonizing the antitumor effects, thus providing selective cardioprotection.

In a small pilot study, five children who were being retreated with anthracyclines at the time of relapse were given ICRF187 in addition on a compassionate basis. There was a clear-cut reduction in the severity of cardiotoxicity when compared to a control group for whom ICRF187 was not available and who received similar cumulative doses of anthracyclines.[11]

A RANDOMIZED, CONTROLLED CLINICAL TRIAL IN CHILDREN[12]

From February 1989 to September 1992, 43 consecutive, previously untreated patients less than 25 years of age with Ewings/PNET sarcoma, rhabdomyosarcoma, or other soft-tissue sarcomas were enrolled in NCI protocol 86C169 and were randomized to receive either ICRF187, granulocyte-macrophage colony-stimulating factor (GM-CSF), both, or neither. Patients were excluded if they had inadequate baseline cardiac function (left ventricular ejection fraction [LVEF] <45%, the lower limit of the institutional normal value) or could not be monitored by multi-gated radionuclide angiography (MUGA) scan. Three patients with chest wall tumors requiring radiation for local control were nonrandomly assigned to treatment with

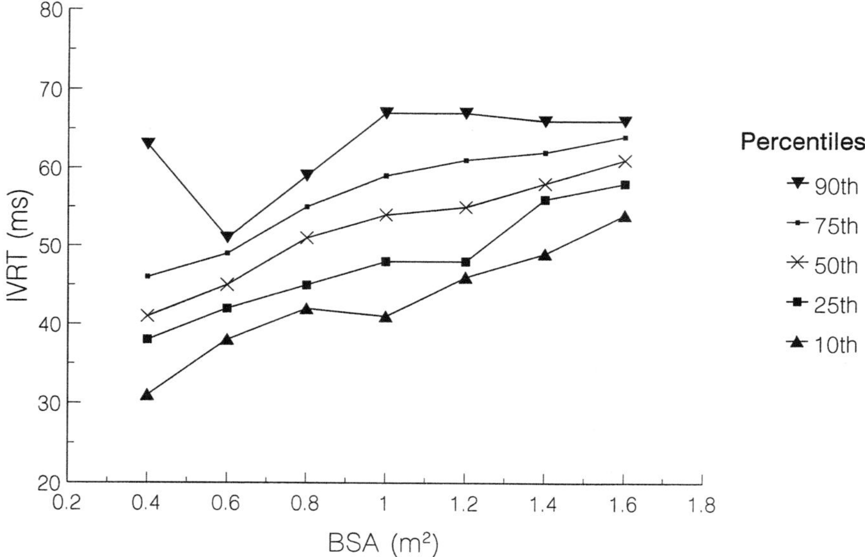

FIGURE 2. Centile values for isovolumic relaxation time (IVRT) relationship to body surface area (BSA).

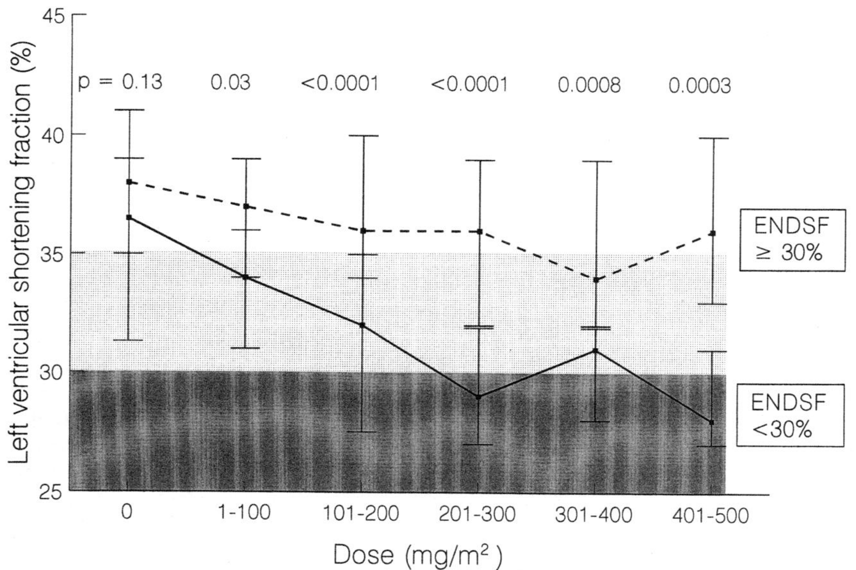

FIGURE 3. Comparison of serial SF values during treatment for those with final SF < 30% and >30%; *bars* show interquartile range.

ICRF187, and 39 patients were randomized to receive ICRF187 or not. Doxorubicin was given at a dose of 70 mg/m^2 per cycle over two days (cycles 1, 3, and 5) with vincristine, cyclophosphamide, and mesna and at a dose of 50 mg/m^2 per cycle on one day during cycles 9, 11, 13, and 15 with the same drugs. ICRF187 was given at 20 times the dose of doxorubicin and was reconstituted in 25 ml of saline within 4 hours of administration i.v. over 15 minutes immediately before the doxorubicin dose, which was also given i.v. over 15 minutes.

MUGA scans were performed at baseline and 6 to 12 weeks following the 210, 310, 360, and 410 mg/m^2 doses of doxorubicin. Dose-limiting cardiotoxicity was defined as a reduction in LVEF to <45%, a decrease of greater than 20% from baseline, or evidence of clinical congestive heart failure. Tumor response to therapy was evaluated after two and four cycles of treatment, and noncardiac toxicities were graded in accordance with standard common toxicity criteria. Three patients were excluded from evaluation of hepatic toxicity (one with hepatic metastases, two with transfusion-acquired hepatitis C) and two from evaluation of hematologic toxicity (treated with recombinant interferon alpha for hepatitis C and treated with recombinant human erythropoietin). Pharmacokinetic monitoring was performed on the first 11 patients.

Cardiac Evaluation

Thirty-three of 38 randomized patients were assessable for cardiac protection. A linear regression analysis of the change from baseline LVEF with increasing cumulative doses of doxorubicin showed a mean decrease in LVEF per 100 mg/m^2 for the control group of 2.7% compared with 1% for those receiving ICRF187 ($p = 0.02$) (FIG. 4). Of 15 patients who received the full intended dose of doxorubicin (410 mg/m^2) and had follow-up MUGA scans, the mean ± SEM LVEF of five control patients was 44 ± 2.8% compared with 53.9 ± 2.2% in the 10 ICRF187 patients ($p = 0.03$). Control patients developed dose-limiting cardiotoxicity significantly sooner than did ICRF187-treated patients, the cumulative proportion after 210, 360, and 410 mg/m^2 of doxorubicin being 5, 7, and 10% in the controls compared to 0, 2, and 4% in those receiving ICRF187 ($p < 0.01$).

Tumor Response

Eighty-one percent of control patients achieved an objective response (3 CRs and 10 PRs) by week 12 compared with 80% in the ICRF187 group (4CRs and 12PRs). At a median follow-up of 39 months, event-free survival time is 17 months in both groups with an estimated probability of event-free survival at 24 months of 39% (95% CI, 20% to 61% in the control group) and 43% (95% CI, 24%–64%) in the ICRF187 group with no significant difference for event-free survival or survival in the two groups (FIG. 5). Noncardiac toxicities showed an inconsistent trend towards an increased hematologic toxicity in the ICRF187 group, though this was of minimal clinical significance; and transient, primarily grade 1 AST elevations were observed with greater frequency in the group that received ICRF187. There were no differences in the incidence of dose modifications for hepatotoxicity or hematologic toxicity and no significant differences were seen in the incidence of mucositis or infection.

Thus, ICRF187 was shown to protect against the development of short-term doxorubicin-induced cardiotoxicity and, within the limited power of this study, no evidence of an adverse impact on the antitumor activity of doxorubicin was seen. The smaller decline in LVEF per 100 mg/m^2 of doxorubicin in the ICRF187 group allowed them to receive a significantly higher median cumulative dose of doxorubicin. No differences in pharmacokinetic criteria were observed between those patients who developed some evidence of cardiotoxicity despite ICRF187 and those who did not.

The number of patients treated was sufficient to detect a modest (28%) difference in antitumor activity of doxorubicin, but no such evidence of tumor protection was found. Follow-up studies will be required to evaluate whether the short-term ICRF187-associated cardioprotection documented will translate into a reduced risk of late cardiac complications.

FUTURE STUDIES

A number of important questions remain to be answered.[13] If early cardiotoxicity is prevented, will this translate into a reduction of morbidity and mortality in long-term survivors? It is certainly our clinical impression that patients treated in recent years since the introduction of a careful monitoring program have a much reduced incidence of late cardiotoxicity, but this remains to be formally demonstrated. Likewise, long-term results of randomized clinical trials of ICRF187 will be needed

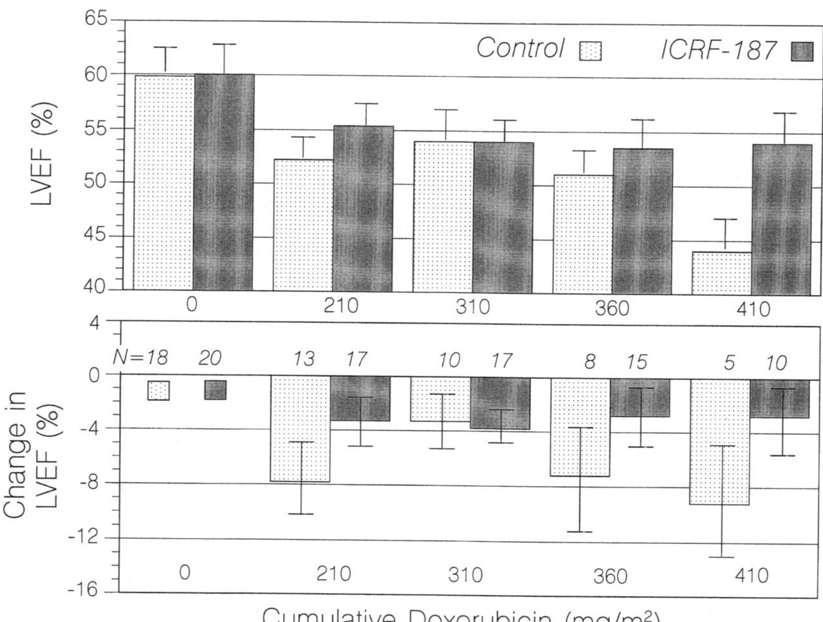

FIGURE 4. (a) Effect of ICRF187 on fall in left ventricular ejection fraction (LVEF) by cumulative doxorubicin dose. **(b)** Effect of ICRF187 on percent change in LVEF in cumulative doxorubicin dose.

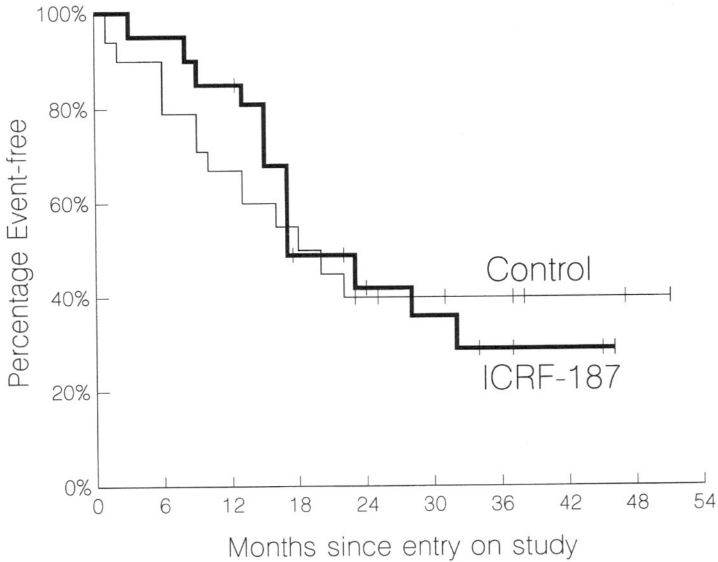

FIGURE 5. Event-free survival for patients randomized according to whether or not they received ICRF187.

to be sure that the anti-tumor efficacy of treatment is not compromised and that additional toxicity is not caused, in particular the potential for an increase in second malignancies.[14] For those patients who have already sustained myocardial injury, would the introduction of afterload-reducing agents such as angiotensin-converting enzyme inhibitors help to preserve residual myocardial function? These and many other questions remain to be addressed if the long-term goal of cure at least cost for children with cancer is to become a reality.

REFERENCES

1. ARCAMONE, F., A. DI MARCO, M. GAETANI, *et al.* 1961. Isolamento ed attivita antitumorle di un antibiotico da Streptomyces sp Giorn. Microbiol **9:** 83–90.
2. DUBOST, M., P. GANTER, R. MARAL, *et al.* 1963. Un novel antibiotique a proprietes antitumorales: La rubidomycine. CR Acad. Sci. (D) (Paris) **257:** 1813–1815.
3. GILLADOGA, A. C., C. MANUEL, C. TAN, *et al.* 1976. The cardiotoxicity of adriamycin and daunomycin in children. Cancer **37:** 1070–1076.
4. GOTTLIEB, J. A., E. A. LEFRAK, O'BRYAN & M. A. BURGESS. 1973. Fatal adriamycin cardiomyopathy (CMY): Prevention by dose limitation. Proc. Am. Assoc. Cancer Res. Abstr. **14:** 88.
5. MYERS, C. E., W. P. MCGUIRE, R. H. LISS, *et al.* 1997. Adriamycin: The role of lipid peroxidation in cardiac toxicity and tumour response. Science **197:** 165–167.
6. VON HOFF, D. D., M. W. LAYARD, P. BASA, *et al.* 1979. Risk factors for doxorubicin-induced congestive heart failure. Ann. Intern. Med. **91:** 701–717.
7. LIPSHULTZ, S. E., S. D. COLAN, R. D. GELBER, *et al.* 1991. Late cardiac effects of adriamycin therapy for childhood acute lymphoblastic leukemia. N. Engl. J. Med. **324:** 808–815.

8. STEINHERZ, L., P. STEINHERZ, C. T. C. TAN, G. HELLER, & M. L. MURPHY 1991. Cardiac toxicity 4–20 years after completing anthracycline therapy. JAMA **226:** 1672–1677.
9. BU'LOCK, F. A., M. G. MOTT, A. OAKHILL & R. P. MARTIN. 1995. Left ventricular diastolic function after anthracycline chemotherapy in childhood: Relationships with systolic function, symptoms and pathophysiology. Br. Heart J. **73:** 340–350.
10. BU'LOCK, F. A., M. G. MOTT, A. OAKHILL & R. P. MARTIN. 1996. Early identification of anthracycline cardiomyopathy: Possibilities and implications. Arch. Dis. Child. **75:** 1–7.
11. BU'LOCK, F. A., H. M. GABRIEL, A. OAKHILL, M. G. MOTT & R. P. MARTIN. 1993. Cardioprotection by ICRF187 against high dose anthracycline toxicity in children with malignant disease. Br. Heart J. **70:** 185–188.
12. WEXLER, LEONARD J., ANDRICH, MARY P., VENZON, DAVID, BERG, STACEY L., WEAVER-MCCLURE, LINDA, CHEN, CLARA C., DILSIZIAN, VASKEN, AVILA, NILO, JAROSINSKI, PAUL, BALIS, FRANK M., POPLACK DAVID G. & HOROWITZ, MARC E. 1996. Randomised trial of the cardioprotective agent ICRF187 in pediatric sarcoma patients treated with doxorubicin. J. Clin. Oncol. **14**(2): 362–372.
13. LIPSHULTZ, STEVEN E. 1991. Dexrazoxane for protection against cardiotoxic effects of anthracyclines in children (Editorial). J. Clin. Oncol. 328–331.
14. PEDERSEN-BJERGAARD, J. 1992. The dioxopiperazine derivatives, their leukemogenic potential and other biological effects. Leuk. Res. **16:** 1057–1059.

Secondary Malignancies or Late Recurrences?

J. ANŽIČ[a] AND B. JEREB[b]

[a]*University Department of Pediatrics*
Clinical Centre
Ljubljana, Slovenia

[b]*Institute of Oncology*
Zaloška 2 1005 Ljubljana, Slovenia

BACKGROUND

There have been numerous reports on the occurrence of secondary tumors after treatment of a malignant disease in childhood.[1-10] The results, generally expressed in terms of overall risk for such an event, are very difficult to compare, since the series differ in their composition, time of follow-up, and definition of the necessary interval between primary and secondary tumors.[4] Some authors assume that for the first five years from diagnosis of the primary tumor the risk is negligible; others report on secondary tumors within one year after diagnosis of the first primary malignancy or after just three years.[5] Also, the calculation of the time interval between the first and second malignancy can be performed differently by different authors; it can be calculated either from the date of diagnosis or from the date of completion of treatment. Some reports are limited to one kind of primary tumor and others to one kind of secondary tumors.[6,8] Sometimes, it is unclear whether a secondary tumor or a recurrent tumor is in question; in lymphomas the problem of a secondary tumor or changing histology might arise.

MATERIALS AND METHODS

In this study we aimed to analyze a population-based series of patients registered for malignant disease at the Cancer Registry of Slovenia. Between 1971 and 1993, 1195 patients, 0 to 15 years old at the time of diagnosis, were registered in Slovenia for malignant disease; 977 were treated at the Institute of Oncology in Ljubljana. The 521 survivors were all followed.

RESULTS

Twenty-six patients developed secondary tumors between one year and 29 years after the diagnosis of the first malignancy (TABLE 1): four developed leukemia, five non-Hodgkin's lymphoma, seven brain tumors, five carcinomas, one embryonal rhabdomyosarcoma, one Hodgkin's disease, one Ewing's sarcoma, and two osteogenic sarcomas. Among patients with leukemia or lymphoma, the second malignancy tended to be mainly leukemia or non-Hodgkin's lymphoma; among those with Hodgkin's disease and solid tumors, solid tumors were most common. No secondary tumors were observed among the survivors of nephroblastoma, neuroblastoma, and

TABLE 1. Clinical Data on 26 Patients with Secondary Tumors[a]

		First Tumor		Second Tumor			
Pt. No.	Age (Sex)	Diagnosis	Treatment	Age	Diagnosis	Interval (yrs/months)	Follow-up
1	14 (F)	Glioblastoma (cerebellum)	S + RT 55 Gy	43	Astrocytoma	29/4	DOD
2	12 (F)	ALL	ChT + RT 20 Gy	20	AML	7/2	DOD
3	3 (M)	Ependymoma	S + RT 45 Gy	11	Astrocytoma (optic nerve)	7/8	DOD[b]
4	8 (F)	ALL	ChT + RT 10 Gy	13	NHL (abdomen) (RT + 20)	4/7	AWD[c]
5	3 (M)	ALL	ChT + RT 24 Gy	12	Glioblastoma	9/11	DOD
6	10 (F)	HD	ChT + RT (1F) 30 Gy (MOPP)	24	Ca uterine cervix	13/8	NED[d]
7	5 (F)	ALL	ChT + RT 24 Gy	14	OS (mandible)	8/5	NED
8	15 (F)	ERMS (clitoris)	S + ChT (T2)	14	Adenoca (paranasal sinus)	12/9	NED
9	13 (M)	HD	ChT + RT (1F) 24 Gy (MOPP)	20	ERMS	6/5	DOC
10	12 (F)	Ewing Sa (tibia)	ChT + (T6 + T2) + RT 60 Gy	25	Ewing Sa (tibia)	12/4	NED
11	5 (M)	NHL	S + ChT (VCR, Pred.)	13	NHL	7/5	DOD
12	9 (M)	Ewing Sa (femur)	ChT (T6) + RT 60 Gy	15	OS (femur)	6/2	NED
13	2 (F)	ERMS (perineum)	S + ChT (CIVADIC)	3	ALL	1/3	DOD[e]
14	4 (M)	ALL	ChT + RT 18 Gy	8	HD	3/8	NED
15	6 (M)	AML	ChT (PATCO) + RT 24 Gy	16	NHL (testis)	10/1	NED
16	5 (M)	ALL	ChT + RT 18 Gy	14	ALL	8/10	NED
17	16 (M)	ALL	ChT (BFM 83) + 18 Gy	24	NHL (pelvis)	8/8	NED
18	4 (M)	ALL	ChT (BFM 83) + 18 Gy	13	meningioma	9/6	NED
19	9 (F)	Germinoma (suprasellar)	S + RT 30 Gy + ChT	13	germcell mixed	5/0	DOC
20	10 (F)	Germinoma (suprasellar)	S + RT 32 Gy + ChT	14	germcell mixed	1/8	DOD
21	12 (F)	HD	ChT + RT (1F) 30 Gy (MOPP)	31	AML	18/2	DOD
22	5 (M)	HD	ChT + RT 37 Gy (MOPP)	26	Ca thyroid	21/0	NED
23	4 (F)	ALL	ChT + RT 24 Gy	14	NHL (uterus)	9/9	NED
24	14 (F)	Ca breast	S + RT 40 Gy	45	Ca salivary gland	31/6	NED[f]
25	4 (F)	ALL	ChT + RT 30 Gy (BFM 86)	12	Glioblastoma	7/7	AWD
26	13 (F)	HD	ChT + RT 30 Gy (Velbe)	32	Ca breast	19/1	NED

[a] Abbreviations: ALL, acute lymphocytic leukemia; HD, Hodgkin's disease; NHL, non-Hodgkin's lymphoma; AML, acute myelocytic leukemia; OS, osteogenic sarcomas; Sa, sarcoma; Ca, carcinoma; DOD, died of disease; AWD, alive with disease; NED, no evidence of disease; DOC, died of complications; S, surgery; Rt, radiation therapy; CRT, chemotherapy.
[b] Third tumor 1989, malignant astrocytoma.
[c] Meningioma: 1994, 1995, 1996, one child.
[d] Two children.
[e] High-risk ALL, three CNS recurrences.
[f] Mo. Cowen.

retinoblastoma. Sixteen of 26 patients are alive, 14 without tumors 1 to 11 years after diagnosis of the second malignancy.

DISCUSSION AND CONCLUSIONS

In two of the three secondary bone tumors, the origin of the secondary tumor was at the primary site of Ewing's sarcoma. (The time intervals were 6 and 12 years between first and second malignancy.) In six acute leukemia patients, non-Hodgkin's lymphoma or a different type of acute leukemia was diagnosed as the second malignancy 7 to 10 years after the first diagnosis. In these eight patients and in two patients (No. 19, 20) with brain tumors from among the 26 cases, the question may be raised whether these were in fact second malignancies or late recurrences. At the first diagnosis of malignancy, it is necessary to explore all diagnostic possibilities (markers) to make the distinction between recurrence and secondary tumor as reliable as possible.

REFERENCES

1. BRESLOW, N., J. R. TAKASHIMA, J. A. WHITTAN, J. MOKSNESS, G. J. D'ANGIO & D. M. GREEN. 1995. Second malignant neoplasms following treatment for Wilms' tumor: A report from the National Wilms' Tumor Study Group. J. Clin. Oncol. **13:** 1851–1859.
2. DE VALTHAIRE, F., O. SCHWEISGUTH, C. RODARY, et al. 1989. Long-term risk of second malignant neoplasm after a cancer in childhood. Br. J. Cancer **59:** 448–452.
3. GREEN, D. M., M. A. ZEVON, P. A. REESE, et al. 1994. Second malignant tumors following treatment during childhood and adolescence for cancer. Med. Pediatr. Oncol. **22:** 1–10.
4. HARTLEY, A. L., J. M. BIRCH, V. BLAIR, P. M. JONES, H. R. GATTAMANENI & A. M. KELSEY. 1994. Second primary neoplasms in a population-based series of patients diagnosed with renal tumours in childhood. Med. Pediatr. Oncol. **22:** 318–324.
5. HAWKINS, M. M., G. J. DRAPER & J. E. KINGSTON. 1987. Incidence of second primary tumors among childhood cancer survivors. Br. J. Cancer **56:** 339–347.
6. KREISSMAN, S. G., R. D. GELBER, H. J. COHEN, L. A. CLAVELL, P. LEAVITT & S. E. SALLAN. 1992. Incidence of secondary acute myelogenous leukemia after treatment of childhood acute lymphoblastic leukemia. Cancer **70:** 2208–2213.
7. MEADOWS, A. T., E. BAUM, F. FOSSATI-BELLANI, et al. 1985. Second malignant neoplasms in children: An update from the late effects study group. J. Clin. Oncol. **3:** 532–538.
8. NYGAARD, R., S. GARWICZ, T. HALDORSEN, et al. 1991. Second malignant neoplasms in patients treated for childhood leukemia. Acta Pathol. Scand. **80:** 1220–1228.
9. ROSSO, P., B. TERRACINI, T. R. FEARS, et al. 1994. Second malignant tumors after elective end of therapy for a first cancer in childhood: A multicenter study in Italy. Int. J. Cancer **59:** 451–456.
10. TUCKER, M. A., G. J. D'ANGIO, J. D. BOICE, et al. 1987. Bone sarcomas linked to radiotherapy and chemotherapy in children. N. Engl. J. Med. **317:** 588–593.

Pediatric Oncology Outreach to Hungary: A Registry of Childhood Malignancies[a]

ENIKÖ APJOK,[b] JOSEPH D. BORSI,[b] ROZÁLIA KOÓS,[b]
DEZSŐ SCHULER,[b] ROBERT E. BOLINGER,[c]
CATHY D. BOYSEN,[c] ARIEL B. BAKER,[c]
AND FREDERICK F. HOLMES[c]

[b]Second Department of Pediatrics
Semmelweis University Medical School
Tuzolto u 7-9
1095 Budapest, Hungary

[c]The University of Kansas Medical Center
39th and Rainbow Boulevard
Kansas City, Kansas 66160

The Pediatric Oncology Outreach to Hungary (POOH) project is a "Partnership in Healthcare" sponsored by the U.S. Agency for International Development (USAID), and the support is shared among the University of Kansas Medical Center (KUMC), Kansas City, Kansas; the Second Department of Pediatrics of the Semmelweis University Medical School (SDP), Budapest, Hungary; and the National Institute of Neurosurgery (NINS), Budapest, Hungary.

Funding for the POOH project was awarded by USAID in September, 1991. In 1992, programming was completed for an institution-based pediatric tumor registry. This menu-driven registry is simple enough in design to allow its use in any country worldwide. The registry program requires an IBM-compatible computer with a DOS operating system of 5.0 or higher and one MEG of central memory.

The POOH Pediatric Tumor Registry maintains two types of user files in its data base. They are *data files* containing clinical information related to patients and *reference files* containing stable information such as diagnosis dictionaries and institutional data.

Clinical information in the data files is divided into five parts. The first part, the *patient master file* (PTMAST), contains demographic information pertaining to a single patient and requires a unique identification number for each patient. Second, the *patient diagnosis file* (DXPT) contains one record for each tumor diagnosis on each patient. This file links the diagnosis with the patient. Both the morphological code and topographic code are included, as well as date of diagnosis. The third part of the data files, the *patient follow-up file* (FUP), contains follow-up data linked to the patient's diagnosis record. Fourth, the *disease-specific file* (DISPEC) contains

[a] This work was jointly sponsored by the Pediatric Oncology Outreach to Hungary project of the University of Kansas Medical Center, Kansas City, Kansas; the Second Department of Pediatrics of the Semmelweis University Medical School, Budapest, Hungary; and the National Institute of Neurosurgery, Budapest, Hungary, through a grant from the U.S. Agency for International Development.

records with a standard data format and must be linked to an existing record in the patient/diagnosis file. The fifth part of the data files is the *joint commission evaluation file* (JCE), containing information that can be used to assist in quality assurance measures.

The reference files are divided into seven sections. A *tumor morphology dictionary* (TMDICT) contains a list of tumor morphology diagnoses along with topographic locations. A *tumor topographic dictionary* (TPDICT) contains topographic codes from the International Classification of Diseases for Oncology (ICDO). The *reporting institution file* (REPINS) contains a record for each reporting institution, in the event that more than one institution submit data to the same registry. The REPINS contains demographic information and a contact person for each institution entering data into the registry. Four final sections, the *outline file* (OUTLINE), the *labels file* (LBLS), the *general terms file* (GENTERMS), and the *specific terms file* (SPECTRUM) are for programmer use.

The POOH Pediatric Tumor Registry contains statistical survival analysis capabilities as well, providing survival function estimates and survival mean and quantile estimates, with a step graph of the survival curve function.

The Pediatric Oncology Outreach to Hungary project is entering data on all patients diagnosed with leukemia and malignant solid tumors who have been treated and followed since 1987. The number of new patients diagnosed with malignancies at SDP and NINS fluctuates between 43 and 76 cases per year. Follow-up information is entered at least annually. Childhood leukemia and cancer frequencies and age distributions are generated annually. Survival rates can be compared from 1987 to present, which helps to evaluate current treatment protocols.

Body Cell Mass as a Reference for Nutritional Support in Pediatric Cancer Patients[a]

PAUL R. SCHLOERB[b] AND HEDVIG E. BODÁNSZKY[c]

[b]Department of Surgery and Nutritional Support Service
University of Kansas Medical Center
39th and Rainbow Boulevard
Kansas City, Kansas, 66160

[c]Second Department of Pediatrics
Semmelweis University Medical School
Budapest, Hungary

Body weight may not be a good reference for nutritional support. In cancer patients with excess body water such as ascites or edema, body weight, including this excess water, is a poor reference for nutritional support. Similarly, the variability of body fat limits the reliability of body weight as a reference. It is the purpose of this paper to point out the deficiencies of body weight as a reference and to describe a method for calculation and use of total body water (TBW) and body cell mass (BCM) for parenteral nutritional support prescriptions in children.

As shown in FIGURE 1, two individuals, weighing the same, 40 kg, have markedly different body compositions. With total body water (TBW) 70% of body weight in the boy and only 35% in the obese girl, there are corresponding major differences in body cell mass (BCM). With ascites, edema, or generalized anasarca, body weight becomes less reliable as a reference as BCM becomes an even smaller proportion of the total. A close relationship between TBW and oxygen consumption has been documented.[1] Body cell mass, a close function of total body water (TBW), is the oxygen- and glucose-consuming, energy-producing portion of the body.[2] BCM determines energy expenditure and, therefore, nutrient requirements. Methods for measurement of TBW for calculation of BCM, such as isotope dilution, bioimpedance analysis, infrared absorption, or DEXA, are too time-consuming, cumbersome, or expensive for routine use. With data based on many observations, equations developed for calculation of TBW and BCM seem preferable.

In 74 children, from birth to 15 years of age, Friis-Hansen measured TBW by dilution of deuterium oxide (heavy water)[3] and documented age, height, weight, and gender for each child. The Harris-Benedict equations for calculation of basal energy expenditure (BEE) in adults were published in 1919.[4] The BEE equation for males is the following:

$$BEE(kcals) = 66.47 + 13.75 \times Wt(kg) + 5 \times Ht(cm) - 6.76 \times age(yrs)$$

Using this equation and the published values of Friis-Hansen for age, height, and

[a] This research was supported by the University of Kansas Endowment Association (PRS).

234

weight, BEE was calculated for each subject (kjoules = kcal $\times$ 4.18). These values were correlated with TBW and with BCM, calculated as 80% of TBW,[5] as shown in FIGURE 2. The relationship between BCM and BEE is as follows:

$$BCM = 9.69 \times e^{(-kjoules/-4827)} - 11.19$$

$$r = 0.996 \ (p < 0.001)$$

With calculation of BCM from this equation and usual activity factors, nutrient requirements of normal children were calculated. Approximate overall requirements for parenteral nutrition are: protein, 25%; carbohydrate, 50%; and fat, 25% of calculated energy requirements.[5] As functions of age (in years) for children with normal weight and height for age, requirements are: kjoules/(kg · day) = 209 − 5.6 $\times$ age; H_2O in ml/(kg · day) = 100 − 4.67 $\times$ age; protein in (g/(kg · day) = 3 − 0.1 $\times$ age; CHO in g/(kg · day) = 7.5 − 0.133 $\times$ age; fat in g/(kg · day) = 1.4 − 0.046 $\times$ age. With an activity factor of 1.25, nutrient requirements, expressed per kilogram BCM, are: protein, 5 g/kg; carbohydrate, 13 g/kg; fat, 2.4 g/kg; H_2O, 50 ml/kg; energy, 365 kj/kg.

With excess body water, a clinical estimate of "dry" weight is necessary for these calculations.[7] BCM, a close function of TBW, calculated from an energy equation, appears to be a useful reference for defining nutritional support in children.

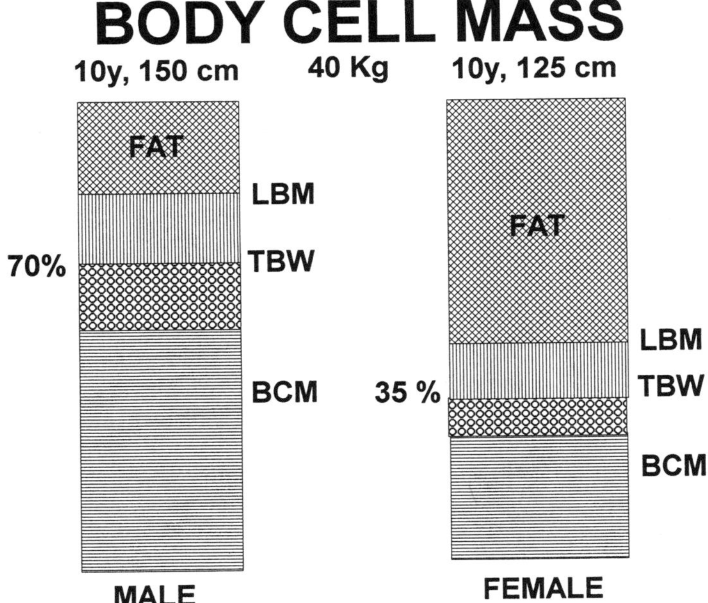

FIGURE 1. Two individuals weighing the same (40 kg), but with different body compositions.

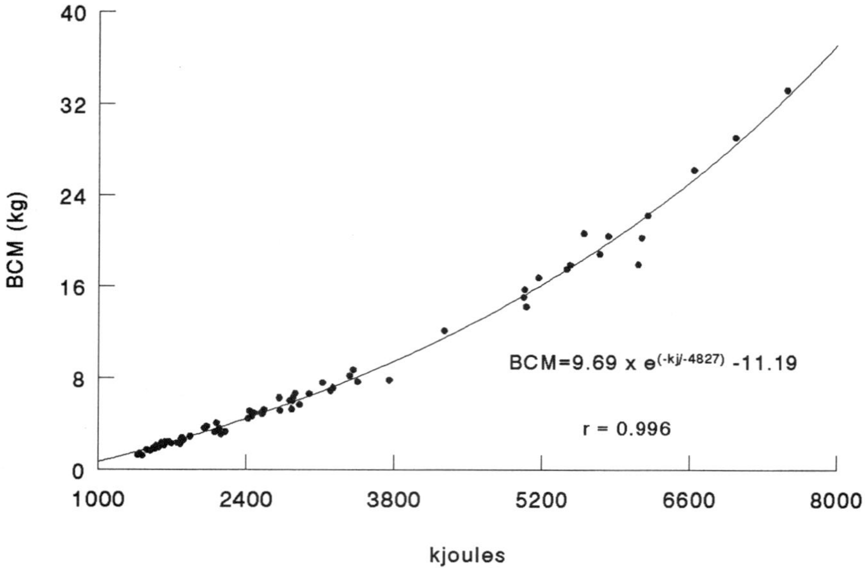

FIGURE 2. Body cell mass (as TBW × 0.8),[5] derived from TBW as measured by deuterium oxide dilution,[3] plotted against resting energy expenditure, and calculated by the male Harris-Benedict equation.[4] KJ = kcal × 4.18.

REFERENCES

1. SCHLOERB, P. R., B. J. FRIIS-HANSEN, I. S. EDELMAN, A. K. SOLOMON & F. D. MOORE. 1950. The measurement of total body water in the human subject by deuterium oxide dilution, with a consideration of the dynamics of deuterium distribution. J. Clin. Invest. **29:** 1296–1310.
2. MOORE, F. D., *et al.* 1963. The Body Cell Mass and Its Supporting Environment: Body Composition in Health and Disease. WB Saunders. Philadelphia, PA.
3. FRIIS-HANSEN, B. 1956. Changes in Body Water Compartments During Growth. Ejnar Munksgaard. Copenhagen.
4. HARRIS, J. A. & F. G. BENEDICT. 1919. A biometric study of basal metabolism in man. Publication #279 of the Carnegie Institution. Washington, DC.
5. BRUCE, A., *et al.* 1980. Body composition. Prediction of normal body potassium, body water, and body fat in adults on the basis of a body height, body weight, and age. Scand. J. Clin. Lab. Invest. **40:** 461–473.
6. HEIRD, W. C. 1987. Chapter 54. *In* Pediatric Nutrition, Theory and Practice. Butterworth. Stoneham, MA.
7. SCHLOERB, P. R. 1984. Computer-assisted nutritional support in surgery. Contemp. Surg. **25:** 53–57.

Neurosurgical Management of Cerebral Astrocytomas in Children

A. V. CIUREA,[a] G. VASILESCU,[a] L. NUTEANU,[a]
MARIA SILVEANU-VLADU,[b] IOANA TEODORESCU,[c]
AND M. LISIEVICI[d]

[a]*Pediatric Neurosurgery Department*
[b]*Pediatric Anesthesiology Department*
[c]*Neuroradiology Department*
[d]*Pathology Department*
Neurosurgical Clinic "Prof. Dr. D. Bagdasar" Hospital
10, Sos. Berceni
Bucharest 4, Romania

Cerebral astrocytomas (CAs) are the most frequently occurring CNS tumor of childhood.[1-4] Many studies refer to median and paramedian deep-seated CAs in children, meaning CAs located in basal ganglia and thalamus as well as the brain stem.[5]

MATERIALS AND METHODS

The authors report 261 cases of CA investigated and operated over a period of 10 years (1986–1995). The age limits of the study were 4 months to 16 years, with a mean age of 9.13 years. The sex ratio was 1.21 with males predominating.

RESULTS

Site: The localization of the tumors was as follows: supratentorial, 115 cases (43.90%); subtentorial, 145 cases (55.04%); and supra- and subtentorial, 1 case (0.34%) (TABLE 1).

Clinical findings: Presenting symptoms in supratentorial CAs were dominated by raised intracranial pressure (ICP) (67.53%), seizures (74.91%), hemiparesis (71.24%), mental changes (42.57%), and speech disturbances (23.47%); in subtentorial CAs the findings noted were raised ICP (84.21%), cerebellar syndromes (62.56%), cranial nerve signs (71.8%), and mental changes (12.45%).

EEG studies: In all cases of supratentorial CAs presenting with seizures, an exact localization of the epileptic focus was searched for before surgery.

Imaging studies: CT scan and especially MRI were fairly suggestive for the neuroimagistic diagnosis.

Pathology: The results of our series for supratentorial CAs (115 cases), according to WHO classifications, were: grade I, 10 cases (8.69%); grade II, 31 cases (26.95%); grade III, 51 cases (44.34%); and grade IV, 23 cases (20%). For subtentorial CAs (145 cases), the results were as follows: grade I, 52 cases (35.86%) and grade II, 93 cases (64.13%).

TABLE 1. Localization of Cerebral Astrocytomas (261 Cases)

Site of CA	No. of Cases	Percent
Supratentorial astrocytomas	**115**	**44.06**
Lobar site	85	32.57
Paramedian structures	12	4.59
Median line structures	18	6.90
Subtentorial astrocytomas	**145**	**55.55**
Cerebellar hemispheres	58	22.22
Cerebellar vermis and fourth ventricle	81	31.03
Brain stem	6	2.30
Supratentorial and subtentorial	**1**	**0.38**

Treatment: Multimodal treatment is necessary in CA, the main request being total microsurgical removal as well as radiotherapy and chemotherapy, depending on the degree of malignancy. The tumoral removal was total in 201 cases (77%) and subtotal in 37 cases (14.2%), involving deep and infiltrative tumors. Bioptic sampling only was done in 23 cases (8.8%) involving basal ganglia and brain stem CAs. Immediate and early operative mortality (including the complications occurring in the first month) was 2.68% (7 cases).

Follow-up and quality of life: The patients were followed, both clinically and by CT, until January 1996 for this study. The quality of life was assessed in all 45 selected patients with long survival (more than 5 years) by adopting a protocol that included repeated CT studies and neuropsychological evaluation (TABLE 2). Recurrences occurred in 56 cases (21.45%). The treatment of recurrences for group I (37 cases) was second surgery; for group II (19 cases), having grade III and IV lesions, chemotherapy and radiotherapy were used (no surgery).

DISCUSSION

CAs represent an important proportion (261 cases or 36.5%) of the total number of intracranial tumors in the same age group (715 cases).

Localization: CAs in children occur mainly in the posterior fossa.[2,6] Our data are similar (55.5%). Brain stem CAs are rather rare (6 cases, 2.29%) in our series in comparison with other studies[2] because the surgical approach of the brain stem is dangerous and limited. In relation to the masses developed in the basal ganglia and thalamus (3 cases, 1.14%) and in the chiasmatic/hypothalamic area (3 cases, 1.14%), the present study refers only to those that were operative and pathologically

TABLE 2. Quality of Life: 45 Selected CA Patients (5-Year Survival)

Neurophychological Protocol	Results
Neurological deficit	+/−
Relapse-free interval	+
Intellectual activity	+
IQ	>77
Mental disturbances	+/−
Behavioral skills problems	+/−

verified. According to the literature, the median line CAs in children occur, in decreasing order, in the chiasmatic/hypothalamic area, brain stem, thalamus, and the pineal region.[5]

Clinical findings: Raised ICP dominates the early signs of CA, followed by seizures, cranial nerve signs, hemiparesis and mental changes. Usually, the growth of this type of tumor is supported rather well in children, with clinical deterioration appearing after a head trauma or intercurrent infection. CSF pathway dissemination was not found in any of the subjects.

Pathology: Supratentorial CA grade II and mostly grade III types are frequent, a fact that explains the recurrences. We have to mention the important number of grade IV tumors (glioblastomas), which are very aggressive (23 cases). In fact, in our series, the grade III and grade IV types represent about two-thirds of all supratentorial CAs. The pathological examination of subtentorial CAs revealed a much more benign aspect.

Treatment: The main therapeutic aim is complete microsurgical removal. Subtotal removal was obtained in infiltrative, diffuse CA. In all chiasmatic/hypothalamic, basal ganglia and thalamus, brain stem, and third ventricle masses with extension to the basal ganglia, only enlarged biopsy was performed.

Operative mortality: We consider that the operative mortality (early and first month postoperative) is rather low (7 cases or 2.68%). We have not had any case of strictly intraoperative death.

Neurological sequelae: Six months after surgery, all neurological complaints disappeared, and the general condition of the children was very good.[7]

Prognostic survival rate (SR): In CA, the outcome depends mainly on the patient's age, general condition at admission, early diagnosis, tumoral volume and localization, total resection possibilities, and mainly the histopathologic type of the tumor. In all grade I and II astrocytomas, the outcome was favorable. Longer survivals were noted in cerebral hemisphere CAs: 0.711 at 1 year and 0.612 at 5 years. The longest survivals in our series were in cerebellar CA: at 1 year they had a 0.825 probability of survival and at 5 years, 0.685. The degree of malignity and the reaction to the adjuvant treatment (radiotherapy and chemotherapy) applied are in direct correlation with the outcome of chiasmatic/hypothalamic, basal ganglia, and brain stem masses, which nevertheless remains severe. The shortest survival was observed with brain stem tumors, with an average survival of 12 months. In total, cumulative probability of survival of children with intracranial CA at 1 year was 0.714 and at 5 years was 0.655.

Quality of life: The quality of life, which generally is maintained within acceptable parameters, was assessed in 45 patients with long survival (more than 5 years), especially cerebellar CA.

Recurrences: All operated cases were followed over a period between 6 months to 10 years. The best means for diagnosing and detecting CNS spreading in recurrences are MRI and PET.[8] Clinical signs of recurrences occurred in 56 cases (21.45%). In 37 (66.07%) of these, a surgical indication was retained; for the other 19 cases, which involved grade III and IV lesions, chemotherapy and radiotherapy were instituted.

CONCLUSIONS

The outcome in cerebellar CA was the most favorable of all intracranial tumors in children.[1,6,7] In supratentorial CA grade II with subtotal removal, we routinely recommend megavoltage radiotherapy to prevent recurrences. In our series, an unexpectedly high prevalence of grade III (51 cases or 44.34%) and IV (23 cases or 20%) supratentorial CA, with corresponding consequences for the outcome, was observed. Supratentorial median line CAs, in spite of generally low malignancy, are extremely aggressive no matter what therapeutic means are used. Furthermore, brain stem CAs remain difficult to approach, and conservative treatment (radiotherapy and chemotherapy) manages to slow down their evolution only for a limited period of time.

SUMMARY

A study of astrocytomas occurring in children (mean age 9.13 years) is reported. Two hundred sixty-one cases operated within 10 years (1986–1995) were reviewed. The subtentorial localization was preponderant and was seen in 145 cases (55.04%). The pathological results for supratentorial astrocytomas show an important number of high-grade, malignant astrocytomas—74 cases (64.3%). Surgery was the most important aspect of the treatment. The microscopic tumoral removal was total in 201 cases (77%), subtotal in 37 cases (14.2%); bioptic sampling was done in only 23 cases (8.8%). In grade III and IV astrocytomas, radiotherapy (high-voltage) and chemotherapy were routinely used. The patients were followed-up, both clinically and by CT, between 6 months and 10 years. Neurological evaluation (6 months post-operative) shows a preponderance of minimal sequelae (71%). Clinical signs of recurrences occurred in 56 cases (21.45%). In 37 cases (66.07%), a surgical indication was retained. Cumulative probability of survival of children with intracranial astrocytomas at 1 year was 0.714 and at 5 years was 0.655.

REFERENCES

1. CUSHING, H. 1931. Experience with cerebellar astrocytomas. A critical review of seventy-six cases. Surg. Gynecol. Obstet. **52:** 129–204.
2. DOHRMANN, G. J., *et al.* 1985. Astrocytomas in childhood: A population-based study. Surg. Neurol. **23:** 64–68.
3. MERCURI, S., *et al.* Hemispheric supratentorial astrocytomas in children. Long term results in 29 cases. J. Neurosurg. **55:** 170–173.
4. SCHNEIDER, J. H., Jr., *et al.* 1991. Benign cerebellar astrocytomas of childhood. Neurosurgery **30:** 1.
5. HOFFMAN, H. J., *et al.* 1993. Management and outcome of low-grade astrocytomas of the midline in children: A retrospective review. Neurosurgery **33:** 964–971.
6. KOOS, W. T. & M. H. MILLER. 1971. Intracranial tumors of children. Thieme, Stuttgart: 9.
7. ABDOLLAHZADEH, M., *et al.* 1994. Benign cerebellar astrocytoma in childhood: Experience at the Hospital for Sick Children 1980–1992. Child. Nerv. Syst. **10:** 380–383.
8. MAMELAK, A. N., *et al.* 1994. Treatment options and prognosis for multicentric juvenile pilocytic astrocytoma. J. Neurosurg. **81:** 24–30.

Pineal Region Teratomas in Children

J. ŠTEŇO,[a] I. BÍZIK, AND P. BIKSADSKÝ

Department of Neurosurgery
Comenius University
833 05 Bratislava, Slovak Republic

INTRODUCTION

Teratomas originating from an embryonic cell precursor undergoing embryonic differentiation may develop as mature, immature, or malignant teratomas.[2] Mature teratomas appear to prevail.[1,4] Immature teratomas may be locally invasive. Mixtures of mature and immature forms do occur and can sometimes make clear distinction between the two varieties difficult.[3] We present our own experience with three children treated for pineal region teratomas with different degrees of immaturity.

PATIENTS AND METHODS

Over three years (1994–1996), 66 patients of up to 15 years of age were operated on in our department for brain tumors. In five of them, the tumor was located in the pineal region (three teratomas, one pineocytoma/pineoblastoma, and one pineoblastoma).

In all three patients with teratomas (two males, 7 and 15 years old, and one female, 9 years of age), the tumor manifested itself by the syndrome of increased intracranial pressure and by oculomotor disturbances (Parinaud's syndrome).

All the tumors were of mixed density and poorly delineated on CT scans. MRI revealed an unhomogenous mass enhanced by gadolinium. The tumor occupied the pineal region and the posterior part of the cavity of the third ventricle (FIG. 1A). In two cases the tumor extended; in one it extended anteriorly, filling the whole cavity of the third ventricle, and in the other one laterally, toward the trigone of the lateral ventricle (FIG. 2A).

Pronounced signs and symptoms of increased intracranial pressure necessitated a CSF shunt before direct attack of the tumor in the female patient. In the other two cases, exposure of the tumor was performed as the initial operation. An infratentorial supracerebellar approach was used in two cases with a midline location of the tumor. In the patient with a lateral extension of the tumor, it was attacked through van Wagenen's posterior transventricular route. Despite some adherence of the tumor to surrounding structures, that is, to the lateral walls of the third ventricle and to Rosenthal or internal cerebral veins, it could be dissected free and removed totally in all patients, as confirmed by contrast-enhanced MRI (FIGS. 1B and 2B).

Histological examination revealed a differentiated, immature teratoma in two patients and a predominantly mature teratoma with "less mature" areas and mitoses in the third patient.

[a] Address for correspondence: J. Šteňo, Department of Neurosurgery, Comenius University, Derer's Hospital, Limbová 5, 833 05 Bratislava, Slovak Republic; Tel./fax: 00421-7-372 787.

External radiotherapy (30 Gy to the whole head, 20 Gy to the tumor bed) was applied in the first, and two courses of chemotherapy BEP/VIP (bleomycine, etoposide, cisplatin/vinblastine, iphosphamide, cisplatin) were administered in the second patient with an immature teratoma. The patient with a mixed teratoma received both external radiotherapy (30 Gy to the whole brain, 20 Gy to the tumor bed, 30 Gy to the spine) and chemotherapy BEP/VIP.

The patients with immature teratomas have been followed for 26 and 25 months. They are alive and well with negative serum germ cell markers (alpha fetoprotein [AFP] human chorionic gonadotropin [hCG]).

The patient with a mixed teratoma died 11 months after surgery from an acute rise of intracranial pressure caused by a recurrent tumor. The high value of serum AFP four months after the operation (229 IU/l) was interpreted as a relapse of hepatitis. Two weeks before death, the patient had normal hCG levels, an insignificantly higher value of AFP (18,1 IU/l), and was found neurologically intact. Neuroradiological examination was not indicated.

DISCUSSION

Although mature teratomas of the pineal region usually prevail,[1,4] in each of our three cases the tumor showed some degree of immaturity. One of them can

A   B

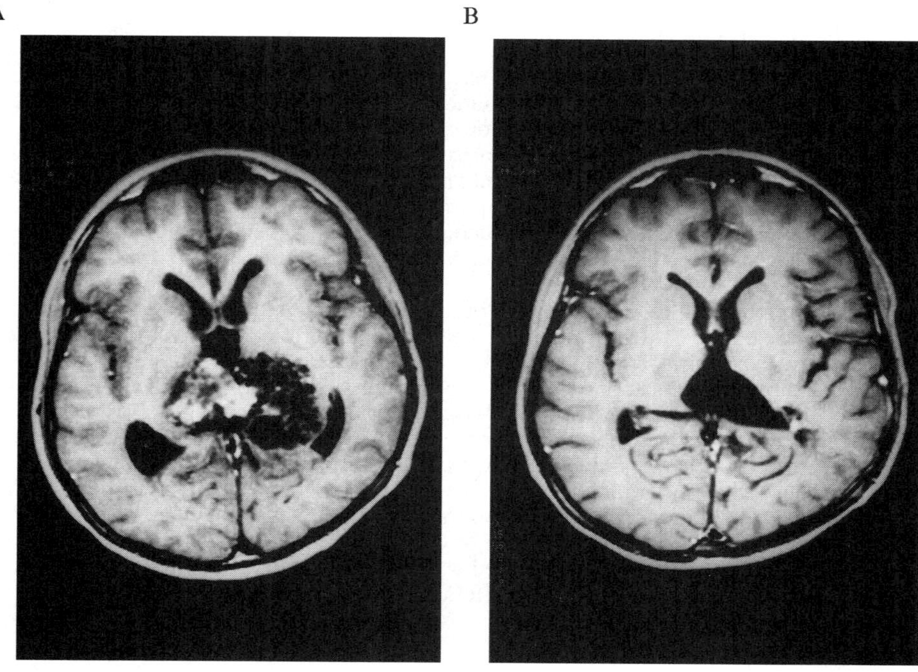

FIGURE 1. Contrast-enhanced MR of immature teratoma of pineal region before (**a**) and after (**b**) radical removal.

A
B

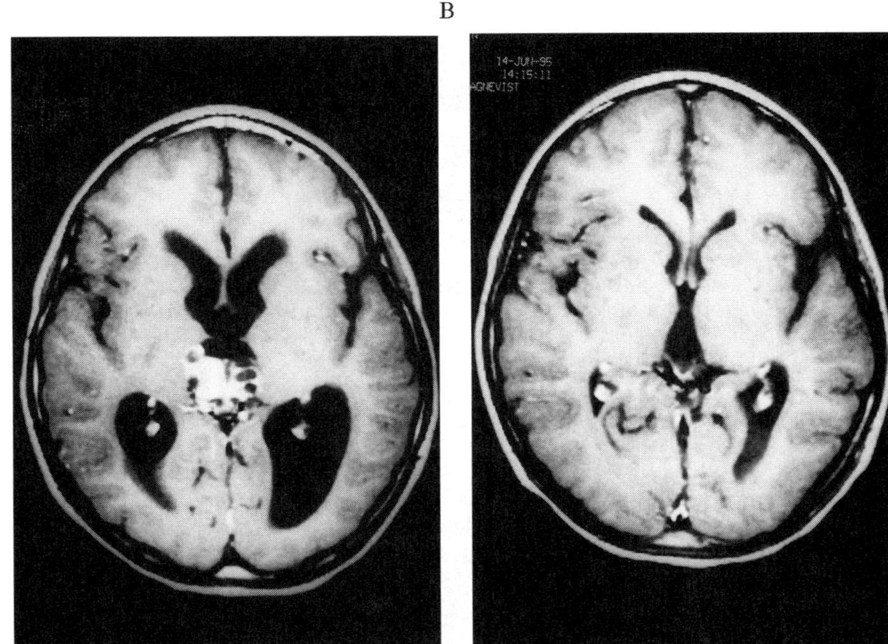

FIGURE 2. Contrast-enhanced MR of immature teratoma extending to the trigone of left lateral ventricle (**a**) radically removed through a posterior transventricular approach (**b**).

be classified as a mixture of mature and immature forms of teratoma.[3] Different histological patterns of tumor tissue at different sites of the lesion justifies direct microsurgical attack of the tumor when deciding between stereotactic biopsy and exposure of the tumor.[5] Direct visual control through an appropriate surgical approach allows radical removal of the tumor despite some local adherence of the immature teratoma to the lateral walls of the third ventricle and to surrounding large veins, even in cases with giant-sized lesions.

Recurrence of the teratoma with some component of immature tissue may occur even after its radical removal and postoperative radio- and chemotherapy. Regular examinations of the levels of tumor markers and neuroradiological investigations are mandatory.

SUMMARY

Immature or mixed mature/immature teratomas of the pineal region were removed radically in three children. After a period of 26 and 25 months, two patients with immature teratomas are alive and well, whereas in the third patient a mixed teratoma locally recurred and caused a fatal rise of intracranial pressure 11 months after the operation. Regular examinations of tumor markers and neuroradiological

investigations are mandatory after radical surgery and adjuvant radio- and chemotherapy.

REFERENCES

1. BRUCE, J. N. & B. M. STEIN. 1995. Surgical management of pineal region tumors. Acta
Neurochirurgica (Wien) **134:** 130–145.
2. CARMEL, P. W. 1996. Brain tumors of disordered embryogenesis. *In* J. R. Youmans, Ed.
Neurological surgery, 4th edit. Vol. 4: 2761–2781. W. B. Saunders Co.
3. DIRKS, P. B., J. T. RUTKA, L. E. BECKER & H. J. HOFFMAN. 1996. Intracranial germ cell
tumors: Classification, diagnosis, and management. *In* Neurological Surgery, 4th edit.
J. R. Youmans, Ed. Vol. 4: 2530–2541. W. B. Saunders Co.
4. FUKUI, K., T. MATSUSHIMA, K. FUJI, J. NISHINO, I. TAKESHITA & T. TOSHIMA. 1991. Pineal
and third ventricle tumors in CT and MR eras. Acta Neurochirurg. Suppl. **53:** 127–136.
5. HOFFMAN, H. J., M. YOSHIDA, L. E. BECKER, E. B. HENDRICH & R. P. HUMPHREYS. 1983.
Pineal region tumors in childhood. Concepts Pediatr. Neurosurg. **4:** 360–386.

Index of Contributors